Student Solutions Manual

Introductory Algebra

Review, Reference, and Practice

Student Solutions Manual

Introductory Algebra

Review, Reference, and Practice

K. Elayn Martin-Gay
University of New Orleans

PRENTICE HALL, Upper Saddle River, NJ 07458

Director of Product Development: *Christine B. Hoag*
Development Editor: *Elaine Page*
Production Editor: *Barbara A. Till*
Supplement Cover Design: *Maureen Eide*
Special Projects Manager: *Barbara A. Murray*
Manufacturing Buyer: *Alan Fischer*

©1999 by Prentice-Hall, Inc.
Simon & Schuster / A Viacom Company
Upper Saddle River, New Jersey 07458

Printed in the United States of America

10 9 8 7 6 5 4 3

0-13-011449-9

Prentice-Hall International (UK) Limited, *London*
Prentice-Hall of Australia Pty. Limited, *Sydney*
Prentice-Hall Canada Inc., *Toronto*
Prentice-Hall Hispanoamericana, S.A., *Mexico*
Prentice-Hall of India Private Limited, *New Delhi*
Prentice-Hall of Japan, Inc., *Tokyo*
Simon & Schuster Asia Pte. Ltd., *Singapore*
Editora Prentice-Hall do Brasil, Ltda., *Rio de Janeiro*

Contents

Student Solutions Manual

Introductory Algebra

Review, Reference, and Practice

Chapter 1

Exercise Set 1.1

1. $4 < 10$

3. $7 > 3$

5. $6.26 = 6.26$

7. $0 < 7$

9. $32 < 212$

11. $11 \le 11$ True

13. $10 > 11$ False

15. $3 + 8 \ge 3(8)$ False

17. $7 > 0$ True

19. $30 \le 45$

21. $8 < 12$

23. $5 \ge 4$

25. $15 \ne -2$

27. 90

29. $70 \le 90$

31. 0; whole, integers, rational, real

33. -2; integers, rational, real

35. 6; natural, whole, integers, rational, real

37. $\dfrac{2}{3}$; rational, real

39. False

41. True

43. True

45. True

47. $-10 > -100$

49. $32 > 5.2$

51. $\dfrac{18}{3} < \dfrac{24}{3}$

53. $-51 < -50$

55. $|-5| > -4$

57. $|-1| = |1|$

59. $|-2| < |-3|$

61. $|0| < |-8|$

63. $0 > -26.7$

65. Sun

67. Sun

69. $5 < 6$ True

71. $-5 < -6$ False

73. $|-5| < |-6|$ True

75. $|-5| \ge |5|$ True

77. $-3 > 2$ False

79. $|8| = |-8|$ True

81. $|0| > |-4|$ False

83. $20 \le 25$

85. $6 > 0$

87. $-12 < -10$

89. Answers may vary.

Exercise Set 1.2

1. $\dfrac{3}{8}$

3. $\dfrac{5}{7}$

5. $20 = 2 \cdot 2 \cdot 5$

7. $75 = 3 \cdot 5 \cdot 5$

9. $45 = 3 \cdot 3 \cdot 5$

11. $\dfrac{2}{4} = \dfrac{2}{2 \cdot 2} = \dfrac{1}{2}$

13. $\dfrac{10}{15} = \dfrac{2 \cdot 5}{3 \cdot 5} = \dfrac{2}{3}$

15. $\dfrac{3}{7}$ is in lowest terms.

17. $\dfrac{18}{30} = \dfrac{2 \cdot 3 \cdot 3}{2 \cdot 3 \cdot 5} = \dfrac{3}{5}$

19. $\dfrac{1}{2} \cdot \dfrac{3}{4} = \dfrac{3}{8}$

21. $\dfrac{2}{3} \cdot \dfrac{3}{4} = \dfrac{2 \cdot 3}{3 \cdot 2 \cdot 2} = \dfrac{1}{2}$

23. $\dfrac{1}{2} + \dfrac{7}{12} = \dfrac{1}{2} \cdot \dfrac{12}{7} = \dfrac{1 \cdot 2 \cdot 2 \cdot 3}{2 \cdot 7} = \dfrac{6}{7}$

25. $\dfrac{3}{4} + \dfrac{1}{20} = \dfrac{3}{4} \cdot \dfrac{20}{1} = \dfrac{3 \cdot 2 \cdot 2 \cdot 5}{2 \cdot 2} = 15$

27. $\dfrac{7}{10} \cdot \dfrac{5}{21} = \dfrac{7 \cdot 5}{2 \cdot 5 \cdot 3 \cdot 7} = \dfrac{1}{6}$

29. $2\dfrac{7}{9} \cdot \dfrac{1}{3} = \dfrac{25}{9} \cdot \dfrac{1}{3} = \dfrac{5 \cdot 5 \cdot 1}{3 \cdot 3 \cdot 3} = \dfrac{25}{27}$

31. $l \cdot w = \dfrac{11}{12} \cdot \dfrac{3}{5} = \dfrac{11 \cdot 3}{2 \cdot 2 \cdot 3 \cdot 5} = \dfrac{11}{20}$ sq miles

33. $\dfrac{4}{5} - \dfrac{1}{5} = \dfrac{4-1}{5} = \dfrac{3}{5}$

35. $\dfrac{4}{5} + \dfrac{1}{5} = \dfrac{4+1}{5} = \dfrac{5}{5} = 1$

37. $\dfrac{17}{21} - \dfrac{10}{21} = \dfrac{17-10}{21} = \dfrac{7}{21} = \dfrac{7}{7 \cdot 3} = \dfrac{1}{3}$

39. $\dfrac{23}{105} + \dfrac{4}{105} = \dfrac{23+4}{105} = \dfrac{27}{105} = \dfrac{3 \cdot 3 \cdot 3}{3 \cdot 5 \cdot 7} = \dfrac{9}{35}$

41. $\dfrac{7}{10} \cdot \dfrac{3}{3} = \dfrac{21}{30}$

43. $\dfrac{2}{9} \cdot \dfrac{2}{2} = \dfrac{4}{18}$

45. $\dfrac{4}{5} \cdot \dfrac{4}{4} = \dfrac{16}{20}$

47. $\dfrac{2}{3} + \dfrac{3}{7}$ LCD = 21

$\dfrac{2}{3} \cdot \dfrac{7}{7} = \dfrac{14}{21}$

$\dfrac{3}{7} \cdot \dfrac{3}{3} = \dfrac{9}{21}$

$\dfrac{14}{21} + \dfrac{9}{21} = \dfrac{14+9}{21} = \dfrac{23}{21}$

49. $2\dfrac{13}{15} - 1\dfrac{1}{5} = \dfrac{43}{15} - \dfrac{6}{5}$ LCD = 15

$\dfrac{6}{5} \cdot \dfrac{3}{3} = \dfrac{18}{15}$

$\dfrac{43}{15} - \dfrac{18}{15} = \dfrac{43-18}{15} = \dfrac{25}{15} = 1\dfrac{2}{3}$

51. $\dfrac{5}{22} - \dfrac{5}{33}$ LCD = 66

$\dfrac{5}{22} \cdot \dfrac{3}{3} = \dfrac{15}{66}$

$\dfrac{5}{33} \cdot \dfrac{2}{2} = \dfrac{10}{66}$

$\dfrac{15}{66} - \dfrac{10}{66} = \dfrac{5}{66}$

53. $\dfrac{12}{5} - 1$ LCD = 5

$1 \cdot \dfrac{5}{5} = \dfrac{5}{5}$

$\dfrac{12}{5} - \dfrac{5}{5} = \dfrac{7}{5}$

55. $1 - \dfrac{3}{10} - \dfrac{5}{10} = \dfrac{10}{10} - \dfrac{3}{10} - \dfrac{5}{10}$

$= \dfrac{10-3-5}{10} = \dfrac{2}{10}$

57. $1 - \dfrac{1}{4} - \dfrac{3}{8} = \dfrac{8}{8} - \dfrac{2}{8} - \dfrac{3}{8}$

$= \dfrac{8-2-3}{8} = \dfrac{3}{8}$

59. $\dfrac{10}{21} + \dfrac{5}{21} = \dfrac{10+5}{21} = \dfrac{15}{21}$

$\dfrac{15}{21} = \dfrac{3 \cdot 5}{3 \cdot 7} = \dfrac{5}{7}$

61. $\dfrac{10}{3} - \dfrac{5}{21}$ LCD $= 21$

$\dfrac{10}{3} \cdot \dfrac{7}{7} = \dfrac{70}{21}$

$\dfrac{70}{21} - \dfrac{5}{21} = \dfrac{65}{21}$

63. $\dfrac{2}{3} \cdot \dfrac{3}{5} = \dfrac{2 \cdot 3}{3 \cdot 5} = \dfrac{2}{5}$

65. $\dfrac{3}{4} + \dfrac{7}{12} = \dfrac{3}{4} \cdot \dfrac{12}{7} = \dfrac{3 \cdot 2 \cdot 2 \cdot 3}{2 \cdot 2 \cdot 7} = \dfrac{9}{7}$

67. $\dfrac{5}{12} + \dfrac{4}{12} = \dfrac{4+5}{12} = \dfrac{9}{12} = \dfrac{3 \cdot 3}{2 \cdot 2 \cdot 3} = \dfrac{3}{4}$

69. $5 + \dfrac{2}{3}$ LCD $= 3$

$\dfrac{5}{1} \cdot \dfrac{3}{3} = \dfrac{15}{3}$

$\dfrac{15}{3} + \dfrac{2}{3} = \dfrac{17}{3}$

71. $\dfrac{7}{8} + 3\dfrac{1}{4} = \dfrac{7}{8} + \dfrac{13}{4}$

$= \dfrac{7}{8} \cdot \dfrac{4}{13} = \dfrac{7 \cdot 2 \cdot 2}{2 \cdot 2 \cdot 2 \cdot 13} = \dfrac{7}{26}$

73. $\dfrac{7}{18} + \dfrac{14}{36} = \dfrac{7}{18} \cdot \dfrac{36}{14}$

$= \dfrac{7 \cdot 2 \cdot 2 \cdot 3 \cdot 3}{2 \cdot 3 \cdot 3 \cdot 2 \cdot 7} = 1$

75. $\dfrac{23}{105} - \dfrac{2}{105} = \dfrac{23-2}{105} = \dfrac{21}{105}$

$\dfrac{21}{105} = \dfrac{3 \cdot 7}{3 \cdot 5 \cdot 7} = \dfrac{1}{5}$

77. $1\dfrac{1}{2} + 3\dfrac{2}{3} = \dfrac{3}{2} + \dfrac{11}{3}$ LCD $= 6$

$\dfrac{3}{2} \cdot \dfrac{3}{3} = \dfrac{9}{6}$

$\dfrac{11}{3} \cdot \dfrac{2}{2} = \dfrac{22}{6}$

$\dfrac{9}{6} + \dfrac{22}{6} = \dfrac{31}{6}$

79. $\dfrac{2}{3} - \dfrac{5}{9} + \dfrac{5}{6}$ LCD $= 18$

$\dfrac{2}{3} \cdot \dfrac{6}{6} = \dfrac{12}{18}$

$\dfrac{5}{9} \cdot \dfrac{2}{2} = \dfrac{10}{18}$

$\dfrac{5}{6} \cdot \dfrac{3}{3} = \dfrac{15}{18}$

$\dfrac{12}{18} - \dfrac{10}{18} + \dfrac{15}{18} = \dfrac{12-10+15}{18} = \dfrac{17}{18}$

81. $5 + 4\dfrac{1}{8} + 15\dfrac{3}{4} + 10\dfrac{1}{2} + 15\dfrac{3}{4} + 4\dfrac{1}{8}$

$= 5 + \dfrac{33}{8} + \dfrac{63}{4} + \dfrac{21}{2} + \dfrac{63}{4} + \dfrac{33}{8}$

$= \dfrac{40}{8} + \dfrac{33}{8} + \dfrac{126}{8} + \dfrac{84}{8} + \dfrac{126}{8} + \dfrac{33}{8}$

$= \dfrac{40+33+126+84+126+33}{8}$

$= \dfrac{442}{8} = 55\dfrac{1}{4}$ feet

83. $20 - 11\dfrac{1}{2} = 20 - \dfrac{23}{2} = \dfrac{40}{2} - \dfrac{23}{2}$

$= \dfrac{40-23}{2} = \dfrac{17}{2} = 8\dfrac{1}{2}$ lbs.

85. Answers may vary.

87. $5\dfrac{1}{2} - 2\dfrac{1}{8} = \dfrac{11}{2} - \dfrac{17}{8} = \dfrac{44}{8} - \dfrac{17}{8}$

$= \dfrac{44-17}{8} = \dfrac{27}{8} = 3\dfrac{3}{8}$ miles

89. $\dfrac{3}{4}$

91. $1 - \dfrac{3}{4} - \dfrac{1}{200} = \dfrac{200}{200} - \dfrac{150}{200} - \dfrac{1}{200}$

$= \dfrac{200-150-1}{200} = \dfrac{49}{200}$

Exercise Set 1.3

1. $3^5 = 3 \cdot 3 \cdot 3 \cdot 3 \cdot 3 = 243$

3. $3^3 = 3 \cdot 3 \cdot 3 = 27$

5. $1^5 = 1 \cdot 1 \cdot 1 \cdot 1 \cdot 1 = 1$

7. $5^1 = 5$

9. $\left(\dfrac{1}{5}\right)^3 = \dfrac{1}{5}\cdot\dfrac{1}{5}\cdot\dfrac{1}{5} = \dfrac{1}{125}$

11. $\left(\dfrac{2}{3}\right)^4 = \dfrac{2}{3}\cdot\dfrac{2}{3}\cdot\dfrac{2}{3}\cdot\dfrac{2}{3} = \dfrac{16}{81}$

13. $7^2 = 7\cdot 7 = 49$

15. $2^5 = 2\cdot 2\cdot 2\cdot 2\cdot 2 = 32$

17. $0^3 = 0\cdot 0\cdot 0 = 0$

19. $4^2 = 4\cdot 4 = 16$

21. $(1.2)^2 = 1.2\cdot 1.2 = 1.44$

23. $5\cdot 5 = 5^2$ square meters

25.
	$5 + 6\cdot 2$
Multiply	$= 5 + 12$
Add	$= 17$

27.
	$4\cdot 8 - 6\cdot 2$
Multiply	$= 32 - 12$
Subtract	$= 20$

29.
	$2(8-3)$
Parentheses	$= 2(5)$
Multiply	$= 10$

31.
	$2 + (5-2) + 4^2$
Parentheses	$= 2 + 3 + 4^2$
Exponent	$= 2 + 3 + 16$
Add	$= 21$

33.
	$5\cdot 3^2$
Exponent	$= 5\cdot 9$
Multiply	$= 45$

35.
	$\dfrac{1}{4}\cdot\dfrac{2}{3} - \dfrac{1}{6}$
Multiply	$= \dfrac{2}{12} - \dfrac{1}{6}$
Simplify	$= \dfrac{1}{6} - \dfrac{1}{6}$
Subtract	$= 0$

37. $\dfrac{6-4}{9-2} = \dfrac{2}{7}$

39.
	$2[5 + 2(8-3)]$
Parentheses	$= 2[5 + 2(5)]$
Multiply inside square brackets	$= 2[5 + 10]$
Add inside square brackets	$= 2[15]$
Multiply	$= 30$

41.
	$\dfrac{3 + 3(5+3)}{3^2 + 1}$
Parentheses	$= \dfrac{3 + 3(8)}{3^2 + 1}$
Exponent	$= \dfrac{3 + 3(8)}{9 + 1}$
Multiply	$= \dfrac{3 + 24}{9 + 1}$
Add	$= \dfrac{27}{10}$

43.
	$\dfrac{6 + |8-2| + 3^2}{18 - 3}$
Simplify inside absolute value	$= \dfrac{6 + |6| + 3^2}{18 - 3}$
Absolute value	$= \dfrac{6 + 6 + 3^2}{18 - 3}$
Exponent	$= \dfrac{6 + 6 + 9}{18 - 3}$
Add and subtract	$= \dfrac{21}{15}$
Reduce	$\dfrac{21}{15} = \dfrac{3\cdot 7}{3\cdot 5} = \dfrac{7}{5}$

45. No; multiplication comes before addition in order of operations.

47. $10 + 12 = 11\cdot 2$

49. $5 + 6 > 10$

51. $3\cdot 5 > 12$

53. $8^1 = 8(1)$

55. $2^3 \; ? \; 2\cdot 3$
$8 \; ? \; 6$
$8 > 6$

57. $\left(\dfrac{2}{3}\right)^3$? $\dfrac{1}{9}+\dfrac{1}{27}$

$\dfrac{2}{3}\cdot\dfrac{2}{3}\cdot\dfrac{2}{3}$? $\dfrac{1}{9}\cdot\dfrac{3}{3}+\dfrac{1}{27}$

$\dfrac{8}{27}$? $\dfrac{3}{27}+\dfrac{1}{27}$

$\dfrac{8}{27}>\dfrac{4}{27}$

59. $(0.6)^2$? 36

$0.6\cdot 0.6$? 36

$0.36<36$

61. Answers may vary.

63.

$\dfrac{19-3\cdot 5}{6-4}$

Multiply $=\dfrac{19-15}{6-4}$

Subtract $=\dfrac{4}{2}$

Reduce $=\dfrac{4}{2}=\dfrac{2\cdot 2}{2}=2$

65.

$\dfrac{|6-2|+3}{8+2\cdot 5}$

Simplify inside absolute value $=\dfrac{|4|+3}{8+2\cdot 5}$

Absolute value $=\dfrac{4+3}{8+2\cdot 5}$

Multiply $=\dfrac{4+3}{8+10}$

Add $=\dfrac{7}{18}$

67.

$\dfrac{3(2+5)}{6+2}$

Parentheses $=\dfrac{3(7)}{6+2}$

Multiply $=\dfrac{21}{6+2}$

Add $=\dfrac{21}{8}$

69.

$5[3(0.2+0.1)+4]$

Parentheses $=5[3(0.3)+4]$

Multiply inside square brackets $=5[0.9+4]$

Add $=5[4.9]$

Multiply $=24.5$

71. $(20-4)\cdot 4+2$

73. $3^2=20+7$

75. $1+2=9+3$

77. $9\le 11\cdot 2$

79. $3\ne 4+2$

Exercise Set 1.4

1. $3x-2$
$=3(1)-2$
$=3-2$
$=1$

3. $|2x+3y|$
$=|2(1)+3(3)|$
$=|2+9|$
$=|11|$
$=11$

5. $xy+z$
$=(1)(3)+5$
$=3+5$
$=8$

7. $5y^2=5\cdot 3^2=5\cdot 9=45$

9. $\left|y^2+z^2\right|=\left|3^2+5^2\right|=|9+25|=|34|=34$

11.

t	$16t^2$
1	16
2	64
3	144
4	256

13. $x+15$

15. $x-5$

17. $3x+22$

19. $3x-6=9$
$3(5)-6\overset{?}{=}9$
$15-6\overset{?}{=}9$
$9=9$
Solution

21. $2x + 6 = 5x - 1$
$2 \cdot 0 + 6 \overset{?}{=} 5 \cdot 0 - 1$
$0 + 6 \overset{?}{=} 0 - 1$
$6 \neq -1$
Not a solution

23. $x^2 + 2x + 1 = 0$
$1^2 + 2 \cdot 1 + 1 \overset{?}{=} 0$
$1 + 2 + 1 \overset{?}{=} 0$
$4 \neq 0$
Not a solution

25. $2x - 5 = 5$
$2 \cdot 8 - 5 \overset{?}{=} 5$
$16 - 5 \overset{?}{=} 5$
$11 \neq 5$
Not a solution

27. $x + 6 = x + 6$
$2 + 6 \overset{?}{=} 2 + 6$
$8 = 8$
Solution

29. $x = 5x + 15$
$0 \overset{?}{=} 5 \cdot 0 + 15$
$0 \overset{?}{=} 0 + 15$
$0 \neq 15$
Not a solution

31. $5 + x = 20$

33. $13 - 3x = 13$

35. $\dfrac{12}{x} = \dfrac{1}{2}$

37. If $x = 2$ and $y = 6$, then
$3x + 8y = 3(2) + 8(6) = 6 + 48 = 54.$

39. If $x = 2$ and $y = 6$, then
$\dfrac{4x}{3y} = \dfrac{4(2)}{3(6)} = \dfrac{8}{18} = \dfrac{4}{9}.$

41. If $x = 2$ and $y = 6$, then
$\dfrac{y}{x} + \dfrac{y}{x} = \dfrac{6}{2} + \dfrac{6}{2} = 3 + 3 = 6.$

43. If $x = 2$ and $y = 6$, then
$3x^2 + y = 3(2)^2 + 6 = 3(4) + 6$
$= 12 + 6 = 18.$

45. If $x = 2$, $y = 6$, and $z = 3$, then
$x + yz = 2 + 6(3) = 2 + 18 = 20.$

47. If $x = 12$, $y = 8$, and $z = 4$, then
$\dfrac{x}{z} + 3y = \dfrac{12}{4} + 3(8) = 3 + 24 = 27.$

49. If $z = 4$, then
$3z + z^2 = 3(4) + 4^2 = 12 + 16 = 28.$

51. If $x = 12$ and $y = 8$, then
$x^2 - 3y + x = (12)^2 - 3(8) + 12$
$= 144 - 24 + 12 = 132.$

53. If $x = 12$, $y = 8$, and $z = 4$, then
$\dfrac{2x + z}{3y - z} = \dfrac{2(12) + 4}{3(8) - 4} = \dfrac{24 + 4}{24 - 4} = \dfrac{28}{20} = \dfrac{7}{5}.$

55. If $x = 12$, $y = 8$, and $z = 4$, then
$\dfrac{x^2 + z}{y^2 + 2z} = \dfrac{(12)^2 + 4}{(8)^2 + 2(4)} = \dfrac{144 + 4}{64 + 8}$
$= \dfrac{148}{72} = \dfrac{37}{18}.$

57. Answers may vary.

59. $\dfrac{x}{13}$

61. $2x - 10 = 18$

63. $20 - 30x$

65. $0.02x = 1.76$

67. $19 - x = 3x$

69. $P = 2l + 2w$
$P = 2(8) + 2(6)$
$P = 16 + 12$
$P = 28$ meters

71. $A = lw$
$A = (120)(100)$
$A = 12,000$ sq ft

73. $R = \dfrac{I}{PT}$
$R = \dfrac{126.75}{(650)(3)}$
$R = \dfrac{126.75}{1950} = .065 = 6.5\%$

75. $I = B + RS$
$I = 300 + (.08)(500)$

$I = 300 + 40$

$I = \$340$

77. $7(x-4)^2 - 5(y-1)^2 = 7(5-4)^2 - 5(1-1)^2$

$= 7(1)^2 - 5(0)^2 = 7 \cdot 1 - 5 \cdot 0 = 7 - 0 = 7$

79. $\dfrac{3xy}{6} - \dfrac{y^3}{12} = \dfrac{3 \cdot 5 \cdot 1}{6} - \dfrac{1^3}{12} = \dfrac{15}{6} - \dfrac{1}{12}$

$= \dfrac{30}{12} - \dfrac{1}{12} = \dfrac{29}{12}$

81. $\dfrac{3[(x+1)+(2y+5)]}{(x+1)^2} = \dfrac{3[(5+1)+(2 \cdot 1 + 5)]}{(5+1)^2}$

$= \dfrac{3[(5+1)+(2+5)]}{(5+1)^2} = \dfrac{3[6+7]}{6^2}$

$= \dfrac{3[13]}{36} = \dfrac{39}{36} = \dfrac{13}{12}$

Exercise Set 1.5

1. $6 + 3 = 9$

3. $-6 + (-8) = -14$

5. $-8 + (-7) = -15$

7. $-52 + 36 = -16$

9. $6 + (-4) + 9 = 2 + 9 = 11$

11. $2\dfrac{3}{4} + \left(-\dfrac{1}{8}\right)$

$= \dfrac{11}{4} + \left(-\dfrac{1}{8}\right)$

$= \dfrac{11}{4} \cdot \dfrac{2}{2} + \left(-\dfrac{1}{8}\right)$

$= \dfrac{22}{8} + \left(-\dfrac{1}{8}\right)$

$= \dfrac{21}{8} = 2\dfrac{5}{8}$

13. 6

The additive inverse is –6.

15. –2

The additive inverse is 2.

17. 0

The additive inverse is 0.

19. $|-6|$

$|-6| = 6$

The additive inverse is –6.

21. Answers may vary.

23. $-|-2|$

$= -(+2)$ The absolute value of –2 is 2.

$= -2$ The additive inverse of 2 is –2.

25. $-|0|$

$= -0$ The absolute value of 0 is 0.

$= 0$ The additive inverse of 0 is 0.

27. $-\left|-\dfrac{2}{3}\right|$

$= -\left(+\dfrac{2}{3}\right)$ The absolute value of $-\dfrac{2}{3}$ is $\dfrac{2}{3}$.

$= -\dfrac{2}{3}$ The additive inverse of $\dfrac{2}{3}$ is $-\dfrac{2}{3}$.

29. $-3 + (-5) = -8$

31. $-9 + (-3) = -12$

33. $+9 + (-3) = 6$

35. $-7 + (+3) = -4$

37. $-3 + (+10) = 7$

39. $8 + 4 = 12$

41. $-15 + 9 + (-2)$

$= -6 + (-2)$

$= -8$

43. $-21 + (-16) + (-22)$

$= -37 + (-22)$

$= -59$

45. $|-8| + (-16)$

$= 8 + (-16)$

$= -8$

47. $-\dfrac{7}{16} + \dfrac{1}{4}$

$= -\dfrac{7}{16} + \dfrac{1}{4} \cdot \dfrac{4}{4}$

$= -\dfrac{7}{16} + \dfrac{4}{16}$

$= -\dfrac{3}{16}$

49. $-\dfrac{7}{10}+\left(-\dfrac{3}{5}\right)$

$=-\dfrac{7}{10}+\left(-\dfrac{3}{5}\cdot\dfrac{2}{2}\right)$

$=-\dfrac{7}{10}+\left(-\dfrac{6}{10}\right)$

$=-\dfrac{13}{10}$

51. $27+(-46)=-19$

53. $-18+49=31$

55. $126+(-67)=59$

57. $6.3+(-8.4)=-2.1$

59. $117+(-79)=38$

61. $-214+(-86)=-300$

63. $-23+16+(-2)$
$=-7+(-2)$
$=-9$

65. $-9.6+(-3.5)=-13.1$

67. $|5+(-10)|$
$=|-5|$
$=5$

69. $[-2+(-7)]+[-11+(-4)]$
$=[-9]+[-15]$
$=-24$

71. $|7+(-10)|+|-16|$
$=|-3|+16$
$=3+16$
$=19$

73. Tuesday

75. $7°$

77. $\dfrac{-4°+3°+7°-2°+1°}{5}=\dfrac{5°}{5}=1°$

79. $-4+49=45°$

81. $-512\text{ ft}+658\text{ ft}=146\text{ ft}$

83. $-1\dfrac{1}{4}+\dfrac{7}{8}$

$=-\dfrac{5}{4}+\dfrac{7}{8}$

$=-\dfrac{5}{4}\cdot\dfrac{2}{2}+\dfrac{7}{8}$

$=-\dfrac{3}{8}$

85. $x+9=5$
$-4+9\overset{?}{=}5$
$5=5$
Solution

87. $2y+(-3)=-7$
$2(-1)+(-3)\overset{?}{=}-7$
$-2+(-3)\overset{?}{=}-7$
$-5\neq-7$
Not a solution

89. Answers may vary.

Exercise Set 1.6

1. $-6-4$
$=-6+(-4)$
$=-10$

3. $4-9$
$=4+(-9)$
$=-5$

5. $16-(-3)$
$=16+(+3)$
$=19$

7. $\dfrac{1}{2}-\dfrac{1}{3}=\dfrac{1}{2}\cdot\dfrac{3}{3}-\dfrac{1}{3}\cdot\dfrac{2}{2}=\dfrac{3}{6}-\dfrac{2}{6}=\dfrac{3-2}{6}=\dfrac{1}{6}$

9. $-16-(-18)=-16+18=2$

11. $8-(-5)=8+5=13$

13. $-6-(-1)=-6+1=-5$

15. $-6-(2-11)$
$=-6-[2+(-11)]$
$=-6-[-9]$
$=-6+(+9)$
$=3$

17. $3^3-8\cdot9$
$=27-72$
$=27+(-72)$
$=-45$

19. $2 - 3(8 - 6)$
$= 2 - [8 + (-6)]$
$= 2 - 3[2]$
$= 2 - 6$
$= 2 + (-6)$
$= -4$

21. $|-3| + 2^2 + [-4 - (-6)]$
$= 3 + 2^2 + [-4 + (+6)]$
$= 3 + 2^2 + [2]$
$= 3 + 4 + 2$
$= 9$

23. Sometimes positive and sometimes negative

25. $x - y$
$= -6 - (-3)$
$= -6 + (+3)$
$= -3$

27. $x + t - 12$
$= -6 + (2) - 12$
$= -6 + 2 - 12$
$= -4 - 12$
$= -4 + (-12)$
$= -16$

29. $\dfrac{x - (-12)}{y + 6}$
$= \dfrac{-6 - (-12)}{(-3) + 6}$
$= \dfrac{-6 + (+12)}{-3 + 6}$
$= \dfrac{6}{3} = 2$

31. $x - t^2$
$= -6 - (2)^2$
$= -6 - (4)$
$= -6 + (-4)$
$= -10$

33. $\dfrac{4t - y}{3t}$
$= \dfrac{4(2) - (-3)}{3(2)}$
$= \dfrac{8 - (-3)}{6}$

$= \dfrac{8 + (+3)}{6}$
$= \dfrac{11}{6}$

35. $-6 - 5$
$= -6 + (-5)$
$= -11$

37. $7 - (-4)$
$= 7 + (+4)$
$= 11$

39. $-6 - (-11)$
$= -6 + (+11)$
$= 5$

41. $6\dfrac{2}{5} - \dfrac{7}{10}$
$= \dfrac{32}{5} - \dfrac{7}{10}$
$= \dfrac{32}{5} \cdot \dfrac{2}{2} - \dfrac{7}{10}$
$= \dfrac{64}{10} - \dfrac{7}{10}$
$= \dfrac{64}{10} + \left(-\dfrac{7}{10}\right)$
$= \dfrac{57}{10} = 5\dfrac{7}{10}$

43. $-\dfrac{3}{4} - \dfrac{1}{9}$
$= -\dfrac{3}{4} \cdot \dfrac{9}{9} - \dfrac{1}{9} \cdot \dfrac{4}{4}$
$= -\dfrac{27}{36} - \dfrac{4}{36}$
$= -\dfrac{27}{36} + \left(-\dfrac{4}{36}\right)$
$= -\dfrac{31}{36}$

45. $-2.6 - (-6.7)$
$= -2.6 + (+6.7)$
$= 4.1$

47. $4 - (-6) - 9$
$= 4 + (+6) + (-9)$
$= 10 + (-9)$
$= 1$

49. $-11 - (-8) - 4$
$= -11 + (+8) + (-4)$

$= -3 + (-4)$
$= -7$

51. $16 - (-21)$
$= 16 + (+21)$
$= 37$

53. $-44 - (-27)$
$= -44 + (+27)$
$= -17$

55. $4\dfrac{2}{3} - \dfrac{5}{12}$

$= \dfrac{14}{3} - \dfrac{5}{12}$

$= \dfrac{14}{3} \cdot \dfrac{4}{4} - \dfrac{5}{12}$

$= \dfrac{56}{12} - \dfrac{5}{12}$

$= \dfrac{56}{12} + \left(-\dfrac{5}{12}\right)$

$= \dfrac{51}{12} = 4\dfrac{3}{12} = 4\dfrac{1}{4}$

57. $-\dfrac{1}{10} - \left(-\dfrac{7}{8}\right)$

$= -\dfrac{1}{10} \cdot \dfrac{4}{4} - \left(-\dfrac{7}{8} \cdot \dfrac{5}{5}\right)$

$= -\dfrac{4}{40} - \left(-\dfrac{35}{40}\right)$

$= -\dfrac{4}{40} + \left(+\dfrac{35}{40}\right)$

$= \dfrac{31}{40}$

59. $9.7 - 16.1$
$= 9.7 + (-16.1)$
$= -6.4$

61. $4.3 - (-0.87)$
$= 4.3 + (+0.87)$
$= 5.17$

63. $4 - (3[9 - (-8)] - 11)$
$= 4 - (3[9 + (+8)] - 11)$
$= 4 - (3[17] - 11)$
$= 4 - (51 - 11)$
$= 4 - [51 + (-11)]$
$= 4 - [40]$
$= 4 + (-40)$
$= -36$

65. $2 - 3 \cdot 5^2$
$= 2 - 3 \cdot 25$
$= 2 - 75$
$= 2 + (-75)$
$= -73$

67. $(2 - 3) + 5^2$
$= (-1) + 25$
$= 24$

69. $2 - (3 \cdot 5)^2$
$= 2 - (15)^2$
$= 2 - (225)$
$= 2 + (-225)$
$= -223$

71. $-5 + [(4 - 15) - (-6) - 8]$
$= -5 + [(-11) - (-6) - 8]$
$= -5 + [-11 + (+6) + (-8)]$
$= -5 + [-5 + (-8)]$
$= -5 + [-13]$
$= -18$

73. $-4 - 9$
$= -4 + (-9)$
$= -13$

75. $11 - (-14)$
$= 11 + (+14)$
$= 25$

77. $5 - 12 = 7$ feet below sea level

79. $15 - 24 = -9$

81. $14 - 77$
$= 14 + (-77)$
$= -63$ or 63 B.C.

83. -250 ft $+ 120$ ft $- 178$ ft
$= -130$ ft $- 178$ ft
$= -130$ ft $+ (-178$ ft$)$
$= -308$ ft

85. Monday

87. If $x = -5$ and $y = 4$, then
$x - y = -5 - 4 = -9$.

89. If $x = -5$, $y = 4$, and $t = 10$, then
$|x| + 2t - 8y = |-5| + 2(10) - 8(4)$
$= 5 + 20 - 32 = 25 - 32 = -7$.

91. If $x = -5$ and $y = 4$, then
$\dfrac{9 - x}{y + 6} = \dfrac{9 - (-5)}{4 + 6} = \dfrac{14}{10} = \dfrac{7}{5}$.

93. If $x = -5$ and $y = 4$, then

$y^2 - x = 4^2 - (-5) = 16 + 5 = 21$.

95. If $x = -5$ and $t = 10$, then

$$\frac{|x - (-10)|}{2t} = \frac{|-5 - (-10)|}{2(10)}$$

$$= \frac{|-5 + 10|}{20} = \frac{|5|}{20} = \frac{5}{20} = \frac{1}{4}.$$

97. Answers may vary.

99. $x - 9 = 5$
$-4 - 9 \overset{?}{=} 5$
$-13 \neq 5$
Not a solution

101. $-x + 6 = -x - 1$
$-(-2) + 6 \overset{?}{=} -(-2) - 1$
$2 + 6 \overset{?}{=} 2 - 1$
$8 \neq 1$
Not a solution

103. $-x - 13 = -15$
$-2 - 13 \overset{?}{=} -15$
$-15 = -15$
Solution

Exercise Set 1.7

1. $(-3)(+4) = -12$

3. $-6(-7) = 42$

5. $(-2)(-5)(0)$
$= 10(0)$
$= 0$

7. $2(-9) = -18$

9. $\left(-\dfrac{3}{4}\right)\left(\dfrac{8}{9}\right)$

$= \dfrac{3}{2 \cdot 2} \cdot \dfrac{2 \cdot 2 \cdot 2}{3 \cdot 3}$

$= -\dfrac{2}{3}$

11. $\left(-1\dfrac{1}{5}\right)\left(-1\dfrac{2}{3}\right)$

$= -\dfrac{6}{5} \cdot -\dfrac{5}{3}$

$= -\dfrac{2 \cdot 3}{5} \cdot -\dfrac{5}{3.}$

$= +2$

13. $(-1)(2)(-3)(-5)$
$= (-2)(-3)(-5)$
$= (+6)(-5)$
$= -30$

15. $(2)(-1)(-3)(5)(3)$
$= (-2)(-3)(5)(3)$
$= (+6)(5)(3)$
$= (30)(3)$
$= (90)$

17. $(-4)^2$
$(-4)^2 = (-4)(-4)$
$= 16$

19. True

21. False

23. $\dfrac{1}{9}$

25. $\dfrac{3}{2}$

27. $-\dfrac{1}{14}$

29. $1, -1$

31. $\dfrac{18}{-2} = -9$

33. $\dfrac{-12}{-4} = 3$

35. $\dfrac{-45}{-9} = 5$

37. $\dfrac{0}{-3} = 0$

39. $-\dfrac{3}{0}$
Undefined—division by zero is not defined.

41. $\dfrac{-6^2 + 4}{-2}$

$= \dfrac{-(6)^2 + 4}{-2}$

$= \dfrac{-(36) + 4}{-2}$

$$= \frac{-36+4}{-2}$$

$$= \frac{-32}{-2}$$

$$= 16$$

43. $\dfrac{8+(-4)^2}{4-12}$

$$= \frac{8+(16)}{4+(-12)}$$

$$= \frac{24}{-8}$$

$$= -3$$

45. $\dfrac{22+(3)(-2)}{-5-2}$

$$= \frac{22+(-6)}{-5+(-2)}$$

$$= \frac{16}{-7} = -\frac{16}{7}$$

47. $3x+2y$

$$= 3(-5)+2(-3)$$

$$= -15+(-6)$$

$$= -21$$

49. $2x^2-y^2$

$$= 2(-5)^2-(-3)^2$$

$$= 2(25)-(9)$$

$$= 50+(-9)$$

$$= 41$$

51. x^3+3y

$$= (-5)^3+3(-3)$$

$$= -125+(-9)$$

$$= -134$$

53. $\dfrac{2x-5}{y-2}$

$$= \frac{2(-5)-5}{-3-2}$$

$$= \frac{-10+(-5)}{-3-2}$$

$$= \frac{-15}{-5}$$

$$= 3$$

55. $\dfrac{6-y}{x-4}$

$$= \frac{6-(-3)}{-5-4}$$

$$= \frac{6+(+3)}{-5+(-4)}$$

$$= \frac{9}{-9}$$

$$= -1$$

57. $(-6)(-2)=12$

59. $(-7)(2)=-14$

61. $\dfrac{18}{-3}=-6$

63. $\dfrac{-6}{0}$

undefined—division by zero is not defined

65. $-\dfrac{15}{-3}$

$$= -(-5)$$

$$= 5$$

67. $\dfrac{0}{-7}=0$

69. $(-6)(3)(-2)(-1)$

$$= (-18)(-2)(-1)$$

$$= (36)(-1)$$

$$= -36$$

71. $(-5)^3=(-5)(-5)(-5)=-125$

73. $(-4)^2=(-4)(-4)=16$

75. $-4^2=-(4)(4)=-(16)=-16$

77. $\dfrac{-3-5^2}{2(-7)}$

$$= \frac{-3-25}{-14}$$

$$= \frac{-3+(-25)}{-14}$$

$$= \frac{-28}{-14}$$

$$= 2$$

79. $\dfrac{6-2(-3)}{4-3(-2)}$

$= \dfrac{6-(-6)}{4-(-6)}$

$= \dfrac{6+(+6)}{4+(+6)}$

$= \dfrac{12}{10} = \dfrac{2\cdot2\cdot3}{2\cdot5} = \dfrac{6}{5}$

81. $\dfrac{-3-2(-9)}{-15-3(-4)}$

$= \dfrac{-3-(-18)}{-15-(-12)}$

$= \dfrac{-3+(+18)}{-15+(+12)}$

$= \dfrac{15}{-3}$

$= -5$

83. $-3(2-8)$
$= -3[2+(-8)]$
$= -3[-6]$
$= 18$

85. $6(3-8)$
$= 6[3+(-8)]$
$= 6[-5]$
$= -30$

87. $-3[(2-8)-(-6-8)]$
$= -3[(2+(-8))-(-6+(-8))]$
$= -3[(-6)-(-14)]$
$= -3[-6+(+14)]$
$= -3[8]$
$= -24$

89. $\left(\dfrac{2}{5}\right)\left(-1\dfrac{1}{4}\right)$

$= \left(\dfrac{2}{5}\right)\left(-\dfrac{5}{4}\right)$

$= \dfrac{2}{5}\cdot\left(-\dfrac{5}{2\cdot2}\right)$

$= -\dfrac{1}{2}$

91. $(1.82)(-4.6) = -8.372$

93. $-22.4 \div (-1.6) = 14$

95. positive

97. Not possible to determine

99. negative

101. $-2+\dfrac{-15}{3} = -2-5 = -7$

103. $2[-5+(-3)] = 2[-8] = -16$

105. $-5x = -35$
$-5(7) \overset{?}{=} -35$
$-35 = -35$
Solution

107. $\dfrac{x}{-10} = 2$

$\dfrac{-20}{-10} \overset{?}{=} 2$

$2 = 2$
Solution

109. $-3x-5 = -20$
$-3(5)-5 \overset{?}{=} -20$
$-15-5 \overset{?}{=} -20$
$-20 = -20$
Solution

111. $38-1.5x = 20$
$x = 12$
$38-1.5(12) = 20$
$38-18 = 20$
$20 = 20$
Twelve days from Monday, September 20 is Saturday, October 2nd.

113. Answers may vary.

Exercise Set 1.8

1. $3\cdot5 = 5\cdot3$
commutative property of multiplication

3. $2+(8+5) = (2+8)+5$
associative property of addition

5. $9(3+7) = 9(3)+9(7)$
distributive property

7. $(4\cdot8)\cdot9 = 4\cdot(8\cdot9)$
associative property of multiplication

9. $0+6 = 6$
identity property of addition

11. $-4(3+7) = -4\cdot3+(-4)\cdot7$
distributive property

13. $-4 \cdot (8 \cdot 3) = (-4 \cdot 8) \cdot 3$
    associative property of multiplication

15. Answers may vary.

17. $3(6 + x)$
    $= 3(6) + 3(x)$
    $= 18 + 3x$

19. $-2(y - z)$
    $= -2(y) + (-2)(-z)$
    $= -2y + 2z$

21. $-7(3y - 5)$
    $= -7(3y) + (-7)(-5)$
    $= -21y + 35$

23. $5(x + 4m + 2)$
    $= 5(x) + 5(4m) + 5(2)$
    $= 5x + 20m + 10$

25. $-4(1 - 2m + n)$
    $= -4(1) + (-4)(-2m) + (-4)(n)$
    $= -4 + 8m - 4n$

27. $-(5x + 2)$
    $= -1(5x + 2)$
    $= -1(5x) + (-1)(2)$
    $= -5x - 2$

29. $-(r - 3 - 7p)$
    $= -1(r - 3 - 7p)$
    $= -1(r) + (-1)(-3) + (-1)(-7p)$
    $= -r + 3 + 7p$

31. $4 \cdot 1 + 4 \cdot y = 4(1 + y)$

33. $11x + 11y = 11(x + y)$

35. $(-1) \cdot 5 + (-1) \cdot x = -1(5 + x)$

37. 16
    The additive inverse is -16.

39. -8
    The additive inverse is 8.

41. $|-9| = 9$
    The additive inverse is -9.

43. $\dfrac{2}{3}$
    The additive inverse is $-\dfrac{2}{3}$.

45. $-(-1.2) = 1.2$
    The additive inverse is -1.2.

47. $-|-2| = -2$
    The additive inverse is 2.

49. $\dfrac{2}{3}$
    The multiplicative inverse is $\dfrac{3}{2}$.

51. $-\dfrac{5}{6}$
    The multiplicative inverse is $-\dfrac{6}{5}$.

53. 6
    The multiplicative inverse is $\dfrac{1}{6}$.

55. -2
    The multiplicative inverse is $-\dfrac{1}{2}$.

57. $-\left|-\dfrac{3}{5}\right| = -\left(\dfrac{3}{5}\right) = -\dfrac{3}{5}$
    The multiplicative inverse is $-\dfrac{5}{3}$.

59. $3\dfrac{5}{6} = \dfrac{23}{6}$
    The multiplicative inverse is $\dfrac{6}{23}$.

61. x: $-x$, $\dfrac{1}{x}$

63. $-3z$: $3z$, $-\dfrac{1}{3z}$

65. $a + b$: $-a - b$, $\dfrac{1}{a+b}$

67. $\dfrac{2}{3} \cdot \dfrac{3}{2}$; multiplicative inverse property
    $\dfrac{2}{3} \cdot \dfrac{3}{2} = 1$

69. $(-4)(-3)$; commutative property of multiplication
    $(-3)(-4)$

71. $3 + (8 + 9)$; associative property of addition
    $(3 + 8) + 9$

73. $y + 0$; additive identity property
$y + 0 = y$

75. $x(a + b)$; distributive property
$xa + xb$

77. $a(b + c)$; commutative property of multiplication
$(b + c)a$

79. $7(2 + x) + 5 = 14 + 7x + 5$ distributive property

                $= 7x + 14 + 5$ commutative property of addition

                $= 7x + 19$

81. $\triangle + (\square + \bigcirc) = (\square + \bigcirc) + \triangle$ commutative property of addition
      $= (\bigcirc + \square) + \triangle$ commutative property of addition
      $= \bigcirc + (\square + \triangle)$ associative property of addition

83. Answers may vary.

Exercise Set 1.9

1. $7\cent$

3. Alaska Village Electric Coop.

5. Answers may vary.

7. \$130 million

9. $176 - 110 = \$66$ million

11. The Lion King

13. \$125

15. 50 miles

17. 85 heartbeats per min

19. 95 heartbeats per min

21. Philadelphia Flyers

23. 8 years

25. $11 - 5 = 6$ years

27. 1988; \$633

29. \$533

31. latitude 30° North, longitude 90° West

33. Answers may vary.

Chapter 1 - Review

1. $8 < 10$

2. $7 > 2$

3. $-4 > -5$

4. $\dfrac{12}{2} > -8$

5. $|-7| < |-8|$

6. $|-9| > -9$

7. $-|-1| = -1$

8. $|-14| = -(-14)$

9. $1.2 > 1.02$

10. $-\dfrac{3}{2} < -\dfrac{3}{4}$

11. $4 \geq -3$

12. $6 \neq 5$

13. $0.03 < 0.3$

14. $50 > 40$

15. **a.** $\{1, 3\}$

     b. $\{0, 1, 3\}$

   c. $\{-6, 0, 1, 3\}$

   d. $\{-6, 0, 1, 1\frac{1}{2}, 3, 9.62\}$

   e. $\{\pi\}$

   f. $\{-6, 0, 1, 1\frac{1}{2}, 3, \pi, 9.62\}$

16. a. $\{2, 5\}$

   b. $\{2, 5\}$

   c. $\{-3, 2, 5\}$

   d. $\{-3, -1.6, 2, 5, \frac{11}{2}, 15.1\}$

   e. $\{\sqrt{5}, 2\pi\}$

   f. $\{-3, -1.6, 2, 5, \frac{11}{2}, 15.1, \sqrt{5}, 2\pi\}$

17. Friday

18. Wednesday

19. $36 = 2 \cdot 2 \cdot 3 \cdot 3$

20. $120 = 2 \cdot 2 \cdot 2 \cdot 3 \cdot 5$

21. $\dfrac{8}{15} \cdot \dfrac{27}{30}$

   $= \dfrac{2 \cdot 2 \cdot 2}{3 \cdot 5} \cdot \dfrac{3 \cdot 3 \cdot 3}{2 \cdot 3 \cdot 5} = \dfrac{12}{25}$

22. $\dfrac{7}{8} \div \dfrac{21}{32} = \dfrac{7}{8} \cdot \dfrac{32}{21}$

   $= \dfrac{1}{1} \cdot \dfrac{4}{3} = \dfrac{4}{3}$

23. $\dfrac{7}{15} + \dfrac{5}{6}$

   $= \dfrac{7}{15} \cdot \dfrac{2}{2} + \dfrac{5}{6} \cdot \dfrac{5}{5}$

   $= \dfrac{14}{30} + \dfrac{25}{30}$

   $= \dfrac{39}{30} = \dfrac{3 \cdot 13}{2 \cdot 3 \cdot 5} = \dfrac{13}{10}$

24. $\dfrac{3}{4} - \dfrac{3}{20} = \dfrac{15}{20} - \dfrac{3}{20}$

   $= \dfrac{12}{20} = \dfrac{3}{5}$

25. $2\dfrac{3}{4} + 6\dfrac{5}{8}$

   $= \dfrac{11}{4} \cdot \dfrac{2}{2} + \dfrac{53}{8}$

   $= \dfrac{22}{8} + \dfrac{53}{8}$

   $= \dfrac{75}{8} = 9\dfrac{3}{8}$

26. $7\dfrac{1}{6} - 2\dfrac{2}{3} = \dfrac{43}{6} - \dfrac{8}{3}$

   $= \dfrac{43}{6} - \dfrac{16}{6} = \dfrac{27}{6} = \dfrac{9}{2} = 4\dfrac{1}{2}$

27. $5 \div \dfrac{1}{3}$

   $= 5 \cdot \dfrac{3}{1} = 15$

28. $2 \cdot 8\dfrac{3}{4} = 2 \cdot \dfrac{35}{4} = \dfrac{35}{2} = 17\dfrac{1}{2}$

29. $1 - \dfrac{1}{6} - \dfrac{1}{4} = 1 \cdot \dfrac{12}{12} - \dfrac{1}{6} \cdot \dfrac{2}{2} - \dfrac{1}{4} \cdot \dfrac{3}{3}$

   $= \dfrac{12}{12} - \dfrac{2}{12} - \dfrac{3}{12} = \dfrac{12 - 2 - 3}{12} = \dfrac{7}{12}$

30. $\text{Area} = 1\dfrac{1}{3} \cdot \dfrac{7}{8} = \dfrac{4}{3} \cdot \dfrac{7}{8} = \dfrac{7}{6} = 1\dfrac{1}{6}$ sq m

   $\text{Perimeter} = \dfrac{7}{8} + \dfrac{7}{8} + 1\dfrac{1}{3} + 1\dfrac{1}{3}$

   $= \dfrac{7}{8} + \dfrac{7}{8} + \dfrac{4}{3} + \dfrac{4}{3}$

   $= \dfrac{7}{8} \cdot \dfrac{3}{3} + \dfrac{7}{8} \cdot \dfrac{3}{3} + \dfrac{4}{3} \cdot \dfrac{8}{8} + \dfrac{4}{3} \cdot \dfrac{8}{8}$

   $= \dfrac{21}{24} + \dfrac{21}{24} + \dfrac{32}{24} + \dfrac{32}{24}$

   $= \dfrac{106}{24} = 4\dfrac{5}{12}$ meters

31. Area $= \dfrac{3}{11} \cdot \dfrac{3}{11} + \dfrac{5}{11} \cdot \dfrac{5}{11}$

$= \dfrac{9}{121} + \dfrac{25}{121} = \dfrac{34}{121}$ sq in.

Perimeter $= \dfrac{5}{11} + \dfrac{5}{11} + \dfrac{3}{11} + \dfrac{2}{11} + \dfrac{3}{11} + \dfrac{3}{11} + \dfrac{5}{11}$

$= \dfrac{26}{11} = 2\dfrac{4}{11}$ in.

32. $7\dfrac{1}{2} - 6\dfrac{1}{8} = \dfrac{15}{2} - \dfrac{49}{8}$

$= \dfrac{15}{2} \cdot \dfrac{4}{4} - \dfrac{49}{8}$

$= \dfrac{60}{8} - \dfrac{49}{8}$

$= \dfrac{11}{8} = 1\dfrac{3}{8}$ ft

33. $6 \cdot 3^2 + 2 \cdot 8$
$= 6 \cdot 9 + 2 \cdot 8$
$= 54 + 16$
$= 70$

34. $68 - 5 \cdot 2^3 = 68 - 5 \cdot 8$
$= 68 - 40 = 28$

35. $3(1 + 2 \cdot 5) + 4$
$= 3(1 + 10) + 4$
$= 3(11) + 4$
$= 33 + 4 = 37$

36. $8 + 3(2 \cdot 6 - 1) = 8 + 3(12 - 1)$
$= 8 + 3(11)$
$= 8 + 33 = 41$

37. $\dfrac{4 + |6 - 2| + 8^2}{4 + 6 \cdot 4}$

$= \dfrac{4 + 4 + 64}{4 + 24}$

$= \dfrac{72}{28}$

$= \dfrac{2 \cdot 2 \cdot 2 \cdot 3 \cdot 3}{2 \cdot 2 \cdot 7}$

$= \dfrac{18}{7}$

38. $5[3(2 - 5) - 5] = 5[3(-3) - 5]$
$= 5[-9 - 5]$
$= 5[-14] = -70$

39. $20 - 12 = 2 \cdot 4$

40. $\dfrac{9}{2} > -5$

41. $2x + 3y$
$= 2(6) + 3(2)$
$= 12 + 6 = 18$

42. If $x = 6$, $y = 2$ and $z = 8$, then $x(y + 2z)$
$= 6[2 + 2(8)]$
$= 6[2 + 16]$
$= 6[18] = 108$

43. $\dfrac{x}{y} + \dfrac{z}{2y}$

$= \dfrac{6}{2} + \dfrac{8}{2(2)}$

$= 3 + \dfrac{8}{4}$

$= 3 + 2 = 5$

44. If $x = 6$ and $y = 2$, then
$x^2 - 3y^2 = 6^2 - 3(2)^2$
$= 36 - 3(4)$
$= 36 - 12 = 24$

45. $180 - 37 - 80 = 63°$

46. $7x - 3 = 18$
$7 \cdot 3 - 3 \stackrel{?}{=} 18$
$21 - 3 \stackrel{?}{=} 18$
$18 = 18$
Solution

47. $3x^2 + 4 = x - 1$
$3 \cdot 1^2 + 4 \stackrel{?}{=} 1 - 1$
$3 \cdot 1 + 4 \stackrel{?}{=} 0$
$3 + 4 \stackrel{?}{=} 0$
$7 \neq 0$
Not a solution

48. $-9 \Rightarrow 9$

49. $\dfrac{2}{3} \Rightarrow -\dfrac{2}{3}$

50. $|-2| = 2 \Rightarrow -2$

51. $-|-7| = -7 \Rightarrow 7$

52. $-15 + 4 = -11$

53. $-6 + (-11) = -17$

54. $\dfrac{1}{16} + \left(-\dfrac{1}{4}\right) = \dfrac{1}{16} - \dfrac{1}{4} \cdot \dfrac{4}{4}$

$= \dfrac{1}{16} - \dfrac{4}{16} = -\dfrac{3}{16}$

55. $-8 + |-3| = -8 + 3 = -5$

56. $-4.6 + (-9.3) = -13.9$

57. $-2.8 + 6.7 = 3.9$

58. $6 - 20 = -14$

59. $-3.1 - 8.4 = -11.5$

60. $-6 - (-11) = -6 + (+11) = 5$

61. $4 - 15 = -11$

62. $-21 - 16 + 3(8 - 2) = -21 - 16 + 3(6)$
$= -21 - 16 + 18 = -19$

63. $\dfrac{11 - (-9) + 6(8 - 2)}{2 + 3 \cdot 4} = \dfrac{11 + 9 + 6(6)}{2 + 12}$

$= \dfrac{11 + 9 + 36}{14}$

$= \dfrac{56}{14} = 4$

64. $2x^2 - y + z = 2(3)^2 - (-6) + (-9)$
$= 2(9) - (-6) + (-9)$
$= 18 - (-6) + (-9)$
$= 18 + (+6) + (-9)$
$= 24 + (-9) = 15$

65. If $x = 3$ and $y = -6$, then
$\dfrac{y - x + 5x}{2x} = \dfrac{-6 - 3 + 5(3)}{2(3)}$

$\dfrac{-6 - 3 + 15}{6} = \dfrac{6}{6} = 1$

66. $50 + 1 - 2 + 5 + 1 - 4 = \$51$

67. -6

The multiplicative inverse is $-\dfrac{1}{6}$.

68. $\dfrac{3}{5} \Rightarrow \dfrac{5}{3}$

69. $6(-8) = -48$

70. $(-2)(-14) = 28$

71. $\dfrac{-18}{-6} = 3$

72. $\dfrac{42}{-3} = -14$

73. $-3(-6)(-2) = +18(-2) = -36$

74. $(-4)(-3)(0)(-6) = (12)(0)(-6)$
$= (0)(-6) = 0$

75. $\dfrac{4 \cdot (-3) + (-8)}{2 + (-2)} = \dfrac{-12 + (-8)}{2 + (-2)}$

$= \dfrac{-20}{0}$

undefined—division by zero is not defined

76. $\dfrac{3(-2)^2 - 5}{-14} = \dfrac{3(4) - 5}{-14}$

$= \dfrac{12 - 5}{-14}$

$= \dfrac{7}{-14} = -\dfrac{1}{2}$

77. $-6 + 5 = 5 + (-6)$
commutative property of addition

78. $6 \cdot 1 = 6$
multiplicative identity property

79. $3(8 - 5) = 3(8) + 3(-5)$
distributive property

80. $4 + (-4) = 0$
additive inverse property

81. $2 + (3 + 9) = (2 + 3) + 9$
associative property of addition

82. $2 \cdot 8 = 8 \cdot 2$
commutative property of multiplication

83. $6(8 + 5) = 6(8) + 6(5)$
distributive property

84. $(3 \cdot 8) \cdot 4 = 3 \cdot (8 \cdot 4)$
associative property of multiplication

85. $4 \cdot \dfrac{1}{4} = 1$
multiplicative inverse

86. $8 + 0 = 8$
additive identity property

87. $4(8 + 3) = 4(3 + 8)$
commutative property of addition

88. $1800 million

89. $1100 - 700 = 400$ million

90. 1994

91. Revenue is increasing.

Chapter 1 - Test

1. $|-7| > 5$

2. $(9 + 5) \geq 4$

3. $-13 + 8 = -5$

4. $-13 - (-2) = -13 + (+2) = -11$

5. $6 \cdot 3 - 8 \cdot 4 = 18 - 32 = 18 + (-32) = -14$

6. $(13)(-3) = -39$

7. $(-6)(-2) = 12$

8. $\dfrac{|-16|}{-8} = \dfrac{16}{-8} = -2$

9. $\dfrac{-8}{0}$
undefined—division by zero is not defined

10. $\dfrac{|-6| + 2}{5 - 6} = \dfrac{6 + 2}{5 - 6} = \dfrac{8}{5 + (-6)} = \dfrac{8}{-1} = -8$

11. $\dfrac{1}{2} - \dfrac{5}{6} = \dfrac{1}{2} + \left(-\dfrac{5}{6}\right) = \dfrac{1}{2} \cdot \dfrac{3}{3} + \left(-\dfrac{5}{6}\right)$
$= \dfrac{3}{6} + \left(-\dfrac{5}{6}\right) = -\dfrac{2}{6} = \dfrac{2}{2 \cdot 3} = -\dfrac{1}{3}$

12. $-1\dfrac{1}{8} + 5\dfrac{3}{4} = -\dfrac{9}{8} + \dfrac{23}{4}$
$= -\dfrac{9}{8} + \dfrac{23}{4} \cdot \dfrac{2}{2} = -\dfrac{9}{8} + \dfrac{46}{8}$
$= \dfrac{37}{8} = 4\dfrac{5}{8}$

13. $-\dfrac{3}{5} + \dfrac{15}{8}$
$= -\dfrac{3}{5} \cdot \dfrac{8}{8} + \dfrac{15}{8} \cdot \dfrac{5}{5}$
$= -\dfrac{24}{40} + \dfrac{75}{40}$
$= \dfrac{51}{40}$

14. $3(-4)^2 - 80$
$= 3(16) - 80$
$= 48 - 80$
$= 48 + (-80)$
$= -32$

15. $6[5 + 2(3 - 8) - 3]$
$= 6[5 + 2(3 + (-8)) - 3]$
$= 6[5 + 2(-5) - 3]$
$= 6[5 + (-10) - 3]$
$= 6[5 + (-10) + (-3)]$
$= 6[-5 + (-3)]$
$= 6[-8]$
$= -48$

16. $\dfrac{-12 + 3 \cdot 8}{4}$
$= \dfrac{-12 + 24}{4}$
$= \dfrac{12}{4} = 3$

17. $\dfrac{(-2)(0)(-3)}{-6} = \dfrac{0}{-6} = 0$

18. $-3 > -7$

19. $4 > -8$

20. $|-3| \; ? \; 2$
$3 \; ? \; 2$
$3 > 2$
$|-3| > 2$

21. $|-2| \; ? \; -1 - (-3)$
$2 \; ? \; -1 + (+3)$
$2 \; ? \; 2$
$2 = 2$
$|-2| = -1 - (-3)$

22. a. $\{1, 7\}$

b. $\{0, 1, 7\}$

c. $\{-5, -1, 0, 1, 7\}$

d. $\{-5, -1, 0, \dfrac{1}{4}, 1, 7, 11.6\}$

e. $\{\sqrt{7}, \ 3\pi\}$

f. $\{-5, -1, 7, 0, \dfrac{1}{4}, 1, \ \sqrt{7}, \ 3\pi, 11.6\}$

23. $x^2 + y^2 = (6)^2 + (-2)^2$
$= 36 + 4 = 40$

24. $x + yz = 6 + (-2)(-3)$
$= 6 + 6 = 12$

25. $2 + 3x - y = 2 + 3(6) - (-2)$
$= 2 + 18 + (+2)$
$= 22$

26. $\dfrac{y + z - 1}{x} = \dfrac{-2 + (-3) - 1}{6}$
$= \dfrac{-2 + (-3) + (-1)}{6}$
$= \dfrac{-6}{6} = -1$

27. $8 + (9 + 3) = (8 + 9) + 3$
associative property of addition

28. $6 \cdot 8 = 8 \cdot 6$
commutative property of multiplication

29. $-6(2 + 4) = -6(2) + (-6)(4)$
distributive property

30. $\dfrac{1}{6}(6) = 1$
multiplicative inverse

31. 9
The opposite of –9 is 9.

32. –3
The reciprocal of $-\dfrac{1}{3}$ is –3.

33. Second down

34. Yes

35. $-14 + 31 = 17°$

36. $280 \cdot 1.5 = 420$
loss of $420

37. $8 billion

38. $3 billion

39. $13.5 - 8 = 5.5$
$5.5 billion

40. 1994, since the increase from 1993 to 1994 is
greater than any other year.

Chapter 2

Section 2.1 Mental Math

1. -7

3. 1

5. 17

7. like

9. unlike

11. like

Exercise Set 2.1

1. $7y + 8y$
$= (7 + 8)y$
$= 15y$

3. $8w - w + 6w$
$= (8 - 1 + 6)w$
$= 13w$

5. $3b - 5 - 10b - 4$
$= 3b - 10b - 5 - 4$
$= (3 - 10)b + (-5 - 4)$
$= -7b - 9$

7. $m - 4m + 2m - 6$
$= (1 - 4 + 2)m - 6$
$= -m - 6$

9. $5(y - 4) = 5y - 20$

11. $7(d - 3) + 10$
$= 7d - 21 + 10$
$= 7d - 11$

13. $-(3x - 2y + 1) = -3x + 2y - 1$

15. $5(x + 2) - (3x - 4)$
$= 5x + 10 - 3x + 4$
$= 5x - 3x + 10 + 4$
$= 2x + 14$

17. Answers may vary.

19. $(6x + 7) + (4x - 10)$
$= 6x + 7 + 4x - 10$
$= 6x + 4x + 7 - 10$
$= 10x - 3$

21. $(3x - 8) - (7x + 1)$
$= 3x - 8 - 7x - 1$
$= 3x - 7x - 8 - 1$
$= -4x - 9$

23. $2x - 4$;
Let x represent the unknown number.

Twice a number	decreased	four
$2x$	–	4

25. $\dfrac{3}{4}x + 12$;
Let x represent the unknown number.

Three fourths of a number	increased	twelve
$\dfrac{3}{4}x$	$+$	12

27. $-2 + 12x$;
Let x represent the unknown number

-2	sum of	5 times a number	added to	seven times a number
-2	$+$	$5x$	$+$	$7x$

29. $7x^2 + 8x^2 - 10x^2 = 5x^2$

31. $6x - 5x + x - 3 + 2x$
$= 6x - 5x + x + 2x - 3$
$= 4x - 3$

33. $-5 + 8(x - 6)$
$= -5 + 8x - 48$
$= 8x - 5 - 48$
$= 8x - 53$

35. $5g - 3 - 5 - 5g$
$= 5g - 5g - 3 - 5$
$= -8$

37. $6.2x - 4 + x - 1.2$
$= 6.2x + x - 4 - 1.2$
$= 7.2x - 5.2$

39. $2k - k - 6 = k - 6$

41. $0.5(m + 2) + 0.4m$
$= 0.5m + 1.0 + 0.4m$
$= 0.5m + 0.4m + 1.0$
$= 0.9m + 1.0$

43. $-4(3y - 4) = -12y + 16$

45. $3(2x - 5) - 5(x - 4)$
$= 6x - 15 - 5x + 20$
$= 6x - 5x - 15 + 20$
$= x + 5$

47. $3.4m - 4 - 3.4m - 7$
$= 3.4m - 3.4m - 4 - 7$
$= -11$

49. $6x + 0.5 - 4.3x - 0.4x + 3$
$= 6x - 4.3x - 0.4x + 0.5 + 3$
$= 6.0x - 4.3x - 0.4x + 0.5 + 3.0$
$= 1.3x + 3.5$

51. $-2(3x - 4) + 7x - 6$
$= -6x + 8 + 7x - 6$
$= -6x + 7x + 8 - 6$
$= x + 2$

53. $-9x + 4x + 18 - 10x$
$= -9x + 4x - 10x + 18$
$= -15x + 18$

55. $5k - (3k - 10)$
$= 5k - 3k + 10$
$= 2k + 10$

57. $(3x + 4) - (6x - 1)$
$= 3x + 4 - 6x + 1$
$= 3x - 6x + 4 + 1$
$= -3x + 5$

59. Perimeter $= 5x + (4x - 1) + 5x + (4x - 1)$
$= 5x + 4x - 1 + 5x + 4x - 1$
$= 5x + 4x + 5x + 4x - 1 - 1$
$= (18x - 2)$ feet

61. $(m - 9) - (5m - 6)$
$= m - 9 - 5m + 6$
$= m - 5m - 9 + 6$
$= -4m - 3$

63. $8x + 48$;
Let x represent the unknown number.

Eight	times	Sum of a number and six
8	$\cdot$	$(x + 6)$

$8(x + 6) = 8x + 48$

65. $x - 10$;
Let x represent the unknown number.

Double	Number	minus	Sum of the number and ten
2	$\cdot x$	$-$	$(x + 10)$

$2x - (x + 10) = 2x - x - 10 = x - 10$

67. $\dfrac{7x}{6}$;
Let x represent the number.

Seven	multiply	quotient of a number and six
7	$\cdot$	$\left(\dfrac{x}{6}\right)$

$7\left(\dfrac{x}{6}\right) = \dfrac{7}{1} \cdot \left(\dfrac{x}{6}\right) = \dfrac{7x}{6}$

69. Balanced

71. Balanced

73. $12 \cdot (x + 2) + (3x - 1)$
$= 12x + 24 + 3x - 1$
$= 12x + 3x + 24 - 1$
$= 15x + 23$
$(15x + 23)$ inches

75. $y - x^2$ when $x = -1$ and $y = 3$
$3 - (-1)^2$
$= 3 - (+1)$
$= 3 - 1$
$= 2$

77. If $a = 2$ and $b = -5$, then $a - b^2$
$= 2 - (-5)^2$
$= 2 - 25 = -23$

79. $yz - y^2$ when $y = -5$ and $z = 0$
$(-5)(0) - (-5)^2$
$= 0 - (+25)$
$= 0 - 25$
$= -25$

81. $5b^2c^3 + 8b^3c^2 - 7b^3c^2 = 5b^2c^3 + b^3c^2$

83. $3x - (2x^2 - 6x) + 7x^2$
$= 3x - 2x^2 + 6x + 7x^2$
$= -2x^2 + 7x^2 + 3x + 6x$
$= 5x^2 + 9x$

85. $-(2x^2y + 3z) + 3z - 5x^2y$
$= -2x^2y - 3z + 3z - 5x^2y$
$= -2x^2y - 5x^2y - 3z + 3z$
$= -7x^2y$

Section 2.2 Mental Math

1. $x + 4 = 6$
$x + 4 - 4 = 6 - 4$
$x = 2$

3. $n + 18 = 30$
$n + 18 - 18 = 30 - 18$
$n = 12$

5. $b - 11 = 6$
$b - 11 + 11 = 6 + 11$
$b = 17$

Exercise Set 2.2

1. $x + 11 = -2$
$x + 11 + (-11) = -2 + (-11)$
$x = -13$

3. $5y + 14 = 4y$
$5y + 14 + (-4y) = 4y + (-4y)$
$y + 14 = 0$
$y + 14 + (-14) = 0 + (-14)$
$y = -14$

5. $8x = 7x - 8$
$8x + (-7x) = 7x - 8 + (-7x)$
$x = -8$

7. $x - 2 = -4$
$x - 2 + 2 = -4 + 2$
$x = -2$

9. $\dfrac{1}{2} + f = \dfrac{3}{4}$
$\dfrac{1}{2} - \dfrac{1}{2} + f = \dfrac{3}{4} - \dfrac{1}{2}$
$f = \dfrac{3}{4} - \dfrac{2}{4}$
$f = \dfrac{1}{4}$

11. $3x - 6 = 2x + 5$
$3x - 6 - 2x = 2x + 5 - 2x$
$x - 6 = 5$
$x - 6 + 6 = 5 + 6$
$x = 11$

13. $3t - t - 7 = t - 7$
$2t - 7 = t - 7$
$2t - 7 - t = t - 7 - t$
$t - 7 = -7$
$t - 7 + 7 = -7 + 7$
$t = 0$

15. $7x + 2x = 8x - 3$
$9x = 8x - 3$
$9x - 8x = 8x - 8x - 3$
$x = -3$

17. $2y + 10 = y$
$2y - y + 10 = y - y$
$y + 10 = 0$
$y + 10 - 10 = 0 - 10$
$y = -10$

19. $y + 0.8 = 9.7$
$y + 0.8 - 0.8 = 9.7 - 0.8$
$y = 8.9$

21. $5b - 0.7 = 6b$
$5b - 5b - 0.7 = 6b - 5b$
$-0.7 = b$

23. $5x - 6 = 6x - 5$
$5x - 5x - 6 = 6x - 5x - 5$
$-6 = x - 5$
$-6 + 5 = x - 5 + 5$
$-1 = x$

25. $7t - 12 = 6t$
$7t - 6t - 12 = 6t - 6t$
$t - 12 = 0$
$t - 12 + 2 = 0 + 12$
$t = 12$

27. $y - 5y + 0.6 = 0.8 - 5y$
$-4y + 0.6 = 0.8 - 5y$
$-4y + 5y + 0.6 = 0.8 - 5y + 5y$
$y + 0.6 = 0.8$
$y + 0.6 - 0.6 = 0.8 - 0.6$
$y = 0.2$

29. Answers may vary.

31. $2(x - 4) = x + 3$
$2x - 8 = x + 3$
$2x - x - 8 = x - x + 3$
$x - 8 = 3$
$x - 8 + 8 = 3 + 8$
$x = 11$

33. $7(6 + w) = 6(2 + w)$
$42 + 7w = 12 + 6w$
$42 + 7w - 6w = 12 + 6w - 6w$
$42 + w = 12$
$42 - 42 + w = 12 - 42$
$w = -30$

35. $10 - (2x - 4) = 7 - 3x$
$10 - 2x + 4 = 7 - 3x$
$14 - 2x = 7 - 3x$
$14 - 2x + 3x = 7 - 3x + 3x$
$14 + x = 7$
$14 - 14 + x = 7 - 14$
$x = -7$

37. $-5(n - 2) = 8 - 4n$
$-5n + 10 = 8 - 4n$
$-5n + 5n + 10 = 8 - 4n + 5n$
$10 = 8 + n$
$10 - 8 = 8 - 8 + n$
$2 = n$

39. $-3(x - 4) = -4x$
$-3x + 12 = -4x$
$-3x + 4x + 12 = -4x + 4x$
$x + 12 = 0$
$x + 12 - 12 = 0 - 12$
$x = -12$

41. $3(n - 5) - (6 - 2n) = 4n$
$3n - 15 - 6 + 2n = 4n$
$5n - 21 = 4n$
$5n - 4n - 21 = 4n - 4n$
$n - 21 = 0$
$n - 21 + 21 = 0 + 21$
$n = 21$

43. $-2(t - 1) - 3t = 8 - 4t$
$-2t + 2 - 3t = 8 - 4t$
$-5t + 2 = 8 - 4t$
$-5t + 5t + 2 = 8 - 4t + 5t$
$2 = 8 + t$
$2 - 8 = 8 - 8 + t$
$-6 = t$

45. $4y - 6(y + 4) = 1 - y$
$4y - 6y - 24 = 1 - y$
$-2y - 24 = 1 - y$
$-2y + 2y - 24 = 1 - y + 2y$
$-24 = 1 + y$

$-24 - i = 1 - 1 + y$
$-25 = y$

47. $7(m - 2) - 6(m + 1) = -20$
$7m - 14 - 6m - 6 = -20$
$m - 20 = -20$
$m - 20 + 20 = -20 + 20$
$m = 0$

49. $0.8t + 0.2(t - 0.4) = 1.75$
$0.8t + 0.2t - 0.08 = 1.75$
$1.0t - 0.08 = 1.75$
$t - 0.08 + 0.08 = 1.75 + 0.08$
$t = 1.83$

51. $20 - p$

53. $(10 - x)$ feet

55. $(180 - x)°$

57. $180 - x - (2x + 7)$
$= 180 - x - 2x - 7$
$= (173 - 3x)°$

59. $(n + 284)$ votes

61. $1.23x - 0.06 = 2.6x - 0.1285$
$1.23 \cdot 0.05 - 0.06 \overset{?}{=} 2.6 \cdot 0.05 - 0.1285$
$0.0615 - 0.06 \overset{?}{=} 0.13 - 0.1285$
$0.0015 = 0.0015$
Solution

63. $3(a + 4.6) = 5a + 2.5$
$3(6.3 + 4.6) \overset{?}{=} 5 \cdot 6.3 + 2.5$
$3(10.9) \overset{?}{=} 31.5 + 2.5$
$32.7 \neq 34$
Not a solution

65. $\dfrac{5}{8} \Rightarrow \dfrac{8}{5}$

67. $2 \Rightarrow \dfrac{1}{2}$

69. $-\dfrac{1}{9} \Rightarrow -9$

71. 5 hours

73. 2 hours

75. 6 hours and 7 hours after arrival

Section 2.3 Mental Math

1. $3a = 27$

$$\frac{3a}{3} = \frac{27}{3}$$

$a = 9$

3. $5b = 10$

$$\frac{5b}{5} = \frac{10}{5}$$

$b = 2$

5. $6x = -30$

$$\frac{6x}{6} = \frac{-30}{6}$$

$x = -5$

Exercise Set 2.3

1. $-5x = 20$

$$\frac{-5x}{-5} = \frac{20}{-5}$$

$x = -4$

3. $3x = 0$

$$\frac{3x}{3} = \frac{0}{3}$$

$x = 0$

5. $-x = -12$

$$\frac{-x}{-1} = \frac{-12}{-1}$$

$x = +12$

7. $\frac{2}{3}x = -8$

$$\frac{3}{2} \cdot \frac{2}{3}x = \frac{3}{2} \cdot -8$$

$x = -12$

9. $\frac{1}{6}d = \frac{1}{2}$

$$6 \cdot \frac{1}{6}d = 6 \cdot \frac{1}{2}$$

$d = 3$

11. $\frac{a}{-2} = 1$

$$-2 \cdot \frac{a}{-2} = -2 \cdot 1$$

$a = -2$

13. $\frac{k}{7} = 0$

$$7 \cdot \frac{k}{7} = 7 \cdot 0$$

$k = 0$

15. $2x - 4 = 16$

$2x - 4 + 4 = 16 + 4$

$2x = 20$

$$\frac{2x}{2} = \frac{20}{2}$$

$x = 10$

17. $-5x + 2 = 22$

$-5x + 2 - 2 = 22 - 2$

$-5x = 20$

$$\frac{-5x}{-5} = \frac{20}{-5}$$

$x = -4$

19. $6x + 10 = -20$

$6x + 10 - 10 = -20 - 10$

$6x = -30$

$$\frac{6x}{6} = \frac{-30}{6}$$

$x = -5$

21. $-4y + 10 = -6y - 2$

$-4y + 6y + 10 = -6y + 6y - 2$

$2y + 10 = -2$

$2y + 10 - 10 = -2 - 10$

$2y = -12$

$$\frac{2y}{2} = \frac{-12}{2}$$

$y = -6$

23. $9x - 8 = 10 + 15x$

$9x - 15x - 8 = 10 + 15x - 15x$

$-6x - 8 = 10$

$-6x - 8 + 8 = 10 + 8$

$-6x = 18$

$$\frac{-6x}{-6} = \frac{18}{-6}$$

$x = -3$

25. $2x - 7 = 6x - 27$

$2x - 2x - 7 = 6x - 2x - 27$

$-7 = 4x - 27$

$-7 + 27 = 4x - 27 + 27$

$20 = 4x$

$$\frac{20}{4} = \frac{4x}{4}$$

$5 = x$

27. $6 - 2x + 8 = 10$
$14 - 2x = 10$
$14 - 14 - 2x = 10 - 14$
$-2x = -4$
$\dfrac{-2x}{-2} = \dfrac{-4}{-2}$
$x = +2$

29. $-3a + 6 + 5a = 7a - 8a$
$2a + 6 = -a$
$2a - 2a + 6 = -a - 2a$
$6 = -3a$
$\dfrac{6}{-3} = \dfrac{-3a}{-3}$
$-2 = a$

31. Answers may vary.

33. Answers may vary.

35. $-3w = 18$
$\dfrac{-3w}{-3} = \dfrac{18}{-3}$
$w = -6$

37. $-0.2z = -0.8$
$\dfrac{-0.2z}{-0.2} = \dfrac{-0.8}{-0.2}$
$z = 4$

39. $-h = -\dfrac{3}{4}$

$-1 \cdot -h = -1 \cdot -\dfrac{3}{4}$

$h = \dfrac{3}{4}$

41. $6a + 3 = 3$
$6a + 3 - 3 = 3 - 3$
$6a = 0$
$\dfrac{6a}{6} = \dfrac{0}{6}$
$a = 0$

43. $5 - 0.3k = 5$
$5 - 5 - 0.3k = 5 - 5$
$-0.3k = 0$
$\dfrac{-0.3k}{-0.3} = \dfrac{0}{-0.3}$
$k = 0$

45. $2x + \dfrac{1}{2} = \dfrac{7}{2}$

$2x + \dfrac{1}{2} - \dfrac{1}{2} = \dfrac{7}{2} - \dfrac{1}{2}$

$2x = \dfrac{6}{2}$

$2x = 3$

$\dfrac{2x}{2} = \dfrac{3}{2}$

$x = \dfrac{3}{2}$

47. $\dfrac{x}{3} - 2 = 5$

$\dfrac{x}{3} - 2 + 2 = 5 + 2$

$\dfrac{x}{3} = 7$

$3 \cdot \dfrac{x}{3} = 3 \cdot 7$

$x = 21$

49. $10 = 2x - 1$
$10 + 1 = 2x - 1 + 1$
$11 = 2x$
$\dfrac{11}{2} = \dfrac{2x}{2}$
$\dfrac{11}{2} = x$

51. $4 - 12x = 7$
$4 - 4 - 12x = 7 - 4$
$-12x = 3$
$\dfrac{-12x}{-12} = \dfrac{3}{-12}$
$x = -\dfrac{1}{4}$

53. $-\dfrac{2}{3}x = \dfrac{5}{9}$

$-\dfrac{3}{2} \cdot \dfrac{-2}{3}x = -\dfrac{3}{2} \cdot \dfrac{5}{9}$

$x = -\dfrac{3}{2} \cdot \dfrac{5}{9}$

$x = -\dfrac{5}{6}$

55. $10 = -6n + 16$
$10 - 16 = -6n + 16 - 16$
$-6 = -6n$
$\dfrac{-6}{-6} = \dfrac{-6n}{-6}$
$+1 = n$

57. $z - 5z = 7z - 9 - z$
$-4z = 6z - 9$
$-4z - 6z = 6z - 6z - 9$
$-10z = -9$
$\dfrac{-10z}{-10} = \dfrac{-9}{-10}$
$z = \dfrac{9}{10}$

59. $5x + 20 = 8 - x$
$5x + x + 20 = 8 - x + x$
$6x + 20 = 8$
$6x + 20 - 20 = 8 - 20$
$6x = -12$
$\dfrac{6x}{6} = \dfrac{-12}{6}$
$x = -2$

61. $5y - y = 2y - 14$
$4y = 2y - 14$
$4y - 2y = 2y - 2y - 14$
$2y = -14$
$\dfrac{2y}{2} = \dfrac{-14}{2}$
$y = -7$

63. $6z - 8 - z + 3 = 0$
$5z - 5 = 0$
$5z - 5 + 5 = 0 + 5$
$5z = 5$
$\dfrac{5z}{5} = \dfrac{5}{5}$
$z = 1$

65. $10 - n - 2 = 2n + 2$
$8 - n = 2n + 2$
$8 - n + n = 2n + n + 2$
$8 = 3n + 2$
$8 - 2 = 3n + 2 - 2$
$6 = 3n$
$\dfrac{6}{3} = \dfrac{3n}{3}$
$2 = n$

67. $0.4x - 0.6x - 5 = 1$
$-0.2x - 5 = 1$
$-0.2x - 5 + 5 = 1 + 5$
$-0.2x = 6$

$\dfrac{-0.2x}{-0.2} = \dfrac{6}{-0.2}$
$x = -30$

69. $x + 2$

71. $x + (x + 2) = 2x + 2$

73. $x + (x + 2) = 2x + 2$

75. $x + (x + 2) + (x + 4) = 3x + 6$

77. $-3.6x = 10.62$
$\dfrac{-3.6x}{-3.6} = \dfrac{10.62}{-3.6}$
$x = -2.95$

79. $7x - 5.06 = -4.92$
$7x - 5.06 + 5.06 = -4.92 + 5.06$
$7x = 0.14$
$\dfrac{7x}{7} = \dfrac{0.14}{7}$
$x = 0.02$

81. $5x + 2(x - 6)$
$= 5x + 2x - 12$
$= 7x - 12$

83. $6(2z + 4) + 20$
$= 12z + 24 + 20$
$= 12z + 44$

85. $-(x - 1) + x$
$= -x + 1 + x = 1$

87. $(-3)^2 > -3^2$
$9 > -9$

89. $(-2)^3 = -2^3$
$-8 = -8$

91. $-|-6| < 6$
$-6 < 6$

Exercise Set 2.4

1. $-2(3x - 4) = 2x$
$-6x + 8 = 2x$
$-6x - 2x + 8 = 2x - 2x$
$-8x + 8 = 0$
$-8x + 8 - 8 = 0 - 8$
$-8x = -8$
$\dfrac{-8x}{-8} = \dfrac{-8}{-8}$
$x = 1$

3. $4(2n - 1) = (6n + 4) + 1$

$8n - 4 = 6n + 4 + 1$

$8n - 4 = 6n + 5$

$8n - 6n - 4 = 6n - 6n + 5$

$2n - 4 = 5$

$2n - 4 + 4 = 5 + 4$

$2n = 9$

$\dfrac{2n}{2} = \dfrac{9}{2}$

$n = \dfrac{9}{2}$

5. $5(2x - 1) - 2(3x) = 4$

$10x - 5 - 6x = 4$

$4x - 5 = 4$

$4x - 5 + 5 = 4 + 5$

$4x = 9$

$\dfrac{4x}{4} = \dfrac{9}{4}$

$x = \dfrac{9}{4}$

7. $6(x - 3) + 10 = -8$

$6x - 18 + 10 = -8$

$6x - 8 = -8$

$6x - 8 + 8 = -8 + 8$

$6x = 0$

$\dfrac{6x}{6} = \dfrac{0}{6}$

$x = 0$

9. $\dfrac{3}{4}x - \dfrac{1}{2} = 1$

$4\left(\dfrac{3}{4}x - \dfrac{1}{2}\right) = 4 \cdot 1$

$4\left(\dfrac{3}{4}x\right) - 4\left(\dfrac{1}{2}\right) = 4$

$3x - 2 = 4$

$3x - 2 + 2 = 4 + 2$

$3x = 6$

$\dfrac{3x}{3} = \dfrac{6}{3}$

$x = 2$

11. $x + \dfrac{5}{4} = \dfrac{3}{4}x$

$4\left(x + \dfrac{5}{4}\right) = 4\left(\dfrac{3}{4}x\right)$

$4(x) + 4\left(\dfrac{5}{4}\right) = 4\left(\dfrac{3}{4}x\right)$

$4x + 5 = 3x$

$4x - 3x + 5 = 3x - 3x$

$x + 5 = 0$

$x + 5 - 5 = 0 - 5$

$x = -5$

13. $\dfrac{x}{2} - 1 = \dfrac{x}{5} + 2$

$10\left(\dfrac{x}{2} - 1\right) = 10\left(\dfrac{x}{5} + 2\right)$

$10\left(\dfrac{x}{2}\right) - 10 \cdot 1 = 10\left(\dfrac{x}{5}\right) + 10 \cdot 2$

$5x - 10 = 2x + 20$

$5x - 2x - 10 = 2x - 2x + 20$

$3x - 10 = 20$

$3x - 10 + 10 = 20 + 10$

$3x = 30$

$\dfrac{3x}{3} = \dfrac{30}{3}$

$x = 10$

15. $\dfrac{6(3 - z)}{5} = -z$

$5 \cdot \dfrac{6(3 - z)}{5} = 5(-z)$

$6(3 - z) = -5z$

$18 - 6z = -5z$

$18 - 6z + 6z = -5z + 6z$

$18 = z$

17. $\dfrac{2(x + 1)}{4} = 3x - 2$

$4 \cdot \dfrac{2(x + 1)}{4} = 4(3x - 2)$

$2(x + 1) = 12x - 8$

$2x + 2 = 12x - 8$

$2x - 2x + 2 = 12x - 2x - 8$

$2 = 10x - 8$

$2 + 8 = 10x - 8 + 8$

$10 = 10x$

$\dfrac{10}{10} = \dfrac{10x}{10}$

$1 = x$

19. $0.50x + 0.15(70) = 0.25(142)$

$0.50x + 10.5 = 35.5$

$0.50x + 10.5 - 10.5 = 35.5 - 10.5$

$0.50x = 25$

$\dfrac{0.50x}{0.50} = \dfrac{25}{0.50}$

$x = 50$

21. $0.12(y - 6) + 0.06y = 0.08y - 0.07(10)$
$0.12y - 0.72 + 0.06y = 0.08y - 0.7$
$0.18y - 0.72 = 0.08y - 0.7$
$0.18y - 0.08y - 0.72 = 0.08y - 0.08y - 0.7$
$0.1y - 0.72 = -0.7$
$0.1y - 0.72 + 0.72 = -0.7 + 0.72$
$0.1y = 0.02$
$\dfrac{0.1y}{0.1} = \dfrac{0.02}{0.1}$
$y = 0.2$

23. $5x - 5 = 2(x + 1) + 3x - 7$
$5x - 5 = 2x + 2 + 3x - 7$
$5x - 5 = 5x - 5$
$5x - 5x - 5 = 5x - 5x - 5$
$-5 = -5$
$-5 + 5 = -5 + 5$
$0 = 0$
All real numbers

25. $\dfrac{x}{4} + 1 = \dfrac{x}{4}$

$4\left(\dfrac{x}{4} + 1\right) = 4\left(\dfrac{x}{4}\right)$

$4\left(\dfrac{x}{4}\right) + 4(1) = 4\left(\dfrac{x}{4}\right)$

$x + 4 = x$
$x - x + 4 = x - x$
$4 = 0$
No solution

27. $3x - 7 = 3(x + 1)$
$3x - 7 = 3x + 3$
$3x - 3x - 7 = 3x - 3x + 3$
$-7 = 3$
No solution

29. Answers may vary.

31. Answers may vary.

33. $4x + 3 = 2x + 11$
$4x - 2x + 3 = 2x - 2x + 11$
$2x + 3 = 11$
$2x + 3 - 3 = 11 - 3$
$2x = 8$
$\dfrac{2x}{2} = \dfrac{8}{2}$
$x = 4$

35. $-2y - 10 = 5y + 18$
$-2y + 2y - 10 = 5y + 2y + 18$
$-10 = 7y + 18$
$-10 - 18 = 7y + 18 - 18$
$-28 = 7y$

$\dfrac{-28}{7} = \dfrac{7y}{7}$
$-4 = y$

37. $6x - 1 = 5x + 2$
$6x - 5x - 1 = 5x - 5x + 2$
$x - 1 = 2$
$x - 1 + 1 = 2 + 1$
$x = 3$

39. $2y + 2 = y$
$2y - 2y + 2 = y - 2y$
$2 = -y$
$\dfrac{2}{-1} = \dfrac{-y}{-1}$
$-2 = y$

41. $3(5c - 1) - 2 = 13c + 3$
$15c - 3 - 2 = 13c + 3$
$15c - 13c - 5 = 13c - 13c + 3$
$2c - 5 = 3$
$2c - 5 + 5 = 3 + 5$
$2c = 8$
$\dfrac{2c}{2} = \dfrac{8}{2}$
$c = 4$

43. $x + \dfrac{7}{6} = 2x - \dfrac{7}{6}$

$6\left(x + \dfrac{7}{6}\right) = 6\left(2x - \dfrac{7}{6}\right)$

$6(x) + 6\left(\dfrac{7}{6}\right) = 6(2x) + 6\left(-\dfrac{7}{6}\right)$

$6x + 7 = 12x - 7$
$6x - 12x + 7 = 12x - 12x - 7$
$-6x + 7 = -7$
$-6x + 7 - 7 = -7 - 7$
$-6x = -14$
$\dfrac{-6x}{-6} = \dfrac{-14}{-6}$
$x = \dfrac{14}{6}$
$x = \dfrac{7}{3}$

45. $2(x - 5) = 7 + 2x$
$2x - 10 = 7 + 2x$
$2x - 2x - 10 = 7 + 2x - 2x$
$-10 = 7$
No solution

47. $\dfrac{2(z+3)}{3} = 5 - z$

$3 \cdot \dfrac{2(z+3)}{3} = 3(5-z)$

$2(z+3) = 3(5-z)$

$2z + 6 = 15 - 3z$

$2z + 3z + 6 = 15 - 3z + 3z$

$5z + 6 = 15$

$5z + 6 - 6 = 15 - 6$

$5z = 9$

$\dfrac{5z}{5} = \dfrac{9}{5}$

$z = \dfrac{9}{5}$

49. $\dfrac{4(y-1)}{5} = -3y$

$5 \cdot \dfrac{4(y-1)}{5} = 5(-3y)$

$4(y-1) = 5(-3y)$

$4y - 4 = -15y$

$4y - 4y - 4 = -15y - 4y$

$-4 = -19y$

$\dfrac{-4}{-19} = \dfrac{-19y}{-19}$

$\dfrac{4}{19} = y$

51. $8 - 2(a-1) = 7 + a$

$8 - 2a + 2 = 7 + a$

$-2a + 10 = 7 + a$

$-2a - a + 10 = 7 + a - a$

$-3a + 10 = 7$

$-3a + 10 - 10 = 7 - 10$

$-3a = -3$

$\dfrac{-3a}{-3} = \dfrac{-3}{-3}$

$a = +1$

53. $2(x+3) - 5 = 5x - 3(1+x)$

$2x + 6 - 5 = 5x - 3 - 3x$

$2x + 1 = 2x - 3$

$2x - 2x + 1 = 2x - 2x - 3$

$1 = -3$

No solution

55. $\dfrac{5x-7}{3} = x$

$3 \cdot \dfrac{5x-7}{3} = 3(x)$

$5x - 7 = 3x$

$5x - 5x - 7 = 3x - 5x$

$-7 = -2x$

$\dfrac{-7}{-2} = \dfrac{-2x}{-2}$

$\dfrac{7}{2} = x$

57. $\dfrac{9+5v}{2} = 2v - 4$

$2 \cdot \dfrac{9+5v}{2} = 2(2v-4)$

$9 + 5v = 4v - 8$

$9 + 5v - 4v = 4v - 4v - 8$

$9 + v = -8$

$9 - 9 + v = -8 - 9$

$v = -17$

59. $-3(t-5) + 2t = 5t - 4$

$-3t + 15 + 2t = 5t - 4$

$-t + 15 = 5t - 4$

$-t + t + 15 = 5t + t - 4$

$15 = 6t - 4$

$15 + 4 = 6t - 4 + 4$

$19 = 6t$

$\dfrac{19}{6} = \dfrac{6t}{6}$

$\dfrac{19}{6} = t$

61. $0.02(6t-3) = 0.05(t-2) + 0.02$

$0.12t - 0.06 = 0.05t - 0.10 + 0.02$

$0.12t - 0.06 = 0.05t - 0.08$

$0.12t - 0.05t - 0.06 = 0.05t - 0.05t - 0.08$

$0.07t - 0.06 = -0.08$

$0.07t - 0.06 + 0.06 = -0.08 + 0.06$

$0.07t = -0.02$

$\dfrac{0.07t}{0.07} = \dfrac{-0.02}{0.07}$

$t = \dfrac{-0.02}{0.07} = -\dfrac{2}{7}$

63. $0.06 - 0.01(x+1) = -0.02(2-x)$

$0.06 - 0.01x - 0.01 = -0.04 + 0.02x$

$0.05 - 0.01x = -0.04 + 0.02x$

$0.05 + 0.04 - 0.01x = -0.04 + 0.04 + 0.02x$

$0.09 - 0.01x = 0.02x$

$0.09 - 0.01x + 0.01x = 0.02x + 0.01x$

$0.09 = 0.03x$

$\dfrac{0.09}{0.03} = \dfrac{0.03x}{0.03}$

$3 = x$

65. $\dfrac{3(x-5)}{2} = \dfrac{2(x+5)}{3}$

$6 \cdot \dfrac{3(x-5)}{2} = 6 \cdot \dfrac{2(x+5)}{3}$

$3 \cdot \dfrac{3(x-5)}{1} = 2 \cdot \dfrac{2(x+5)}{1}$

$9(x-5) = 4(x+5)$

$9x - 45 = 4x + 20$

$9x - 4x - 45 = 4x - 4x + 20$

$5x - 45 = 20$

$5x - 45 + 45 = 20 + 45$

$5x = 65$

$\dfrac{5x}{5} = \dfrac{65}{5}$

$x = 13$

67. $1000(7x - 10) = 50(412 + 100x)$

$7000x - 10{,}000 = 20{,}600 + 5000x$

$7000x - 5000x - 10{,}000 = 20{,}600 + 5000x - 5000x$

$2000x - 10{,}000 = 20{,}600$

$2000x - 10{,}000 + 10{,}000 = 20{,}600 + 10{,}000$

$2000x = 30{,}600$

$\dfrac{2000x}{2000} = \dfrac{30{,}600}{2000}$

$x = 15.3$

69. $0.035x + 5.112 = 0.010x + 5.107$

$0.035x - 0.010x + 5.112 = 0.010x - 0.010x + 5.107$

$0.025x + 5.112 = 5.107$

$0.025x + 5.112 - 5.112 = 5.107 - 5.112$

$0.025x = -0.005$

$\dfrac{0.025x}{0.025} = \dfrac{-0.005}{0.025}$

$x = -0.2$

71. Let x represent the number.

$2x + \dfrac{1}{5} = 3x - \dfrac{4}{5}$

$5\left(2x + \dfrac{1}{5}\right) = 5\left(3x - \dfrac{4}{5}\right)$

$5(2x) + 5\left(\dfrac{1}{5}\right) = 5(3x) + 5\left(-\dfrac{4}{5}\right)$

$10x + 1 = 15x - 4$

$10x - 15x + 1 = 15x - 15x - 4$

$-5x + 1 = -4$

$-5x + 1 - 1 = -4 - 1$

$-5x = -5$

$\dfrac{-5x}{-5} = \dfrac{-5}{-5}$

$x = +1$

73. $2x + 7 = x + 6$

$2x - x + 7 = x - x + 6$

$x + 7 = 6$

$x + 7 - 7 = 6 - 7$

$x = -1$

75. $3x - 6 = 2x + 8$

$3x - 2x - 6 = 2x - 2x + 8$

$x - 6 = 8$

$x - 6 + 6 = 8 + 6$

$x = 14$

77. $\dfrac{1}{3}x = \dfrac{5}{6}$

$\dfrac{3}{1} \cdot \dfrac{1}{3}x = \dfrac{5}{6} \cdot \dfrac{3}{1}$

$x = \dfrac{5}{2} = 2\dfrac{1}{2}$

79. $x - 4 = 2x$

$x - x - 4 = 2x - x$

$-4 = x$

81. $\dfrac{x}{4} + \dfrac{1}{2} = \dfrac{3}{4}$

$\dfrac{x}{4} + \dfrac{1}{2} - \dfrac{1}{2} = \dfrac{3}{4} - \dfrac{1}{2}$

$\dfrac{x}{4} = \dfrac{3}{4} - \dfrac{1}{2} \cdot \dfrac{2}{2}$

$\dfrac{x}{4} = \dfrac{3}{4} - \dfrac{2}{4}$

$\dfrac{x}{4} = \dfrac{1}{4}$

$4 \cdot \dfrac{x}{4} = \dfrac{1}{4} \cdot 4$

$x = 1$

83. $x + x + x + 2x + 2x = 28$

$7x = 28$

$\dfrac{7x}{7} = \dfrac{28}{7}$

$x = 4 \text{ cm}$

$2x = 8 \text{ cm}$

85. $10 - 5x = 3x$

$10 - 5x + 5x = 3x + 5x$

$10 = 8x$

$\dfrac{10}{8} = \dfrac{8x}{8}$

$\dfrac{5}{4} = x$

87. Midway

89. $x + 55 = 2x - 90$
$x - x + 55 = 2x - x - 90$
$55 = x - 90$
$55 + 90 = x - 90 + 90$
$145 = x$
145 cities, towns, or villages

91. $\left|2^3 - 3^2\right| - |5 - 7|$
$= |8 - 9| - |5 - 7|$
$= |-1| - |-2|$
$= 1 - 2 = -1$

93. $\dfrac{5}{4 + 3 \cdot 7} = \dfrac{5}{4 + 21} = \dfrac{5}{25} = \dfrac{1}{5}$

95. $x + (2x - 3) + (3x - 5) = 6x - 8$
$(6x - 8)$ meters

97. $x(x - 3) = x^2 + 5x + 7$
$x^2 - 3x = x^2 + 5x + 7$
$x^2 - x^2 - 3x = x^2 - x^2 + 5x + 7$
$-3x = 5x + 7$
$-3x - 5x = 5x - 5x + 7$
$-8x = 7$
$\dfrac{-8x}{-8} = \dfrac{7}{-8}$
$x = -\dfrac{7}{8}$

99. $2z(z + 6) = 2z^2 + 12z - 8$
$2z^2 + 12z = 2z^2 + 12z - 8$
$2z^2 - 2z^2 + 12z = 2z^2 - 2z^2 + 12z - 8$
$12z = 12z - 8$
$12z - 12z = 12z - 12z - 8$
$0 = -8$
No solution

101. $n(3 + n) = n^2 + 4n$
$3n + n^2 = n^2 + 4n$
$3n + n^2 - n^2 = n^2 - n^2 + 4n$
$3n = 4n$
$3n - 3n = 4n - 3n$
$0 = n$

Exercise Set 2.5

1. Let x = salary of the governor of Nebraska
$2x + x = 195,000$
$3x = 195,000$
$\dfrac{3x}{3} = \dfrac{195,000}{3}$
$x = 65,000$
$2x = 130,000$
Nebraska: $65,000
New York: $130,000

3. $x + 2x + 5x = 40$
$8x = 40$
$\dfrac{8x}{8} = \dfrac{40}{8}$
$x = 5$
$2x = 10$
$5x = 25$
1st: 5 in.
2nd: 10 in.
3rd: 25 in.

5. $(2x) \cdot 3 = 5x - \dfrac{3}{4}$
$6x = 5x - \dfrac{3}{4}$
$6x - 5x = 5x - 5x - \dfrac{3}{4}$
$x = -\dfrac{3}{4}$

7. $3(x + 5) = 2x - 1$
$3x + 15 = 2x - 1$
$3x - 2x + 15 = 2x - 2x - 1$
$x + 15 = -1$
$x + 15 - 15 = -1 - 15$
$x = -16$

9. Let x = number of miles
$2(24.95) + 0.29x = 100$
$49.90 + 0.29x = 100$
$0.29x = 50.1$
$x = 172$
172 miles

11. $x + (2x + 2) = 17$
$3x + 2 = 17$
$3x + 2 - 2 = 17 - 2$
$3x = 15$
$\dfrac{3x}{3} = \dfrac{15}{3}$
$x = 5$
$2x + 2 = 12$

Shorter piece: 5 ft
Longer piece: 12 ft

13. $x + 3x = 180$
$4x = 180$
$$\frac{4x}{4} = \frac{180}{4}$$
$x = 45$
$3x = 135$
$45°$ and $135°$

15. Let x = score for Phoenix Suns
$x + (x + 1) = 197$
$2x + 1 = 197$
$2x + 1 - 1 = 197 - 1$
$2x = 196$
$$\frac{2x}{2} = \frac{196}{2}$$
$x = 98$
$x + 1 = 99$
Bulls: 99
Suns: 98

17. Let x = 1st odd integer
$7x = 54 + 5(x + 2)$
$7x = 54 + 5x + 10$
$7x = 64 + 5x$
$2x = 64$
$x = 32$
However, 32 is not odd. Therefore, there are none.

19. $2(x - 8) = 3(x + 3)$
$2x - 16 = 3x + 9$
$2x - 2x - 16 = 3x - 2x + 9$
$-16 = x + 9$
$-16 - 9 = x + 9 - 9$
$-25 = x$

21. Let x = smaller integer
$2(x + 2) = 15 + 3x$
$2x + 4 = 15 + 3x$
$2x - 2x + 4 = 15 + 3x - 2x$
$4 = 15 + x$
$4 - 15 = 15 - 15 + x$
$-11 = x$
$-9 = x + 2$
The integers are -11 and -9.

23. Let x = height
$x + (2x - 19) = 83$
$3x - 19 = 83$
$3x - 19 + 19 = 83 + 19$
$3x = 102$
$$\frac{3x}{3} = \frac{102}{3}$$
$x = 34$
$2x - 19 = 49$

height: 34 in.
diameter: 49 in.

25. $x + 2x = 15{,}000$
$3x = 15{,}000$
$$\frac{3x}{3} = \frac{15{,}000}{3}$$
$x = 5000$
$2x = 10{,}000$
Son: \$5000
Husband: \$10,000

27. Let x = one angle
$2x + 2(2x - 15) = 360$
$2x + 4x - 30 = 360$
$6x - 30 = 360$
$6x - 30 + 30 = 360 + 30$
$6x = 390$
$$\frac{6x}{6} = \frac{390}{6}$$
$x = 65$
$2x - 15 = 115$
The angles are $65°$ and $115°$.

29. Let x = smallest angle
$x + (x + 2) + (x + 4) = 180$
$3x + 6 = 180$
$3x + 6 - 6 = 180 - 6$
$3x = 174$
$$\frac{3x}{3} = \frac{174}{3}$$
$x = 58$
$x + 2 = 60$
$x + 4 = 62$
The angles are $58°$, $60°$, and $62°$.

31. Let x = measure of first angle
$x + 2x + 3x = 180$
$6x = 180$
$$\frac{6x}{6} = \frac{180}{6}$$
$x = 30$
$2x = 60$
$3x = 90$
The angles are $30°$, $60°$, and $90°$.

33. Answers may vary.

35. Answers may vary.

37. $3 + (-7) = 3 - 7 = -4$

39. $4 - 10 = -6$

41. $-5 - (-1) = -5 + 1 = -4$

43. $\dfrac{1}{2}(x-1)=37$

45. $\dfrac{3(x+2)}{5}=0$

Exercise Set 2.6

1. $A=bh$
$45=15h$
$\dfrac{45}{15}=\dfrac{15h}{15}$
$3=h$

3. $S=4lw+2wh$
$102=4(7)(3)+2(3)h$
$102=84+6h$
$102-84=84-84+6h$
$18=6h$
$\dfrac{18}{6}=\dfrac{6h}{6}$
$3=h$

5. $C=2\pi r$
$15.7=2(3.14)r$
$15.7=6.28r$
$\dfrac{15.7}{6.28}=\dfrac{6.28r}{6.28}$
$2.5=r$

7. $I=PRT$
$3750=(25,000)(0.05)T$
$3750=1250T$
$\dfrac{3750}{1250}=\dfrac{1250T}{1250}$
$3=T$

9. $A=\dfrac{1}{2}(B+b)h$
$180=\dfrac{1}{2}(11+7)h$
$180=\dfrac{1}{2}(18)h$
$180=9h$
$\dfrac{180}{9}=\dfrac{9h}{9}$
$20=h$

11. $P=a+b+c$
$30=8+10+c$
$30=18+c$
$30-18=18-18+c$
$12=c$

13. $V=\dfrac{1}{3}\pi r^2 h$
$565.2=\dfrac{1}{3}(3.14)(6)^2 h$
$565.2=\dfrac{1}{3}(3.14)(36)h$
$565.2=37.68h$
$\dfrac{565.2}{37.68}=\dfrac{37.68h}{37.68}$
$15=h$

15. $f=5gh$
$\dfrac{f}{5g}=\dfrac{5gh}{5g}$
$\dfrac{f}{5g}=h$

17. $V=LWH$
$\dfrac{V}{LH}=\dfrac{LWH}{LH}$
$\dfrac{V}{LH}=W$

19. $3x+y=7$
$3x-3x+y=7-3x$
$y=7-3x$

21. $A=p+PRT$
$A-p=p-p+PRT$
$A-p=PRT$
$\dfrac{A-p}{PT}=\dfrac{PRT}{PT}$
$\dfrac{A-p}{PT}=R$

23. $V=\dfrac{1}{3}Ah$
$3(V)=3\left(\dfrac{1}{3}Ah\right)$
$3V=Ah$
$\dfrac{3V}{h}=\dfrac{Ah}{h}$
$\dfrac{3V}{h}=A$

25. $P=a+b+c$
$P-b=a+b-b+c$
$P-b=a+c$
$P-b-c=a+c-c$
$P-b-c=a$

27. $S = 2\pi rh + 2\pi r^2$

$S - 2\pi r^2 = 2\pi rh + 2\pi r^2 - 2\pi r^2$

$S - 2\pi r^2 = 2\pi rh$

$\dfrac{S - 2\pi r^2}{2\pi r} = \dfrac{2\pi rh}{2\pi r}$

$\dfrac{S - 2\pi r^2}{2\pi r} = h$

29. $V = LWH$

$V = (2L)(2W)(2H)$

$V = 8LWH$

It multiplies the volume by 8.

31. $d = rt$

$d = (470)(5.5)$

$d = 2585$

Miles earned = 3000

33. $d = rt$

$93,000,000 = (186,000)t$

$\dfrac{93,000,000}{186,000} = \dfrac{186,000t}{186,000}$

$500 = t$

500 seconds or $8\dfrac{1}{3}$ minutes

35. $C = 2\pi r$

$C = 2(3.14)(4000)$

$C = 25,120$

25,120 miles of rope

37. $F = \dfrac{9}{5}C + 32$

$F = \dfrac{9}{5}(-78.5) + 32$

$F = -141.3 + 32$

$F = -109.3$

$-109.3°$ F

39. $F = \dfrac{9}{5}C + 32$

$14 = \dfrac{9}{5}C + 32$

$-18 = \dfrac{9}{5}C$

$-90 = 9C$

$\dfrac{-90}{9} = \dfrac{9C}{9}$

$-10° = C$

41. $d = rt$

$3000 = r\left(1\dfrac{1}{2}\right)$

$3000 = \dfrac{3}{2}r$

$3000 \cdot \dfrac{2}{3} = \dfrac{3}{2} \cdot \dfrac{2}{3}r$

$2000 = r$

2000 mph

43. $V = lwh$

$V = 10(8)(10)$

$V = 800$ cubic ft

45. $V = lwh$

$V = 8(3)(6) = 144$ cubic feet

$\dfrac{144}{1.5} = 96$ piranhas

47. $d = rt$

$135 = 60t$

$2.25 = t$

2.25 hours

49. $F = \dfrac{9}{5}C + 32$

Find F and C such that $F = C$.

$F = \dfrac{9}{5}F + 32$

$-32 = \dfrac{4}{5}F$

$-40 = F$

$-40°$

51. $d = rt$

$d = 270,000 \cdot 1$

$\dfrac{270,000}{25,120} = 10.8$

53. $V = \dfrac{4}{3}\pi r^3$

$V = \dfrac{4}{3}(3.14)(2000)^2$

$V = 3.3493 \times 10^{10}$

33,349,333,333 cubic miles

55. 15 feet = 180 inches

$V = \pi r^2 h$

$V = (3.14)(1)^2(180)$

$V = 565.2$ cubic inches

57. $F = \dfrac{9}{5}C + 32$

$78 = \dfrac{9}{5}C + 32$

$46 = \dfrac{9}{5}C$

$25\dfrac{5}{9} = C$

$25\dfrac{5}{9}°C$

59. $d = rt$
$25{,}000 = 4000t$
$6.25 = t$
6.25 hours

61. $A = \dfrac{1}{2}bh$

$20 = \dfrac{1}{2}(5)h$

$8 = h$
8 feet

63. Let x = number
$\dfrac{1}{2}(x)(5)$

65. $\dfrac{1}{3}\left(\dfrac{x}{6}\right)$

67. Let x = number
$\dfrac{2x}{3x}$

69. $x - (x + 6)$

Exercise Set 2.7

1. $120\% = 1.20$

3. $22.5\% = 0.225$

5. $0.12\% = 0.0012$

7. $0.75 = 75\%$

9. $2 = 200\%$

11. $\dfrac{1}{8} = 0.125 = 12.5\%$

13. 38%

15. $38\% + 16\% = 54\%$

17. 38% of $360°$
$0.38(360) = 136.8°$

19. Answers may vary.

21. Let x = unknown number
$x = 0.16(70)$
$x = 11.2$

23. Let x = unknown percent
$28.6 = x(52)$
$0.55 = x$
55%

25. Let x = unknown number
$45 = 0.25x$
$180 = x$

27. $0.23(20) = 4.6$

29. Let x = unknown number
$40 = 0.80x$
$50 = x$

31. Let x = unknown percent
$144 = x(480)$
$0.3 = x$
30%

33. Decrease $= 156(0.25) = \$39$
Sale price $= 156 - 39 = \$117$

35. Increase $= 110{,}000 - 95{,}500 = 14{,}500$
$14{,}500 = x(95{,}500)$
$0.152 = x$
15.2% increase

37. Increase $= 447.2(0.448) = 200.3$
Bergehamn's throw $= 447.2 + 200.3 = 647.5$ ft

39. 55.40%

41. $230(0.237) = 54.51$ or 54 people

43. No, many people use several medications.

45. Increase $= 70 - 40 = 30$
$30 = x(40)$
$0.75 = x$
75% increase

47. $121(0.26) = 31.46$
31 men

49.

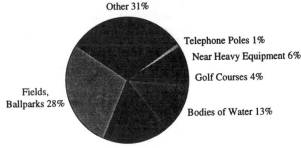

Other 31%

Telephone Poles 1%

Near Heavy Equipment 6%

Golf Courses 4%

Fields, Ballparks 28%

Bodies of Water 13%

Under Trees 17%

51. Increase $= 6.08 - 5.38 = 0.7$
$0.7 = x(5.38)$
$0.13 = x$
13% increase; Yes

53. Increase $= 20.8 - 20.7 = 0.1$
$0.1 = x(20.7)$
$0.0048 = x$
0.48% increase

55. 39%

57. $63 - 58 = 5\%$

59. $230 = x(2400)$
$0.096 = x$
9.6%

61. $35 = x(130)$
$0.269 = x$
26.9%; Yes

63. $4(12) = 48$
$\dfrac{48}{280} = 0.171 = 17.1\%$

65. $2a + b - c = 2(5) + (-1) - 3 = 10 - 1 - 3 = 6$

67. $4ab - 3bc = 4(-5)(-8) - 3(-8)(2) = 160 + 48 = 208$

69. $n^2 - m^2 = (-3)^2 - (-8)^2 = 9 - 64 = -55$

71. $I = PRT$
$I = 2000(0.04)(3)$
$I = \$240$

73. $d = rt$
$d = 57(3.5)$
$d = 199.5$ miles

Exercise Set 2.8

1. Let $x =$ length
then $\dfrac{2}{3}x =$ width
$P = 2L + 2W$
$260 = 2(x) + 2\left(\dfrac{2}{3}x\right)$
$260 = 2x + \dfrac{4}{3}x$
$260 = \dfrac{6}{3}x + \dfrac{4}{3}x$
$260 = \dfrac{10}{3}x$
$\dfrac{3}{10}(260) = \dfrac{3}{10} \cdot \dfrac{10}{3}x$
78 ft $= x$
length $= 78$ ft
width $= \dfrac{2}{3}(78$ ft$) = 52$ ft

3.

$P = a + b + c$
$102 = x + 2x + x + 30$
$102 = 4x + 30$
$102 - 30 = 4x + 30 - 30$
$72 = 4x$
$\dfrac{72}{4} = \dfrac{4x}{4}$
18 ft $= x$
The sides are 18 ft, 36 ft, and 48 ft.

5.

rate · time = distance			
Jet	500	x	$500x$
Propeller	200	$x+2$	$200(x+2)$

$500x = 200(x+2)$
$500x = 200x + 400$
$500x - 200x = 200x - 200x + 400$
$300x = 400$
$\dfrac{300x}{300} = \dfrac{400}{300}$
$x = \dfrac{4}{3}$

Distance $= 500\left(\dfrac{4}{3}\right) = \dfrac{2000}{3} = 666\dfrac{2}{3}$ miles

7.

rate · time = distance			
To Disneyland	50	x	$50x$
From Disneyland	40	$7.2 - x$	$40(7.2 - x)$

$50x = 40(7.2 - x)$
$50x = 288 - 40x$
$50x + 40x = 288 - 40x + 40x$
$\dfrac{90x}{90} = \dfrac{288}{90}$
$90x = 288$
$x = 3.2$ hours
Distance $= 50(3.2) = 160$ miles

9.

Strength of solution	Liters of Solution	Gallons of Acid
100%	x	$1.00x$
40%	2	$.40(2)$
70%	$x+2$	$.70(x+2)$

$1.00x + 0.40(2) = 0.70(x+2)$
$1.00x + 0.80 = 0.70x + 1.40$
$1.00x - 0.70x + 0.80 = 0.70x - 0.70x + 1.40$
$0.30x + 0.80 = 1.40$
$0.30x + 0.80 - 0.80 = 1.40 - 0.80$
$0.30x = 0.60$
$\dfrac{0.30x}{0.30} = \dfrac{0.60}{0.30}$
$x = 2$ gallons

11.

Pounds of Nuts	Cost of One Pound	Total Cost
20	3	60
x	5	$5x$
$20 + x$	3.50	$3.50(20 + x)$

$60 + 5x = 3.50(20 + x)$
$60 + 5x = 70 + 3.50x$
$60 + 5x - 3.50x = 70 + 3.50x - 3.50x$
$60 + 1.50x = 70$
$60 - 60 + 1.50x = 70 - 60$
$1.50x = 10$
$\dfrac{1.50x}{1.50} = \dfrac{10}{1.5}$
$x = 6.\overline{6}$ or $6\dfrac{2}{3}$ lbs

13. No; the mixture will have a percent antifreeze content between 30% and 50%.

15.

Principal · Rate · Time = Interest				
8%	x	0.08	1	$0.08x$
9%	$25,000 - x$	0.09	1	$0.09(25,000 - x)$
Total	25,000			2135

$0.08x + 0.09(25,000 - x) = 2135$
$0.08x + 2250 - 0.09x = 2135$
$-0.01x + 2250 = 2135$
$-0.01x + 2250 - 2250 = 2135 - 2250$
$-0.01x = -115$
$\dfrac{-0.01x}{-0.01} = \dfrac{-115}{-0.01}$
$x = \$11,500$
$\$11,500$ at 8% and $\$13,500$ at 9%.

17. 1st investment at 11% $= x$
2nd investment at 4% loss $= 10,000 - x$
$0.11(x) - 0.04(10,000 - x) = 650$
$11x - 4(10,000 - x) = 65,000$
$11x - 40,000 + 4x$
$15x - 40,000 = 65,000$
$\dfrac{15x}{15} = \dfrac{105,000}{15}$
$x = 7000$
$10,000 - x = 3000$
$\$7,000$ @ 11% profit
$\$3,000$ @ 4% loss

19. Let x = side of square
$x + 5$ = side of triangle
$3(x + 5) = 7 + 4x$
$3x + 15 = 7 + 4x$
$8 = x$
$x + 5 = 13$
Square's side length: 8 in.
Triangle's side length: 13 in.

21.

Principal · Rate · Time = Interest			
x	0.08	1	$0.08x$
$54,000 - x$	0.10	1	$0.10(54,000 - x)$

$0.08x = 0.10(54,000 - x)$
$0.08x = 5400 - 0.10x$
$0.08x + 0.10x = 5400 - 0.10x + 0.10x$
$0.18x = 5400$
$\dfrac{0.18x}{0.18} = \dfrac{5400}{0.18}$
$x = 30,000$
$30,000$ at 8% and $24,000 at 10%

23. 1st investment = x
2nd investment = $20,000 - x$
$0.12(x) - 0.04(20,000 - x) = 0$
$12x - 80,000 + 4x = 0$
$\dfrac{16x}{16} = \dfrac{80,000}{16}$
$x = 5,000$
$20,000 - x = 15,000$
$5,000 at 12%
$15,000 at 4%

25.

Principal · Rate · Time = Interest			
3000	0.06	1	180
x	0.09	1	$0.09x$

$180 + 0.09x = 585$
$180 - 180 + 0.09x = 585 - 180$
$0.09x = 405$
$\dfrac{0.09x}{0.09} = \dfrac{405}{0.09}$
$x = \$4500$

27. amount invested at 9% = x
amount invested at 10% = $2x$
amount invested at 11% = $3x$
$\$2790 = 0.09(x) + 0.10(2x) + 0.11(3x)$
$279,000 = 9x + 20x + 33x$
$\dfrac{279,000}{62} = \dfrac{62x}{62}$
$x = 4,500$

$2x = 9,000$
$3x = 13,500$
$\$4,500 at 9%
$\$9,000 at 10%
$\$13,500 at 11%

29.

	Number of Tickets	Cost of One Ticket	Total Cost
Adults	x	5.75	$5.75x$
Children	$8 - x$	3.00	$3.00(8 - x)$

$5.75x + 3.00(8 - x) = 32.25$
$5.75x + 24.00 - 3.00x = 32.25$
$2.75x + 24.00 = 32.25$
$2.75x + 24.00 - 24.00 = 32.25 - 24.00$
$2.75x = 8.25$
$\dfrac{2.75x}{2.75} = \dfrac{8.25}{2.75}$
$x = 3$
3 adult tickets

31. Rate of 1st hiker = x
Rate of 2nd hiker = $\dfrac{3}{2}x$

$2x + 2\left(\dfrac{3}{2}x\right) = 11$
$2x + 3x = 11$
$\dfrac{5x}{5} = \dfrac{11}{5}$
$x = 2\dfrac{1}{5}$ mph $= 2.2$ mph
$\dfrac{3}{2} \cdot \dfrac{11}{5} = \dfrac{33}{10} = 3\dfrac{3}{10}$ mph $= 3.3$ mph

33.

	Rate	Time
Upstream	5	x
Downstream	11	$4 - x$

$5x = 11(4 - x)$
$5x = 44 - 11x$
$\dfrac{16x}{16} = \dfrac{44}{16}$
$x = 2.75$
Distance:
$5x = 5(2.75) = 11(4 - x) = 11(4 - 2.75) = 11(1.75)$
$= 13.75 \times 2 = 27.5$ miles

35. $R = 60x$
$C = 50x + 5000$
$60x = 50x + 5000$
$10x = 5000$
$x = 500$
$C = 50(500) + 5000 = \$30{,}000$

37. $C = 870 + 70x$
$R = 105x$
$870 + 70x = 105x$
$870 = 35x$
$24.9 = x$
25 monitors

39. $3 + (-7) = -4$

41. $\dfrac{3}{4} - \dfrac{3}{16} = \dfrac{12}{16} - \dfrac{3}{16} = \dfrac{9}{16}$

43. $-5 - (-1) = -5 + 1 = -4$

45. $-5 > -7$

47. $|-5| = 5$
$-(-5) = 5$
$|-5| = -(-5)$

Section 2.9 Mental Math

1. $5x > 10$
$\dfrac{5x}{5} > \dfrac{10}{5}$
$x > 2$

3. $2x \geq 16$
$\dfrac{2x}{2} \geq \dfrac{16}{2}$
$x \geq 8$

Exercise Set 2.9

1. $x \leq -1$

3. $x > \dfrac{1}{2}$

5. $-1 < x < 3$

7. $0 \leq y < 2$

9. $2x < -6$
$\dfrac{2x}{2} < -\dfrac{6}{2}$
$x < -3$

11. $x - 2 \geq -7$
$x - 2 + 2 \geq -7 + 2$
$x \geq -5$

13. $-8x \leq 16$
$\dfrac{-8x}{-8} \geq \dfrac{16}{-8}$
$x \geq -2$

15. $3x - 5 > 2x - 8$
$3x - 5 - 2x > 2x - 2x - 8$
$x - 5 > -8$
$x - 5 + 5 > -8 + 5$
$x > -3$

17. $4x - 1 \leq 5x - 2x$
$4x - 1 \leq 3x$
$4x - 4x - 1 \leq 3x - 4x$
$-1 \leq -1x$
$\dfrac{-1}{-1} \geq \dfrac{-1x}{-1}$
$1 \geq x$

19. $x - 7 < 3(x + 1)$
$x - 7 < 3x + 3$
$x - x - 7 < 3x - x + 3$
$-7 < 2x + 3$
$-7 - 3 < 2x + 3 - 3$
$-10 < 2x$
$\dfrac{-10}{2} < \dfrac{2x}{2}$
$-5 < x$

21. $-6x + 2 \geq 2(5 - x)$
$-6x + 2 \geq 10 - 2x$
$-6x + 2x + 2 \geq 10 - 2x + 2x$
$-4x + 2 \geq 10$
$-4x + 2 - 2 \geq 10 - 2$
$-4x \geq 8$

$$\frac{-4x}{-4} \le \frac{8}{-4}$$
$$x \le -2$$
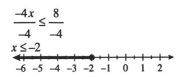

23. $4(3x - 1) \le 5(2x - 4)$
$12x - 4 \le 10x - 20$
$12x - 10x - 4 \le 10x - 10x - 20$
$2x - 4 \le -20$
$2x - 4 + 4 \le -20 + 4$
$2x \le -16$
$$\frac{2x}{2} \le \frac{-16}{2}$$
$x \le -8$

25. $3(x + 2) - 6 > -2(x - 3) + 14$
$3x + 6 - 6 > -2x + 6 + 14$
$3x > -2x + 20$
$3x + 2x > -2x + 2x + 20$
$5x > 20$
$$\frac{5x}{5} > \frac{20}{5}$$
$x > 4$

27. $-3 < 3x < 6$
$$\frac{-3}{3} < \frac{3x}{3} < \frac{6}{3}$$
$-1 < x < 2$

29. $2 \le 3x - 10 \le 5$
$2 + 10 \le 3x - 10 + 10 \le 5 + 10$
$12 \le 3x \le 15$
$$\frac{12}{3} \le \frac{3x}{3} \le \frac{15}{3}$$
$4 \le x \le 5$

31. $-4 < 2(x - 3) < 4$
$-4 < 2x - 6 < 4$
$-4 + 6 < 2x - 6 + 6 < 4 + 6$
$2 < 2x < 10$
$$\frac{2}{2} < \frac{2x}{2} < \frac{10}{2}$$
$1 < x < 5$

33. Answers may vary.

35. $-2x \le -40$
$$\frac{-2x}{-2} \ge \frac{-40}{-2}$$
$x \ge 20$

37. $-9 + x > 7$
$-9 + 9 + x > 7 + 9$
$x > 16$

39. $3x - 7 < 6x + 2$
$3x - 3x - 7 < 6x - 3x + 2$
$-7 < 3x + 2$
$-7 - 2 < 3x + 2 - 2$
$-9 < 3x$
$$\frac{-9}{3} < \frac{3x}{3}$$
$-3 < x$

41. $5x - 7x \le x + 2$
$-2x \le x + 2$
$-2x - x \le x - x + 2$
$-3x \le 2$
$$\frac{-3x}{-3} \ge \frac{2}{-3}$$
$x \ge -\frac{2}{3}$

43. $\frac{3}{4}x > 2$
$$\frac{4}{3}\left(\frac{3}{4}x\right) > \frac{4}{3}(2)$$
$x > \frac{8}{3}$
$x > 2\frac{2}{3}$

45. $3(x - 5) < 2(2x - 1)$
$3x - 15 < 4x - 2$
$3x - 3x - 15 < 4x - 3x - 2$
$-15 < x - 2$
$-15 + 2 < x - 2 + 2$
$-13 < x$

47. $4(2x + 1) > 4$
$8x + 4 > 4$
$8x + 4 - 4 > 4 - 4$
$8x > 0$
$\dfrac{8x}{8} > \dfrac{0}{8}$
$x > 0$

49. $-5x + 4 \le -4(x - 1)$
$-5x + 4 \le -4x + 4$
$-5x + 5x + 4 \le -4x + 5x + 4$
$4 \le x + 4$
$4 - 4 \le x + 4 - 4$
$0 \le x$

51. $-2 < 3x - 5 < 7$
$-2 + 5 < 3x - 5 + 5 < 7 + 5$
$3 < 3x < 12$
$\dfrac{3}{3} < \dfrac{3x}{3} < \dfrac{12}{3}$
$1 < x < 4$

53. $-2(x - 4) - 3x < -(4x + 1) + 2x$
$-2x + 8 - 3x < -4x - 1 + 2x$
$-5x + 8 < -2x - 1$
$-5x + 2x + 8 < -2x + 2x - 1$
$-3x + 8 < -1$
$-3x + 8 - 8 < -1 - 8$
$-3x < -9$
$\dfrac{-3x}{-3} > \dfrac{-9}{-3}$
$x > 3$

55. $-3x + 6 \ge 2x + 6$
$-3x + 3x + 6 \ge 2x + 3x + 6$
$6 \ge 5x + 6$
$6 - 6 \ge 5x + 6 - 6$
$0 \ge 5x$
$\dfrac{0}{5} \ge \dfrac{5x}{5}$
$0 \ge x$

57. $-6 < 3(x - 2) < 8$
$-6 < 3x - 6 < 8$
$-6 + 6 < 3x - 6 + 6 < 8 + 6$
$0 < 3x < 14$

$\dfrac{0}{3} < \dfrac{3x}{3} < \dfrac{14}{3}$
$0 < x < 4\dfrac{2}{3}$

59. Let x = number
$2x + 6 > -14$
$2x + 6 - 6 > -14 - 6$
$2x > -20$
$\dfrac{2x}{2} > \dfrac{-20}{2}$
$x > -10$

61. $P = 2L + 2W$
$2L + 2(15) \le 100$
$2L + 30 \le 100$
$2L + 30 - 30 \le 100 - 30$
$2L \le 70$
$\dfrac{2L}{2} \le \dfrac{70}{2}$
$L \le 35$ cm

63. $-39 \le \dfrac{5}{9}(F - 32) \le 45$
$-70.2 \le F - 32 \le 81$
$-38.2° \le F \le 113°$

65. $2.9 \le 3.14d \le 3.1$
$\dfrac{2.9}{3.14} \le \dfrac{3.14d}{3.14} \le \dfrac{3.1}{3.14}$
$0.924 \le d \le 0.987$

67. Let x = number
$-5 < 2x + 1 < 7$
$-5 - 1 < 2x + 1 - 1 < 7 - 1$
$-6 < 2x < 6$
$\dfrac{-6}{2} < \dfrac{2x}{2} < \dfrac{6}{2}$
$-3 < x < 3$

69.

Principal ·	Rate ·	Time =	Interest
10,000	0.11	1	1100
5,000	x	1	$5000x$

$1100 + 5000x \ge 1600$
$1100 - 1100 + 5000x \ge 1600 - 1100$
$5000x \ge 500$
$\dfrac{5000x}{5000} \ge \dfrac{500}{5000}$

$x \geq 0.10$

$x \geq 10\%$

71. Let x = score of 3rd game

$$\frac{146 + 201 + x}{3} \geq 180$$

$$3\left(\frac{146 + 201 + x}{3}\right) \geq (3)180$$

$146 + 201 + x \geq 540$

$347 + x \geq 540$

$347 - 347 + x \geq 540 - 347$

$x \geq 193$

73. $(2)^3 = 8$

75. $(1)^{12} = 1$

77. $\left(\dfrac{4}{7}\right)^2 = \dfrac{16}{49}$

79. 32 million

81. 1992

83. $x(x + 4) > x^2 - 2x + 6$

$x^2 + 4x > x^2 - 2x + 6$

$x^2 - x^2 + 4x > x^2 - x^2 - 2x + 6$

$4x > -2x + 6$

$4x + 2x > -2x + 2x + 6$

$6x > 6$

$\dfrac{6x}{6} > \dfrac{6}{6}$

$x > 1$

85. $x^2 + 6x - 10 < x(x - 10)$

$x^2 + 6x - 10 < x^2 - 10x$

$x^2 - x^2 + 6x - 10 < x^2 - x^2 - 10x$

$6x - 10 < -10x$

$6x - 6x - 10 < -10x - 6x$

$-10 < -16x$

$\dfrac{-10}{-16} > \dfrac{-16x}{-16}$

$\dfrac{10}{16} > x$

$\dfrac{5}{8} > x$

87. $x(2x - 3) \leq 2x^2 - 5x$

$2x^2 - 3x \leq 2x^2 - 5x$

$2x^2 - 2x^2 - 3x \leq 2x^2 - 2x^2 - 5x$

$-3x \leq -5x$

$-3x + 5x \leq -5x + 5x$

$2x \leq 0$

$\dfrac{2x}{2} \leq \dfrac{0}{2}$

$x \leq 0$

Chapter 2 - Review

1. $5x - x + 2x$

$= 4x + 2x = 6x$

2. $0.2z - 4.6x - 7.4z = -4.6x - 7.2z$

3. $\dfrac{1}{2}x + 3 + \dfrac{7}{2}x - 5$

$= \dfrac{8}{2}x - 2$

$= 4x - 2$

4. $\dfrac{4}{5}y + 1 + \dfrac{6}{5}y + 2$

$= \dfrac{10}{5}y + 3$

$= 2y + 3$

5. $2(n - 4) + n - 10$

$= 2n - 8 + n - 10$

$= 3n - 18$

6. $3(w + 2) - (12 - w)$

$= 3w + 6 - 12 + w$

$= 4w - 6$

7. $x + 5 - (7x - 2)$

$= x + 5 - 7x + 2$

$= -6x + 7$

8. $y - 0.7 - (1.4y - 3)$

$= y - 0.7 - 1.4y + 3$

$= -0.4y + 2.3$

9. Let x = number

$3x - 7$

10. $2(x + 2.8) + 3x$

11. $8x + 4 = 9x$

$8x - 8x + 4 = 9x - 8x$

$4 = x$

12. $5y - 3 = 6y$
$-3 = y$

13. $3x - 5 = 4x + 1$
$3x - 3x - 5 = 4x - 3x + 1$
$-5 = x + 1$
$-5 - 1 = x + 1 - 1$
$-6 = x$

14. $2x - 6 = x - 6$
$x = 0$

15. $4(x + 3) = 3(1 + x)$
$4x + 12 = 3 + 3x$
$4x - 3x + 12 = 3 + 3x - 3x$
$x + 12 = 3$
$x + 12 - 12 = 3 - 12$
$x = -9$

16. $6(3 + n) = 5(n - 1)$
$18 + 6n = 5n - 5$
$n = -23$

17. Let x = number
$10 - x$ = other number

18. $x - 5$

19. $180 - (x + 5) = 180 - x - 5 = 175 - x$ or $(175 - x)°$

20. $\dfrac{3}{4}x = -9$
$\dfrac{4}{3}\left(\dfrac{3}{4}x\right) = \dfrac{4}{3}(-9)$
$x = -12$

21. $\dfrac{x}{6} = \dfrac{2}{3}$
$\dfrac{3x}{3} = \dfrac{12}{3}$
$x = 4$

22. $-3x + 1 = 19$
$-3x + 1 - 1 = 19 - 1$
$-3x = 18$
$\dfrac{-3x}{-3} = \dfrac{18}{-3}$
$x = -6$

23. $5x + 25 = 20$
$\dfrac{5x}{5} = \dfrac{-5}{5}$
$x = -1$

24. $5x - 6 + x = 9 + 4x - 1$
$6x - 6 = 4x + 8$
$6x - 4x - 6 = 4x - 4x + 8$
$2x - 6 = 8$
$2x - 6 + 6 = 8 + 6$
$2x = 14$
$\dfrac{2x}{2} = \dfrac{14}{2}$
$x = 7$

25. $8 - y + 4y = 7 - y - 3$
$8 + 3y = 4 - y$
$\dfrac{4y}{4} = \dfrac{-4}{4}$
$y = -1$

26. $x + (x + 2) + (x + 4) = 3x + 6$

27. $\dfrac{2}{7}x - \dfrac{5}{7} = 1$
$2x - 5 = 7$
$\dfrac{2x}{2} = \dfrac{12}{2}$
$x = 6$

28. $\dfrac{5}{3}x + 4 = \dfrac{2}{3}x$
$\dfrac{5}{3}x - \dfrac{5}{3}x + 4 = \dfrac{2}{3}x - \dfrac{5}{3}x$
$4 = -\dfrac{3}{3}x$
$4 = -x$
$\dfrac{4}{-1} = \dfrac{-x}{-1}$
$-4 = x$

29. $-(5x + 1) = -7x + 3$
$-5x - 1 = -7x + 3$
$\dfrac{2x}{2} = \dfrac{4}{2}$
$x = 2$

30. $-4(2x + 1) = -5x + 5$
$-8x - 4 = -5x + 5$
$-8x + 5x - 4 = -5x + 5x + 5$
$-3x - 4 = 5$
$-3x - 4 + 4 = 5 + 4$
$-3x = 9$
$\dfrac{-3x}{-3} = \dfrac{9}{-3}$
$x = -3$

31. $-6(2x - 5) = -3(9 + 4x)$
$-12x + 30 = -27 - 12x$
$30 = -27$
No solution

32. $3(8y - 1) = 6(5 + 4y)$
$24y - 3 = 30 + 24y$
$24y - 24y - 3 = 30 + 24y - 24y$
$-3 = 30$
No solution

33. $\dfrac{3(2 - z)}{5} = z$
$3(2 - z) = 5z$
$6 - 3z = 5z$
$\dfrac{6}{8} = \dfrac{8z}{8}$
$\dfrac{6}{8} = z$
$z = \dfrac{3}{4}$

34. $\dfrac{4(n + 2)}{5} = -n$
$5\left[\dfrac{4(n + 2)}{5}\right] = 5(-n)$
$4(n + 2) = -5n$
$4n + 8 = -5n$
$4n - 4n + 8 = -5n - 4n$
$8 = -9n$
$\dfrac{8}{-9} = \dfrac{-9n}{-9}$
$\dfrac{8}{-9} = n$

35. $5(2n - 3) - 1 = 4(6 + 2n)$
$10n - 15 - 1 = 24 + 8n$
$10n - 16 = 24 + 8n$
$\dfrac{2n}{2} = \dfrac{40}{2}$
$n = 20$

36. $-2(4y - 3) + 4 = 3(5 - y)$
$-8y + 6 + 4 = 15 - 3y$
$-8y + 10 = 15 - 3y$
$-8y + 3y + 10 = 15 - 3y + 3y$
$-5y + 10 = 15$
$-5y + 10 - 10 = 15 - 10$
$-5y = 5$
$\dfrac{-5y}{-5} = \dfrac{5}{-5}$
$y = -1$

37. $9z - z + 1 = 6(z - 1) + 7$
$8z + 1 = 6z - 6 + 7$
$8z + 1 = 6z + 1$
$\dfrac{2z}{2} = \dfrac{0}{2}$
$z = 0$

38. $5t - 3 - t = 3(t + 4) - 15$
$4t - 3 = 3t + 12 - 15$
$4t - 3 = 3t - 3$
$4t - 3t - 3 = 3t - 3t - 3$
$t - 3 = -3$
$t - 3 + 3 = -3 + 3$
$t = 0$

39. $-n + 10 = 2(3n - 5)$
$-n + 10 = 6n - 10$
$\dfrac{20}{7} = \dfrac{7n}{7}$
$\dfrac{20}{7} = n$

40. $-9 - 5a = 3(6a - 1)$
$-9 - 5a = 18a - 3$
$-9 - 5a + 5a = 18a + 5a - 3$
$-9 = 23a - 3$
$-9 + 3 = 23a - 3 + 3$
$-6 = 23a$
$\dfrac{-6}{23} = \dfrac{23a}{23}$
$\dfrac{-6}{23} = a$

41. $\dfrac{5(c + 1)}{6} = 2c - 3$
$5(c + 1) = 6(2c - 3)$
$5c + 5 = 12c - 18$
$\dfrac{23}{7} = \dfrac{7c}{7}$
$\dfrac{23}{7} = c$

42. $\dfrac{2(8 - a)}{3} = 4 - 4a$
$3\left[\dfrac{2(8 - a)}{3}\right] = 3(4 - 4a)$
$2(8 - a) = 12 - 12a$
$16 - 2a = 12 - 12a$
$16 - 2a + 12a = 12 - 12a + 12a$
$16 + 10a = 12$
$16 - 16 + 10a = 12 - 16$
$10a = -4$

$$\frac{10a}{10} = \frac{-4}{10}$$

$$a = -\frac{4}{10} = -\frac{2}{5}$$

43. $200(70x - 3560) = -179(150x - 19{,}300)$
$14{,}000x - 712{,}000 = -26{,}850x + 3{,}454{,}700$
$14{,}000x + 26{,}850x - 712{,}000$
$\qquad\qquad = -26{,}850x + 26{,}850x + 3{,}454{,}700$
$40{,}850x - 712{,}000 = 3{,}454{,}700$
$40{,}850x - 712{,}000 + 712{,}000$
$\qquad\qquad = 3{,}454{,}700 + 712{,}000$
$40{,}850x = 4{,}166{,}700$
$$\frac{40{,}850x}{40{,}850} = \frac{4{,}166{,}700}{40{,}850}$$
$x = 102$

44. $1.72y - 0.04y = 0.42$
$1.68y = 0.42$
$y = 0.25$

45. $\dfrac{x}{3} = x - 2$
$x = 3(x - 2)$
$x = 3x - 6$
$$\frac{6}{2} = \frac{2x}{2}$$
$3 = x$

46. Let x = number
$2(x + 6) = -x$
$2x + 12 = -x$
$2x - 2x + 12 = -x - 2x$
$12 = -3x$
$$\frac{12}{-3} = \frac{-3x}{-3}$$
$-4 = x$

47. Let x = side of base
$68 + 3x + x = 1380$
$68 + 4x = 1380$
$68 - 68 + 4x = 1380 - 68$
$4x = 1312$
$$\frac{4x}{4} = \frac{1312}{4}$$
$x = 328$
height = $68 + 3(328) = 1052$ ft

48. $x + 2x = 12$
$3x = 12$
$x = 4$
$2x = 8$
4 ft and 8 ft

49. Let x = smaller area code
$34 + 3x + x = 1262$
$34 + 4x = 1262$
$34 - 34 + 4x = 1262 - 34$
$4x = 1228$
$$\frac{4x}{4} = \frac{1228}{4}$$
$x = 307$
$34 + 3x = 955$
The codes are 307 and 955.

50. Let x = smallest integer
$x + (x + 2) + (x + 4) = -114$
$3x + 6 = -114$
$3x = -120$
$x = -40$
$x + 2 = -38$
$x + 4 = -36$
The integers are −40, −38, and −36.

51. $P = 2l + 2w$
$46 = 2(14) + 2w$
$46 = 28 + 2w$
$18 = 2w$
$9 = w$

52. $V = lwh$
$192 = 8(6)h$
$192 = 48h$
$4 = h$

53. $y = mx + b$
$$\frac{y - b}{x} = \frac{mx}{x}$$
$$\frac{y - b}{x} = m$$

54. $r = vst - 5$
$r + 5 = vst - 5 + 5$
$r + 5 = vst$
$$\frac{r + 5}{vt} = \frac{vst}{vt}$$
$$\frac{r + 5}{vt} = s$$

55. $2y - 5x = 7$
$$\frac{2y - 7}{5} = \frac{5x}{5}$$
$$\frac{2y - 7}{5} = x$$

56. $3x - 6y = -2$

$3x - 3x - 6y = -2 - 3x$

$-6y = -2 - 3x$

$\dfrac{-6y}{-6} = \dfrac{-2 - 3x}{-6}$

$y = \dfrac{-2 - 3x}{-6}$

$y = \dfrac{-1(-2 - 3x)}{-1(-6)}$

$y = \dfrac{2 + 3x}{6}$

57. $\dfrac{C}{D} = \dfrac{\pi D}{D}$

$\dfrac{C}{D} = \pi$

58. $C = 2\pi r$

$\dfrac{C}{2r} = \dfrac{2\pi r}{2r}$

$\dfrac{C}{2r} = \pi$

59. $V = lwh$

$900 = (20)(w)(3)$

$\dfrac{900}{60} = \dfrac{60w}{60}$

$15 = w$

15 meters $= w$

60. $C = \dfrac{5}{9}(F - 32)$

$C = \dfrac{5}{9}(104 - 32)$

$C = \dfrac{5}{9}(72)$

$C = 40°$

61. $D = RT$

$\dfrac{10,000}{125} = \dfrac{125T}{125}$

80 minutes $= T$

1 hour 20 minutes $= T$

62. $0.12(250) = 30$

63. $1.10(85) = 93.5$

64. Let $x =$ percent

$9 = x(45)$

$0.2 = x$

20%

65. Let $x =$ percent

$59.5 = x(85)$

$0.7 = x$

70%

66. Let $x =$ unknown number

$137.5 = 1.25x$

$110 = x$

67. Let $x =$ unknown number

$768 = 0.6x$

$1280 = x$

68. $0.126(50,000) = 6300$

69. 6%

70. Eat from the Minibar

71. $0.40(300) = 120$ travelers

72. No; some business travelers may have chosen more than one category.

73. $\dfrac{210 - 180}{210} = \dfrac{30}{210} = 0.143$

14.3%

74.

Principal $\cdot$ rate $\cdot$ time $=$ interest				
Money Market	x	0.085	1	$0.085x$
C.D.	$50000 - x$	0.105	1	$.105(50,000 - x)$

$0.085x + 0.105(50,000 - x) = 4550$

$0.085x + 5250 - 0.105x = 4550$

$-0.02x + 5250 = 4550$

$-0.02x + 5250 - 5250 = 4550 - 5250$

$-0.02x = -700$

$\dfrac{-0.02x}{-0.02} = \dfrac{-700}{-0.02}$

$x = 35,000$

$\$35,000$ in the money market and $\$15,000$ in the C.D.

75. dimes $= x$

quarters $= 2x$

nickels $= 500 - 3x$

$0.10(x) + 0.25(2x) + 0.05[500 - 3x] = 88$

$10x + 25(2x) + 5[500 - 3x] = 8800$

$10x + 50x + 2500 - 15x = 8800$

$45x + 2500 = 8800$

$\dfrac{45x}{45} = \dfrac{6300}{45}$

$x = 140$, so

$500 - 3x = 500 - 3(140) = 80$
80 nickels

76.

	Rate	Time	Distance
Passenger Train	60 mph	x	$60x$
Freight Train	45 mph	$x + \dfrac{3}{2}$	$45\left(x + \dfrac{3}{2}\right)$

$60x = 45\left(x + \dfrac{3}{2}\right)$

$60x = 45x + \dfrac{135}{2}$

$120x = 90x + 135$

$\dfrac{30x}{30} = \dfrac{135}{30}$

$x = 4.5$ hrs

77.

	Rate	Time	Distance
Up	8	x	$8x$
Down	12	$5 - x$	$12(5 - x)$

$8x = 12(5 - x)$
$8x = 60 - 12x$
$8x + 12x = 60 - 12x + 12x$
$20x = 60$
$\dfrac{20x}{20} = \dfrac{60}{20}$
$x = 3$
Round trip distance: 48 miles
$8(3) = 24$
$12(5 - 3) = 24$

78. $x \leq -2$

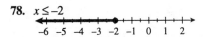

79. $x > 0$

number line from -4 to 4, open circle at 0, shaded right

80. $-1 < x < 1$

number line from -4 to 4, open circles at -1 and 1, shaded between

81. $0.5 \leq y < 1.5$

number line from -4 to 4, closed circle at $.5$, open circle at 1.5, shaded between

82. $-2x \geq -20$

$\dfrac{-2x}{-2} \leq \dfrac{-20}{-2}$

$x \leq 10$

number line from 6 to 14, closed circle at 10, shaded left

83. $-3x > 12$

$\dfrac{-3x}{-3} < \dfrac{12}{-3}$

$x < -4$

number line from -6 to 2, open circle at -4, shaded left

84. $5x - 7 > 8x + 5$

$\dfrac{-12}{3} > \dfrac{3x}{3}$

$-4 > x$

number line from -8 to 0, open circle at -4, shaded left

85. $x + 4 \geq 6x - 16$

$x - x + 4 \geq 6x - x - 16$

$4 \geq 5x - 16$

$4 + 16 \geq 5x - 16 + 16$

$20 \geq 5x$

$\dfrac{20}{5} \geq \dfrac{5x}{5}$

$4 \geq x$

number line from -2 to 6, closed circle at 4, shaded left

86. $2 \leq 3x - 4 < 6$

$\underline{+4 \qquad +4 \qquad +4}$

$\dfrac{6}{3} \leq \dfrac{3x}{3} < \dfrac{10}{3}$

$2 \leq x < \dfrac{10}{3}$

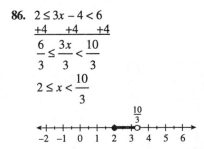

87. $-3 < 4x - 1 < 2$

$-3 + 1 < 4x - 1 + 1 < 2 + 1$

$-2 < 4x < 3$

$\dfrac{-2}{4} < \dfrac{4x}{4} < \dfrac{3}{4}$

$-\dfrac{1}{2} < x < \dfrac{3}{4}$

number line from -4 to 4, open circles at $-\dfrac{1}{2}$ and $\dfrac{3}{4}$, shaded between

88. $-2(x-5) > 2(3x-2)$
$-2x + 10 > 6x - 4$
$\dfrac{14}{8} > \dfrac{8x}{8}$
$\dfrac{7}{4} > x$

89. $4(2x-5) \le 5x-1$
$8x - 20 \le 5x - 1$
$8x - 5x - 20 \le 5x - 5x - 1$
$3x - 20 \le -1$
$3x - 20 + 20 \le -1 + 20$
$3x \le 19$
$\dfrac{3x}{3} \le \dfrac{19}{3}$
$x \le \dfrac{19}{3}$
$x \le 6\dfrac{1}{3}$

90. $175 + 0.05(x) \ge 300$
$17500 + 5x \ge 30{,}000$
$\dfrac{5x}{5} \ge \dfrac{12{,}500}{5}$
$x \ge \$2{,}500$

91. Let x = score on next round
$\dfrac{76 + 82 + 79 + x}{4} < 80$
$4\left[\dfrac{76 + 82 + 79 + x}{4}\right] < 4(80)$
$76 + 82 + 79 + x < 320$
$237 + x < 320$
$237 - 237 + x < 320 - 237$
$x < 83$

Chapter 2 - Test

1. $2y - 6 - y - 4 = y - 10$

2. $2.7x + 6.1 + 3.2x - 4.9 = 5.9x + 1.2$

3. $4(x-2) - 3(2x-6)$
$= 4x - 8 - 6x + 18$
$= -2x + 10$

4. $-5(y+1) + 2(3-5y)$
$= -5y - 5 + 6 - 10y$
$= -15y + 1$

5. $-\dfrac{4}{5}x = 4$
$-\dfrac{5}{4}\left(-\dfrac{4}{5}x\right) = -\dfrac{5}{4}(4)$
$x = -5$

6. $4(n-5) = -(4-2n)$
$4n - 20 = -4 + 2n$
$4n - 2n - 20 = -4 + 2n - 2n$
$2n - 20 = -4$
$2n - 20 + 20 = -4 + 20$
$2n = 16$
$\dfrac{2n}{2} = \dfrac{16}{2}$
$n = 8$

7. $5y - 7 + y = -(y + 3y)$
$5y - 7 + y = -y - 3y$
$6y - 7 = -4y$
$6y - 6y - 7 = -4y - 6y$
$-7 = -10y$
$\dfrac{-7}{-10} = \dfrac{-10y}{-10}$
$\dfrac{-7}{-10} = y$
$\dfrac{7}{10} = y$

8. $4z + 1 - z = 1 + z$
$3z + 1 = 1 + z$
$3z - z + 1 = 1 + z - z$
$2z + 1 = 1$
$2z + 1 - 1 = 1 - 1$
$2z = 0$
$\dfrac{2z}{2} = \dfrac{0}{2}$
$z = 0$

9. $\dfrac{2(x+6)}{3} = x - 5$
$3\left[\dfrac{2(x+6)}{3}\right] = 3(x-5)$
$2(x+6) = 3(x-5)$
$2x + 12 = 3x - 15$
$2x - 2x + 12 = 3x - 2x - 15$
$12 = x - 15$
$12 + 15 = x - 15 + 15$
$27 = x$

10. $\dfrac{4(y-1)}{5} = 2y + 3$

$$5\left[\dfrac{4(y-1)}{5}\right] = 5(2y+3)$$

$4(y-1) = 5(2y+3)$
$4y - 4 = 10y + 15$
$4y - 10y - 4 = 10y - 10y + 15$
$-6y - 4 = 15$
$-6y - 4 + 4 = 15 + 4$
$-6y = 19$
$\dfrac{-6y}{-6} = \dfrac{19}{-6}$
$y = -\dfrac{19}{6}$

11. $\dfrac{1}{2} - x + \dfrac{3}{2} = x - 4$

$-x + \dfrac{4}{2} = x - 4$
$-x + 2 = x - 4$
$-x + x + 2 = x + x - 4$
$2 = 2x - 4$
$2 + 4 = 2x - 4 + 4$
$6 = 2x$
$\dfrac{6}{2} = \dfrac{2x}{2}$
$3 = x$

12. $\dfrac{1}{3}(y+3) = 4y$

$$3\left[\dfrac{1}{3}(y+3)\right] = 3(4y)$$

$y + 3 = 12y$
$y - y + 3 = 12y - y$
$3 = 11y$
$\dfrac{3}{11} = \dfrac{11y}{11}$
$\dfrac{3}{11} = y$

13. $-0.3(x-4) + x = 0.5(3-x)$
$-0.3x + 1.2 + x = 1.5 - 0.5x$
$0.7x + 1.2 = 1.5 - 0.5x$
$0.7x + 0.5x + 1.2 = 1.5 - 0.5x + 0.5x$
$1.2x + 1.2 = 1.5$
$1.2x + 1.2 - 1.2 = 1.5 - 1.2$
$1.2x = 0.3$
$\dfrac{1.2x}{1.2} = \dfrac{0.3}{1.2}$
$x = 0.25$

14. $-4(a+1) - 3a = -7(2a-3)$
$-4a - 4 - 3a = -14a + 21$
$-7a - 4 = -14a + 21$
$-7a + 14a - 4 = -14a + 14a + 21$
$7a - 4 = 21$
$7a - 4 + 4 = 21 + 4$
$7a = 25$
$\dfrac{7a}{7} = \dfrac{25}{7}$
$a = \dfrac{25}{7}$

15. Let x = number

$x + \dfrac{2}{3}x = 35$
$\dfrac{3}{3}x + \dfrac{2}{3}x = 35$
$\dfrac{5}{3}x = 35$
$\dfrac{3}{5}\left(\dfrac{5}{3}x\right) = \dfrac{3}{5}(35)$
$x = 21$

16. Area of deck
$A = lw$
$A = (20\text{ ft})(35\text{ ft})$
$A = 700$ sq ft
Two coats are needed.
Twice the Area $= 2(700$ sq ft$)$
Twice the Area $= 1400$ sq ft

$\dfrac{1400\text{ sq ft}}{200\text{ sq ft / gal}}$

7 gallons

17.

Principal	· rate	· time	= interest
Amoxil	x	0.10	1	0.10x
IBM	2x	0.12	1	0.24x

$0.10x + 0.24x = 2890$
$0.34x = 2890$
$x = 8500$
$2x = 17{,}000$
Amoxil: \$8500
IBM: \$17,000

18.

	rate	· time	= distance
1st train	50	x	50x
2nd train	65	x	65x

$$50x + 65x = 287.5$$
$$115x = 287.5$$
$$\frac{115x}{115} = \frac{287.5}{115}$$
$$x = 2.5 \text{ hours}$$

19. $y = mx + b$
$$-14 = -2x - 2$$
$$-12 = -2x$$
$$6 = x$$

20. $V = \pi r^2 h$
$$\frac{V}{\pi r^2} = \frac{\pi r^2 h}{\pi r^2}$$
$$\frac{V}{\pi r^2} = h$$

21. $3x - 4y = 10$
$$3x - 3x - 4y = 10 - 3x$$
$$-4y = 10 - 3x$$
$$\frac{-4y}{-4} = \frac{10 - 3x}{-4}$$
$$y = \frac{10 - 3x}{-4}$$
$$y = \frac{-1(10 - 3x)}{-1(-4)}$$
$$y = \frac{-10 + 3x}{4}$$
$$y = \frac{3x - 10}{4}$$

22. $3x - 5 > 7x + 3$
$$3x - 7x - 5 > 7x - 7x + 3$$
$$-4x - 5 > 3$$
$$-4x - 5 + 5 > 3 + 5$$
$$-4x > 8$$
$$\frac{-4x}{-4} < \frac{8}{-4}$$
$$x < -2$$

```
◄─┼──┼──┼──○──┼──┼──┼──┼──►
 -6 -5 -4 -3 -2 -1  0  1  2
```

23. $x + 6 > 4x - 6$
$$x - 4x + 6 > 4x - 4x - 6$$
$$-3x + 6 > -6$$
$$-3x + 6 - 6 > -6 - 6$$
$$-3x > -12$$
$$\frac{-3x}{-3} < \frac{-12}{-3}$$
$$x < 4$$

```
◄─┼──┼──┼──┼──○──┼──┼──┼──┼──►
  0  1  2  3  4  5  6  7  8
```

24. $-2 < 3x + 1 < 8$
$$-2 - 1 < 3x + 1 - 1 < 8 - 1$$
$$-3 < 3x < 7$$
$$-\frac{3}{3} < \frac{3x}{3} < \frac{7}{3}$$
$$-1 < x < \frac{7}{3}$$
$$-1 < x < 2\frac{1}{3}$$

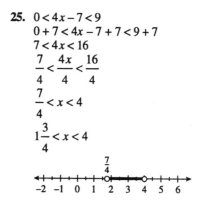

25. $0 < 4x - 7 < 9$
$$0 + 7 < 4x - 7 + 7 < 9 + 7$$
$$7 < 4x < 16$$
$$\frac{7}{4} < \frac{4x}{4} < \frac{16}{4}$$
$$\frac{7}{4} < x < 4$$
$$1\frac{3}{4} < x < 4$$

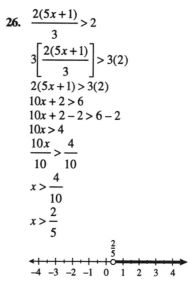

26. $\dfrac{2(5x + 1)}{3} > 2$
$$3\left[\frac{2(5x + 1)}{3}\right] > 3(2)$$
$$2(5x + 1) > 3(2)$$
$$10x + 2 > 6$$
$$10x + 2 - 2 > 6 - 2$$
$$10x > 4$$
$$\frac{10x}{10} > \frac{4}{10}$$
$$x > \frac{4}{10}$$
$$x > \frac{2}{5}$$

```
          2
          5
◄─┼──┼──┼──┼──○──┼──┼──┼──┼──►
 -4 -3 -2 -1  0  1  2  3  4
```

27. 81.3%

28. $0.047(126.2) = \$5.9314 \text{ billion}$

29. $0.067(360) = 24.12°$

Chapter 2 - Cumulative Review

1. **a.** True

   b. True

   c. False

   d. True

2. **a.** $|0| < 2$

   b. $|-5| = 5$

   c. $|-3| > |-2|$

   d. $|5| < |6|$

   e. $|-7| > |6|$

3. $\dfrac{2}{15} \cdot \dfrac{5}{13} = \dfrac{10}{195} = \dfrac{2}{39}$

4. $\dfrac{3 + |4 - 3| + 2^2}{6 - 3} = \dfrac{3 + |1| + 4}{6 - 3} = \dfrac{8}{3}$

5. **a.** $3 + (-7) = 3 - 7 = -4$

   b. $(-2) + (10) = -2 + 10 = 8$

   c. $0.2 + (-0.5) = 0.2 - 0.5 = -0.3$

6. **a.** $-3 + [(-2 - 5) - 2] = -3 + [-7 - 2] = -3 + [-9]$
   $= -3 - 9 = -12$

   b. $2^3 - |10| + [-6 - (-5)] = 8 - 10 + [-6 + 5]$
   $= 8 - 10 - 1 = -3$

7. **a.** $(-1.2)(0.05) = -0.06$

   b. $\dfrac{2}{3} \cdot \left(-\dfrac{7}{10}\right) = -\dfrac{14}{30} = -\dfrac{7}{15}$

8. **a.** $-3 \Rightarrow 3$

   b. $5 \Rightarrow -5$

   c. $0 \Rightarrow 0$

   d. $|-2| = 2 \Rightarrow -2$

9. **a.** $70

   b. 278 miles

10. **a.** unlike

    b. like

    c. like

    d. like

11. $(2x - 3) - (4x - 2)$
    $= 2x - 3 - 4x + 2$
    $= -2x - 1$

12. $x - 7 = 10$
    $x - 7 + 7 = 10 + 7$
    $x = 17$

13. $5x - 2 = 18$
    $5x - 2 + 2 = 18 + 2$
    $5x = 20$
    $\dfrac{5x}{5} = \dfrac{20}{5}$
    $x = 4$

14. $\dfrac{2(a + 3)}{3} = 6a + 2$
    $\dfrac{2a + 6}{3} = 6a + 2$
    $2a + 6 = 18a + 6$
    $0 = 20a$
    $0 = a$

15. Let x = number of Republicans
    $x + (x + 12) = 100$
    $2x + 12 = 100$
    $2x = 88$
    $\dfrac{2x}{2} = \dfrac{88}{2}$
    $x = 44$
    $x + 12 = 56$
    44 Republicans
    56 Democrats

16. $\dfrac{31,680}{400} = 79.2$ years

17. **a.** $35\% = 0.35$

    b. $89.5\% = 0.895$

    c. $150\% = 1.50$

18. Let x = unknown percent
    $63 = x(72)$
    $0.875 = x$
    87.5%

19.

rate · time = distance			
fast	18	x	$18x$
slow	10	x	$10x$

$18x + 10x = 98$
$28x = 98$
$x = 3.5$
$x + x = 7$ hours

20. $2 < x \le 4$

21. $2(x - 3) - 5 \le 3(x + 2) - 18$
$2x - 6 - 5 \le 3x + 6 - 18$
$2x - 11 \le 3x - 12$
$1 \le x$

Chapter 3

Section 3.1 Mental Math

1. Answers may vary; Ex. (5, 5), (7, 3)

3. Answers may vary; Ex. (3, 5), (3, 0)

Exercise Set 3.1

1. quadrant I

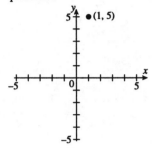

3. no quadrant, x-axis

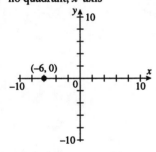

5. quadrant IV

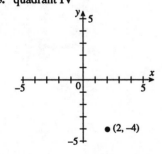

7. no quadrant, x-axis

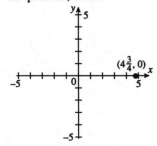

9. no quadrant, origin

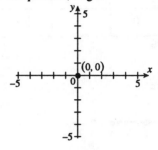

11. no quadrant, y-axis

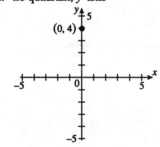

13. when $a = b$

15. $2(4) + 2(9) = 26$ units

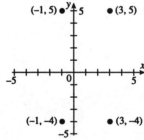

17.

	$2x + y = 7$
(3, 1)	$2(3) + 1 \; ? \; 7$
	$6 + 1 \; ? \; 7$
yes	$7 = 7$
	$2x + y = 7$
(7, 0)	$2(7) + 0 \; ? \; 7$
	$14 + 0 \; ? \; 7$
no	$14 \neq 7$
	$2x + y = 7$
(0, 7)	$2(0) + 7 \; ? \; 7$
	$0 + 7 \; ? \; 7$
yes	$7 = 7$

19.

$y = -5x$

(−1, −5) $-5 \ ? \ -5(-1)$

no $-5 \neq +5$

$y = -5x$

(0, 0) $0 \ ? \ -5(0)$

yes $0 = 0$

$y = -5x$

(2, −10) $-10 \ ? \ -5(2)$

yes $-10 = -10$

21.

$x = 5$

(4, 5) $4 \neq 5$ no

(5, 4) $5 = 5$ yes

(5, 0) $5 = 5$ yes

23. $x + 2y = 9$; (5, 2)

$5 + 2(2) \ ? \ 9$

$5 + 4 \ ? \ 9$

$9 = 9$ yes

$x + 2y = 9$; (0, 9)

$0 + 2(9) \ ? \ 9$

$18 \neq 9$ no

25. $2x - y = 11$; (3, −4)

$2(3) - (-4) \ ? \ 11$

$6 + 4 \ ? \ 11$

$10 \neq 11$ no

$2x - y = 11$; (9, 8)

$2(9) - 8 \ ? \ 11$

$18 - 8 \ ? \ 11$

$10 \neq 11$ no

27. $x = \dfrac{1}{3}y$; (0, 0)

$0 \ ? \ \dfrac{1}{3}(0)$

$0 = 0$ yes

$x = \dfrac{1}{3}y$; (3, 9)

$3 \ ? \ \dfrac{1}{3}(9)$

$3 = 3$ yes

29. $y = -2$; (−2, −2)

$-2 = -2$ yes

$y = -2$; (5, −2)

$-2 = -2$ yes

31.

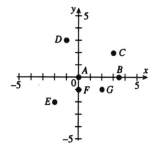

A: (0, 0); B: $\left(3\dfrac{1}{2}, \ 0\right)$; C: (3, 2);

D: (−1, 3); E: (−2, −2); F: (0, −1); G: (2, −1)

33. (+, −); quadrant IV

35. (−, y); quadrant II or III

37. $x - 4y = 4$; (, −2)

$x - 4(-2) = 4$

$x + 8 = 4$

$x = -4$ (−4, −2)

$x - 4y = 4$ (4,)

$4 - 4y = 4$

$-4y = 0$

$y = 0$ (4, 0)

39. $3x + y = 9$; (0,)

$3(0) + y = 9$

$0 + y = 9$

$y = 9$ (0, 9)

$3x + y = 9$ (, 0)

$3x + 0 = 9$

$3x = 9$

$x = 3$ (3, 0)

41. $y = -7$; (11,)

$y = -7$ (11, −7)

$y = -7$; (, −7)

$-7 = -7$

identity, true for all x.

43. $x + 3y = 6$; (0,)

$0 + 3y = 6$

$3y = 6$

$y = 2$ (0, 2)

$x + 3y = 6$; (, 0)

$x + 3(0) = 6$

$x + 0 = 6$

$x = 6$ (6, 0)

$x + 3y = 6;$ (, 1)
$x + 3(1) = 6$
$x + 3 = 6$
$x = 3$ (3, 1)

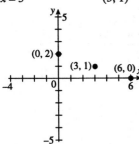

45. $2x - y = 12;$ (0,)
$2(0) - y = 12$
$-y = 12$
$y = -12$ (0, -12)

$2x - y = 12;$ (, -2)
$2x - (-2) = 12$
$2x + 2 = 12$
$2x = 10$
$x = 5$ (5, -2)

$2x - y = 12;$ (-3,)
$2(-3) - y = 12$
$-6 - y = 12$
$-y = 18$
$y = -18$ (-3, -18)

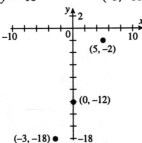

47. $2x + 7y = 5;$ (0,)
$2(0) + 7y = 5$
$7y = 5$
$y = \dfrac{5}{7}$ $\left(0, \dfrac{5}{7}\right)$

$2x + 7y = 5;$ (, 0)
$2x + 7(0) = 5$
$2x = 5$
$x = \dfrac{5}{2}$ $\left(\dfrac{5}{2}, 0\right)$

$2x + 7y = 5;$ (, 1)
$2x + 7(1) = 5$
$2x + 7 = 5$
$2x = -2$
$x = -1$ (-1, 1)

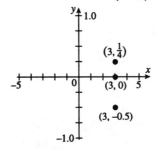

49. $x = 3;$ (, 0)
$x = 3$ (3, 0)

$x = 3;$ (, -0.5)
$x = 3$ (3, -0.5)

$x = 3;$ $\left(\ , \dfrac{1}{4}\right)$
$x = 3$ $\left(3, \dfrac{1}{4}\right)$

51. $x = -5y;$ (, 0)
$x = -5(0)$
$x = 0$ (0, 0)

$x = -5y;$ (, 1)
$x = -5(1)$
$x = -5$ (-5, 1)

$x = -5y;$ (10,)
$10 = -5y$
$-2 = y$ (10, -2)

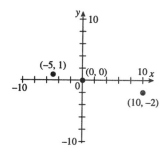

53. Answers may vary.

55. a.

x	100	200	300
y	13,000	21,000	29,000

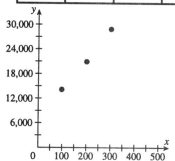

b. $y = 80x + 5000$
$8600 = 80x + 5000$
$3600 = 80x$
$45 = x$

57. Year 1: $6,000 - 5,500 = \$500$ million
Year 2: $7,500 - 6,000 = \$1500$ million
Year 3: $8,500 - 7,500 = \$1000$ million
Year 4: $10,000 - 8,500 = \$1500$ million

59. $x + y = 5$
$x - x + y = 5 - x$
$y = 5 - x$

61. $2x + 4y = 5$
$2x - 2x + 4y = 5 - 2x$
$4y = 5 - 2x$
$\dfrac{4y}{4} = \dfrac{5 - 2x}{4}$
$y = \dfrac{5}{4} - \dfrac{1}{2}x$

63. $10x = -5y$
$\dfrac{10x}{-5} = \dfrac{-5y}{-5}$
$-2x = y$
or
$y = -2x$

65. $x - 3y = 6$
$x - 3y - x = 6 - x$
$-3y = 6 - x$
$\dfrac{-3y}{-3} = \dfrac{6 - x}{-3}$
$y = -2 + \dfrac{1}{3}x$

Exercise Set 3.2

1. Yes; it can be written in the form $Ax + By = C.$

3. Yes; it can be written in the form $Ax + By = C.$

5. No; x is squared.

7. Yes; it can be written in the form $Ax + By = C.$

9. $x + y = 4$
Let $x = 1$, $x = 0$, $x = -2$
$1 + y = 4$ $0 + y = 4$ $-2 + y = 4$
$y = 3$ $y = 4$ $y = 6$
$(1, 3)$ $(0, 4)$ $(-2, 6)$

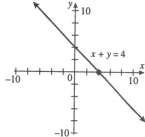

11. $x - y = -2$
Let $x = 2$, $x = 0$, $x = -1$
$2 - y = -2$ $0 - y = -2$ $-1 - y = -2$
$-y = -4$ $-y = -2$ $-y = -1$
$y = 4$ $y = 2$ $y = 1$
$(2, 4)$ $(0, 2)$ $(-1, 1)$

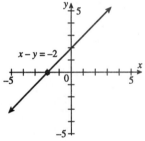

17. $y = x + 5$

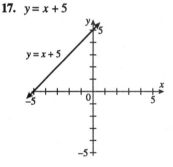

13. $x - 2y = 4$

Let $y = 0$, $y = -2$, $y = 2$
$x - 2(0) = 4$ $x - 2(-2) = 4$ $x - 2(2) = 4$
$x = 4$ $x + 4 = 4$ $x - 4 = 4$
$(4, 0)$ $x = 0$ $x = 8$
  $(0, -2)$ $(8, 2)$

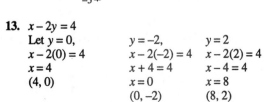

19. $2x + 3y = 6$

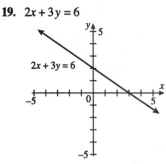

21. Answers may vary.

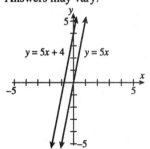

15. $y = 6x + 3$

Let $x = 0$, $x = -1$, $x = 1$
$y = 6(0) + 3$ $y = 6(-1) + 3$ $y = 6(1) + 3$
$y = 0 + 3$ $y = -6 + 3$ $y = 6 + 3$
$y = 3$ $y = -3$ $y = 9$
$(0, 3)$ $(-1, -3)$ $(1, 9)$

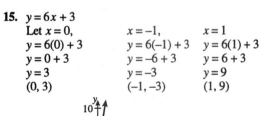

23. Answers may vary.

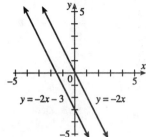

25. Answers may vary.

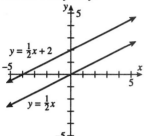

27. $x - 2y = -6$

Let $x = 0$,	$y = 0$,	$y = 1$
$0 - 2y = -6$	$x - 2(0) = -6$	$x - 2(1) = -6$
$-2y = -6$	$x + 0 = -6$	$x - 2 = -6$
$y = 3$	$x = -6$	$x = -4$
$(0, 3)$	$(-6, 0)$	$(-4, 1)$

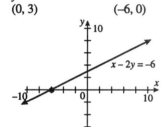

29. $y = 6x$

Let $x = 0$,	$x = 1$,	$x = -1$
$y = 6(0)$	$y = 6(1)$	$y = 6(-1)$
$y = 0$	$y = 6$	$y = -6$
$(0, 0)$	$(1, 6)$	$(-1, -6)$

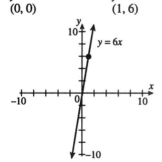

31. $3y - 10 = 5x$

Let $x = 0$,	$y = 0$,	$y = 5$
$3y - 10 = 5(0)$	$3(0) - 10 = 5x$	$3(5) - 10 = 5x$
$3y = 10$	$-10 = 5x$	$5 = 5x$
$y = \dfrac{10}{3}$	$-2 = x$	$1 = x$
$\left(0, \dfrac{10}{3}\right)$	$(-2, 0)$	$(1, 5)$

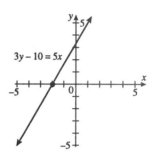

33. $x + 3y = 9$

If $x = 0$, then	If $x = 9$, then	If $x = 3$, then
$0 + 3y = 9$	$9 + 3y = 9$	$3 + 3y = 9$
$3y = 9$	$3y = 0$	$3y = 6$
$y = 3$	$y = 0$	$y = 2$

x	y
0	3
9	0
3	2

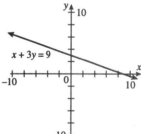

35. $y - x = -1$

If $x = 0$, then	If $x = 1$, then	If $x = 2$, then
$y - 0 = -1$	$y - 1 = -1$	$y - 2 = -1$
$y = -1$	$y = 0$	$y = 1$

x	y
0	-1
1	0
2	1

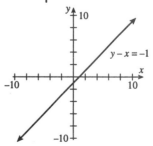

37. $x = -3y$

If $x = 0$, then   If $x = -3$, then   If $x = 6$, then
$0 = -3y$ $-3 = -3y$ $6 = -3y$
$0 = y$ $1 = y$ $-2 = y$

x	y
0	0
−3	1
6	−2

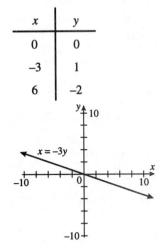

39. $5x - y = 10$

If $x = 0$, then   If $x = 2$, then   If $x = 1$, then
$5(0) - y = 10$ $5(2) - y = 10$ $5(1) - y = 10$
$0 - y = 10$ $10 - y = 10$ $5 - y = 10$
$y = -10$ $y = 0$ $-y = 5$
                                              $y = -5$

x	y
0	−10
2	0
1	−5

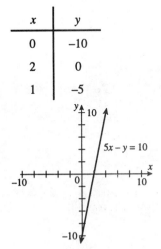

41. $y = \dfrac{1}{2}x + 2$

If $x = 0$, then   If $x = -4$, then   If $x = 4$, then
$y = \dfrac{1}{2}(0) + 2$ $y = \dfrac{1}{2}(-4) + 2$ $y = \dfrac{1}{2}(4) + 2$
$y = 0 + 2$ $y = -2 + 2$ $y = 2 + 2$
$y = 2$ $y = 0$ $y = 4$

x	y
0	2
−4	0
4	4

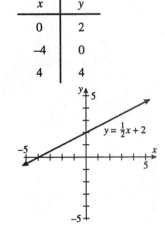

43. $y = x^2$

x	y
0	0
1	1
−1	1
2	4
−2	4

45. C

47. D

49. D

51. A

53. B

55. Answers may vary.

57. Yes; answers may vary.

59. $(4, -1)$

61. $3(x - 2) + 5x = 6x - 16$
$3x - 6 + 5x = 6x - 16$
$8x - 6 = 6x - 16$
$2x - 6 = -16$
$2x = -10$
$x = -5$

63. $3x + \dfrac{2}{5} = \dfrac{1}{10}$

$3x = \dfrac{1}{10} - \dfrac{2}{5}$

$3x = \dfrac{1}{10} - \dfrac{4}{10}$

$3x = -\dfrac{3}{10}$

$x = -\dfrac{1}{10}$

65. $x - y = -3$

x	y
0	3
−3	0

67. $y = 2x$

x	y
0	0
0	0

Exercise Set 3.3

1. $x = -1$; $y = 1$; $(-1, 0)$; $(0, 1)$

3. $x = -2$; $(-2, 0)$

5. $x = -1$; $x = 1$; $y = 1$; $y = -2$;
$(-1, 0)$; $(1, 0)$; $(0, 1)$; $(0, -2)$

7. infinite

9. zero

11.

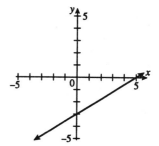

13.

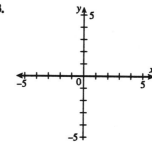

15. $x - y = 3$

Let $x = 0$ $y = 0$
$0 - y = 3$ $x - 0 = 3$
$-y = 3$ $x = 3$
$y = -3$ $(3, 0)$
$(0, -3)$

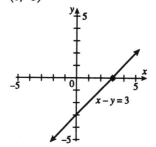

17. $x = 5y$

Let $x = 0$ $y = 0$
$0 = 5y$ $x = 5(0)$
$0 = y$ $x = 0$
$(0, 0)$ $(0, 0)$

The x-intercept and y-intercept are the same so another point must be found.

$x = 5y$
If $y = 1$
$x = 5(1)$
$x = 5$
$(5, 1)$

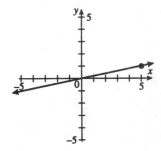

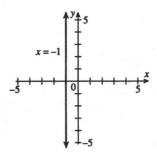

19. $-x + 2y = 6$
Let $x = 0$, $y = 0$
$-0 + 2y = 6$ $-x + 2(0) = 6$
$2y = 6$ $-x = 6$
$y = 3$ $x = -6$
$(0, 3)$ $(-6, 0)$

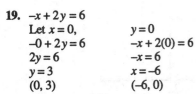

25. $y = 0$
$0x + y = 0$
Let $x = 1$, $x = 2$, $x = 0$
$0(1) + y = 0$ $0(2) + y = 0$ $0(0) + y = 0$
$y = 0$ $y = 0$ $y = 0$
$(1, 0)$ $(2, 0)$ $(0, 0)$

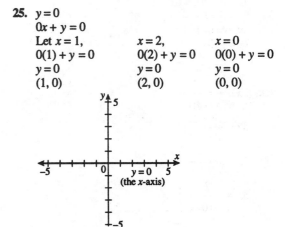

21. $2x - 4y = 8$
Let $x = 0$, $y = 0$
$2(0) - 4y = 8$ $2x - 4(0) = 8$
$-4y = 8$ $2x = 8$
$y = -2$ $x = 4$
$(0, -2)$ $(4, 0)$

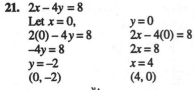

27. $y + 7 = 0$
$y = -7$
$0x + y = -7$
Let $x = -1$, $x = 2$, $x = 0$
$0(-1) + y = -7$ $0(2) + y = -7$ $0(0) + y = -7$
$y = -7$ $y = -7$ $y = -7$
$(-1, -7)$ $(2, -7)$ $(0, -7)$

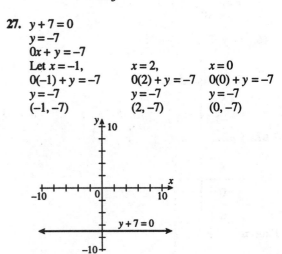

23. $x = -1$
$x + 0y = -1$
Let $y = 1$, $y = 0$, $y = -2$
$x + 0(1) = -1$ $x + 0(0) = -1$ $x + 0(-2) = -1$
$x = -1$ $x = -1$ $x = -1$
$(-1, 1)$ $(-1, 0)$ $(-1, -2)$

29. $x = 1$

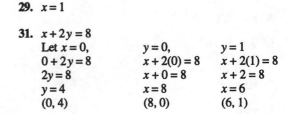

31. $x + 2y = 8$
Let $x = 0$, $y = 0$, $y = 1$
$0 + 2y = 8$ $x + 2(0) = 8$ $x + 2(1) = 8$
$2y = 8$ $x + 0 = 8$ $x + 2 = 8$
$y = 4$ $x = 8$ $x = 6$
$(0, 4)$ $(8, 0)$ $(6, 1)$

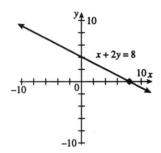

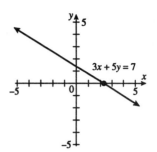

33. $x - 7 = 3y$

Let $x = 0$,	$y = 0$,	$x = 1$
$0 - 7 = 3y$	$x - 7 = 3(0)$	$1 - 7 = 3y$
$-7 = 3y$	$x - 7 = 0$	$-6 = 3y$
$-\dfrac{7}{3} = y$	$x = 7$	$-2 = y$
$\left(0, \ -\dfrac{7}{3}\right)$	$(7, 0)$	$(1, -2)$

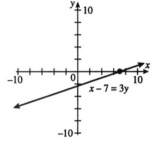

35. $x = -3$
$x + 0y = -3$

Let $y = 1$,	$y = 2$,	$y = 0$
$x + 0(1) = -3$	$x + 0(2) = -3$	$x + 0(0) = -3$
$x = -3$	$x = -3$	$x = -3$
$(-3, 1)$	$(-3, 2)$	$(-3, 0)$

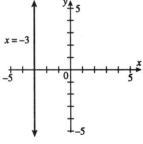

37. $3x + 5y = 7$

Let $x = 0$,	$y = 0$,	$x = 1$
$3(0) + 5y = 7$	$3x + 5(0) = 7$	$3(1) + 5y = 7$
$5y = 7$	$3x = 7$	$5y = 4$
$x = \dfrac{7}{5}$	$x = \dfrac{7}{3}$	$y = \dfrac{4}{5}$
$\left(0, \dfrac{7}{5}\right)$	$\left(\dfrac{7}{3}, 0\right)$	$\left(1, \dfrac{4}{5}\right)$

39. $x = y$

Let $x = 0$,	$y = 2$	$y = -1$
$0 = y$	$x = 2$	$x = -1$
$(0, 0)$	$(2, 2)$	$(-1, -1)$

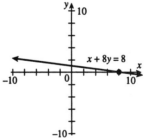

41. $x + 8y = 8$

Let $x = 0$,	$y = 0$,	$x = 3$
$0 + 8y = 8$	$x + 8(0) = 8$	$3 + 8y = 8$
$8y = 8$	$x = 8$	$8y = 5$
$y = 1$	$(8, 0)$	$y = \dfrac{5}{8}$
$(0, 1)$		$\left(3, \ \dfrac{5}{8}\right)$

43. $5 = 6x - y$

Let $x = 0$,	$y = 0$,	$y = 1$
$5 = 6(0) - y$	$5 = 6x - 0$	$5 = 6x - 1$
$5 = -y$	$5 = 6x$	$6 = 6x$
$-5 = y$	$\dfrac{5}{6} = x$	$1 = x$
$(0, -5)$	$\left(\dfrac{5}{6}, 0\right)$	$(1, 1)$

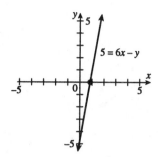

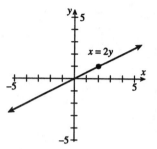

45. $-x + 10y = 11$

Let $x = 0$, $y = 0$, $x = -1$
$-0 + 10y = 11$ $-x + 10(0) = 11$ $-1(-1) + 10y = 11$
$10y = 11$ $-x = 11$ $1 + 10y = 11$
$y = \dfrac{11}{10}$ $x = -11$ $10y = 10$

$\left(0, \dfrac{11}{10}\right)$ $(-11, 0)$ $y = 1$

                                     $(-1, 1)$

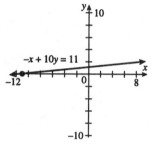

47. $y = 1$
$0x + y = 1$
Let $x = 1$, $x = 2$, $x = 0$
$0(1) + y = 1$ $0(2) + y = 1$ $0(0) + y = 1$
$y = 1$ $y = 1$ $y = 1$
$(1, 1)$ $(2, 1)$ $(0, 1)$

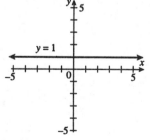

49. $x = 2y$
Let $y = 0$, $y = 1$, $y = -2$
$x = 2(0)$ $x = 2(1)$ $x = 2(-2)$
$x = 0$ $x = 2$ $x = -4$
$(0, 0)$ $(2, 1)$ $(-4, -2)$

51. $x + 3 = 0$
$x = -3$
$x + 0y = -3$
Let $y = 1$, $y = 2$, $y = 0$
$x + 0(1) = -3$ $x + 0(2) = -3$ $x + 0(0) = -3$
$x = -3$ $x = -3$ $x = -3$
$(-3, 1)$ $(-3, 2)$ $(-3, 0)$

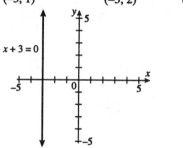

53. $x = 4y - \dfrac{1}{3}$

Let $y = 0$, $y = \dfrac{1}{3}$, $y = -\dfrac{2}{3}$

$x = 4(0) - \dfrac{1}{3}$ $x = 4\left(\dfrac{1}{3}\right) - \dfrac{1}{3}$ $x = 4\left(-\dfrac{2}{3}\right) - \dfrac{1}{3}$

$x = -\dfrac{1}{3}$ $x = \dfrac{4}{3} - \dfrac{1}{3}$ $x = -\dfrac{8}{3} - \dfrac{1}{3}$

$\left(-\dfrac{1}{3}, 0\right)$ $x = \dfrac{3}{3} = 1$ $x = -\dfrac{9}{3} = -3$

                 $\left(1, \dfrac{1}{3}\right)$ $\left(-3, -\dfrac{2}{3}\right)$

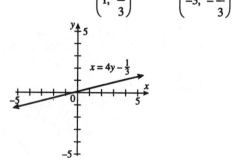

55. $2x + 3y = 6$

Let $x = 0$, $y = 0$, $x = 1$

$2(0) + 3y = 6$ $2x + 3(0) = 6$ $2(1) + 3y = 6$

$3y = 6$ $2x = 6$ $2 + 3y = 6$

$y = 2$ $x = 3$ $3y = 4$

$(0, 2)$ $(3, 0)$ $y = \dfrac{4}{3}$

$$\left(1,\ \dfrac{4}{3}\right)$$

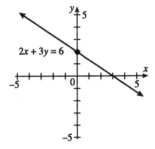

57. C

59. E

61. B

63.

$y = 4000 - 800x$

65. Answers may vary.

67. Answers may vary.

69. $\dfrac{-6 - 3}{2 - 8} = \dfrac{-9}{-6} = \dfrac{3}{2}$

71. $\dfrac{-8 - (-2)}{-3 - (-2)} = \dfrac{-8 + 2}{-3 + 2} = \dfrac{-6}{-1} = 6$

73. $\dfrac{0 - 6}{5 - 0} = \dfrac{-6}{5} = -\dfrac{6}{5}$

Section 3.4 Mental Math

1. upward

3. horizontal

Exercise Set 3.4

1. $(-1, 2)$ and $(2, -2)$

$$m = \dfrac{-2 - 2}{2 - (-1)} = \dfrac{-4}{2 + 1} = -\dfrac{4}{3}$$

3. $(2, 3)$ and $(2, -1)$

$$m = \dfrac{-1 - 3}{2 - 2} = \dfrac{-4}{0}$$

undefined slope

5. $(-3, -2)$ and $(-1, 3)$

$$m = \dfrac{3 - (-2)}{-1 - (-3)} = \dfrac{3 + 2}{-1 + 3} = \dfrac{5}{2}$$

7. $(0, 0)$ and $(7, 8)$

$$m = \dfrac{8 - 0}{7 - 0} = \dfrac{8}{7}$$

9. $(-1, 5)$ and $(6, -2)$

$$m = \dfrac{-2 - 5}{6 - (-1)} = \dfrac{-7}{6 + 1} = \dfrac{-7}{7} = -1$$

11. $(1, 4)$ and $(5, 3)$

$$m = \dfrac{3 - 4}{5 - 1} = \dfrac{-1}{4} = -\dfrac{1}{4}$$

13. $(-4, 3)$ and $(-4, 5)$

$$m = \dfrac{5 - 3}{4 - (-4)} = \dfrac{2}{0}$$

Undefined, division by zero is not defined.

15. $(-2, 8)$ and $(1, 6)$

$$m = \dfrac{6 - 8}{1 - (-2)} = \dfrac{-2}{1 + 2} = -\dfrac{2}{3}$$

17. $(1, 0)$ and $(1, 1)$

$$m = \dfrac{1 - 0}{1 - 1} = \dfrac{1}{0}$$

undefined

19. $(5, -11)$ and $(1, -11)$

$$m = \dfrac{-11 - (-11)}{1 - 5} = \dfrac{-11 + 11}{-4} = \dfrac{0}{4} = 0$$

21. line 1

23. line 2

25. D

27. B

29. E

31. (2, 3) and (6, 4)

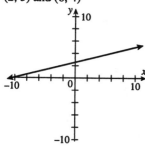

33. (0, –4) and (1, –1)

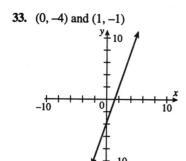

35. (–2, –7) and (–1, –8)

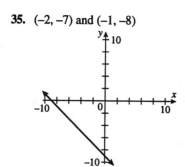

37. (–3, 4) and (2, 1)

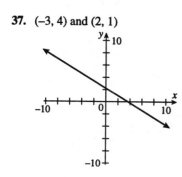

39. $y = 5x - 2$
$m = 5$
$b = -2$

41. $2x + y = 7$
$y = -2x + 7$
$m = -2$
$b = 7$

43. $x = 1$
vertical line
slope is undefined

45. $y = -3$
horizontal line
$m = 0$

47. $2x - 3y = 10$
$-3y = -2x + 10$
$y = \dfrac{-2}{-3}x + \dfrac{10}{-3}$
$y = \dfrac{2}{3}x - \dfrac{10}{3}$
$m = \dfrac{2}{3}$
$b = -\dfrac{10}{3}$

49. $x = 2y$
$2y = x$
$y = \dfrac{1x}{2}$
$y = \dfrac{1}{2}x + 0$
$m = \dfrac{1}{2}$
$b = 0$

51. $\dfrac{\text{rise}}{\text{run}} = \dfrac{6}{10} = \dfrac{3}{5}$

53. $\dfrac{x}{18} = \dfrac{1}{3}$
$18 \cdot \dfrac{x}{18} = \dfrac{1}{3} \cdot 18$
$x = 6$

55. $\dfrac{\text{rise}}{\text{run}} = \dfrac{16}{100} = 0.16$
16%

57. a. $m = \dfrac{0-(-3)}{0-(-3)} = \dfrac{3}{3} = 1$

   b. -1

59. a. $m = \dfrac{5-(-4)}{3-(-8)} = \dfrac{5+4}{3+8} = \dfrac{9}{11}$

   b. $-\dfrac{11}{9}$

61. $m = \dfrac{0-6}{-2-0} = \dfrac{-6}{-2} = 3$

   $m = \dfrac{8-5}{1-0} = \dfrac{3}{1} = 3$

   parallel

63. $m = \dfrac{8-6}{-2-2} = \dfrac{2}{-4} = -\dfrac{1}{2}$

   $m = \dfrac{5-3}{1-0} = \dfrac{2}{1} = 2$

   perpendicular

65. $m = \dfrac{8-6}{7-3} = \dfrac{2}{4} = \dfrac{1}{2}$

   $m = \dfrac{7-6}{2-0} = \dfrac{1}{2}$

   parallel

67. $m = \dfrac{-5-(-3)}{6-2} = \dfrac{-5+3}{4} = \dfrac{-2}{4} = -\dfrac{1}{2}$

   $m = \dfrac{-4-(-2)}{-3-5} = \dfrac{-4+2}{-8} = \dfrac{-2}{-8} = \dfrac{1}{4}$

   neither

69. $m = \dfrac{0-(-3)}{-1-(-4)} = \dfrac{3}{-1+4} = \dfrac{3}{3} = 1$

   $m = \dfrac{0-(-4)}{0-4} = \dfrac{4}{-4} = -1$

   perpendicular

71. Answers may vary.

73. 20 mpg

75. $21.6 - 18.2 = 3.4$ mpg

77. Between 1986 and 1987

79. 1987

81. $m = \dfrac{-6-(-5)}{-2-(-7)} = \dfrac{-6+5}{-2+7} = -\dfrac{1}{5}$

83. $\dfrac{-3-0}{1-0} = \dfrac{-3}{1} = -3$

   $m = \dfrac{1}{3}$

85. $m = 1$

87. $m = \dfrac{9.3-6.7}{-8.3-2.1} = \dfrac{2.6}{-10.4} = -0.25$

89. $m = \dfrac{5.1-0.2}{7.9-2.3} = \dfrac{4.9}{5.6} = 0.875$

91. $x + 3y = 6$

   line passes through $(6, 0)$ and $(0, 2)$

   $m = \dfrac{2-0}{0-6} = \dfrac{2}{-6} = -\dfrac{1}{3}$

   y-intercept $= 2$

   $x + 3y = 6$

   $3y = -x + 6$

   $y = \dfrac{-x+6}{3}$

   $y = -\dfrac{1}{3}x + 2$

   Answers may vary.

93. The line becomes steeper.

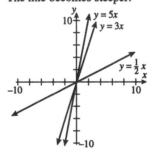

95. $-3x \le -9$

   $\dfrac{-3x}{-3} \ge \dfrac{-9}{-3}$

   $x \ge 3$

97. $\dfrac{x-6}{2} < 3$

$2\left(\dfrac{x-6}{2}\right) < 3 \cdot 2$

$x - 6 < 6$

$x - 6 + 6 < 6 + 6$

$x < 12$

99. $x - y = 6$ or $y = x - 6$
If $x = 0$, then $y = 0 - 6 = -6$
If $x = 6$, then $y = 6 - 6 = 0$
If $x = 7$, then $y = 7 - 6 = 1$

x	y
0	-6
6	0
7	1

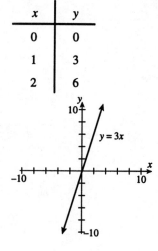

101. $y = 3x$
If $x = 0$, then $y = 3 \cdot 0 = 0$
If $x = 1$, then $y = 3 \cdot 1 = 3$
If $x = 2$, then $y = 3 \cdot 2 = 6$

x	y
0	0
1	3
2	6

103. $x = -2$

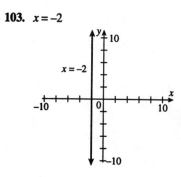

Section 3.5 Mental Math

1. yes

3. yes

5. $0 + 0 > -5$
$0 > -5$
yes

7. $x - y \le -1$
$0 - 0 \le -1$
$0 \le -1$
no

Exercise Set 3.5

1. $x - y > 3$

$0 - 3 > 3$	$2 - (-1) > 3$	$5 - 1 > 3$
$-3 > 3$	$2 + 1 > 3$	$4 > 3$
	$3 > 3$	
no	no	yes

3. $3x - 5y \le -4$

$3(2) - 5(3) \le -4$	$3(-1) - 5(-1) \le -4$	$3(4) - 5(0) \le -4$
$6 - 15 \le -4$	$-3 + 5 \le -4$	$12 - 0 \le -4$
$-9 \le -4$	$2 \le -4$	$12 \le -4$
yes	no	no

5. $x < -y$

$6 < -6$	$0 < -2$	$-5 < -1$
no	no	yes

7. $x + y \le 1$
Find the intercepts.
Let $x = 0$, $0 + y = 1$, $y = 1$, $(0, 1)$
Let $y = 0$, $x + 0 = 1$, $x = 1$, $(1, 0)$
The boundary line is solid.
Choose $(0, 0)$ as a test point.
$x + y \le 1$
$0 + 0 \,?\, 1$
Since $0 < 1$, the side containing $(0, 0)$ is shaded.

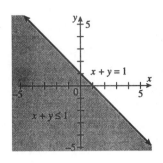

9. $2x + y > -4$
Find the intercepts.
Let $x = 0$, $2(0) + y = -4$, $y = -4$, $(0, -4)$
Let $y = 0$, $2x + 0 = -4$, $2x = -4$, $x = -2$, $(-2, 0)$
The boundary line is dashed.
Choose $(0, 0)$ as a test point.
$2x + y > -4$
$2(0) + 0 \ ? -4$
$0 \ ? -4$
Since $0 > -4$, the side of the line containing $(0, 0)$ is shaded.

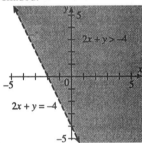

11. $x + 6y \le -6$
Find the intercepts.
Let $x = 0$, $0 + 6y = -6$, $6y = -6$, $y = -1$, $(0, -1)$
Let $y = 0$, $x + 6(0) = -6$, $x = -6$, $(-6, 0)$
The boundary line is solid.
Choose $(0, 0)$ as a test point.
$x + 6y \le -6$
$0 + 6(0) \ ? -6$
$0 \ ? -6$
Since $0 \not< -6$, the side *not* containing $(0, 0)$ is shaded.

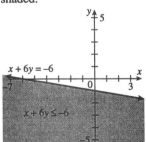

13. $2x + 5y > -10$
Find the intercepts.
Let $x = 0$, $2(0) + 5y = -10$, $5y = -10$, $y = -2$, $(0, -2)$
Let $y = 0$, $2x + 5(0) = -10$, $2x = -10$, $x = -5$, $(-5, 0)$
The boundary line is dashed.
Choose $(0, 0)$ as a test point.
$2x + 5y > -10$
$2(0) + 5(0) \ ? -10$
$0 \ ? -10$
Since $0 > -10$, the side containing $(0, 0)$ is shaded.

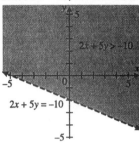

15. $x + 2y \le 3$
Find the intercepts.
Let $x = 0$, $0 + 2y = 3$, $2y = 3$, $y = \dfrac{3}{2}$, $\left(0, \ \dfrac{3}{2}\right)$
Let $y = 0$, $x + 2(0) = 3$, $x = 3$, $(3, 0)$
The boundary line is solid.
Choose $(0, 0)$ as a test point.
$x + 2y \le 3$
$0 + 2(0) \ ? 3$
$0 \ ? 3$
Since $0 \le 3$, the side containing $(0, 0)$ is shaded.

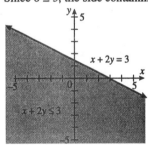

17. $2x + 7y > 5$
Find the intercepts.
Let $x = 0$, $2(0) + 7y = 5$, $7y = 5$, $y = \dfrac{5}{7}$, $\left(0, \ \dfrac{5}{7}\right)$
Let $y = 0$, $2x + 7(0) = 5$, $2x = 5$, $x = \dfrac{5}{2}$, $\left(\dfrac{5}{2}, \ 0\right)$
The boundary line is dashed.
Choose $(0, 0)$ as a test point.
$2x + 7y > 5$
$2(0) + 7(0) \ ? 5$
$0 \ ? 5$

Since $0 \not> 5$, the side *not* containing $(0, 0)$ is shaded.

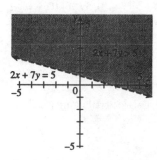

19. $x - 2y \geq 3$

Find the intercepts.

Let $x = 0$, $0 - 2y = 3$, $-2y = 3$, $y = -\dfrac{3}{2}$, $\left(0, -\dfrac{3}{2}\right)$

Let $y = 0$, $x - 2(0) = 3$, $x = 3$, $(3, 0)$

The boundary line is solid.

Choose $(0, 0)$ as a test point.

$x - 2y \geq 3$

$0 - 2(0) ? 3$

$0 ? 3$

Since $0 \not\geq 3$, the side *not* containing $(0, 0)$ is shaded.

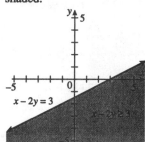

21. $5x + y < 3$

Find the intercepts.

Let $x = 0$, $5(0) + y = 3$, $y = 3$, $(0, 3)$

Let $y = 0$, $5x + 0 = 3$, $5x = 3$, $x = \dfrac{3}{5}$, $\left(\dfrac{3}{5}, 0\right)$

The boundary line is dashed.

Choose $(0, 0)$ as a test point.

$5x + y < 3$

$5(0) + 0 ? 3$

$0 ? 3$

Since $0 < 3$, the side containing $(0, 0)$ is **shaded**.

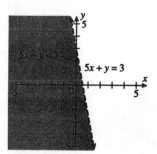

23. $4x + y < 8$

Find the intercepts.

Let $x = 0$, $4(0) + y = 8$, $y = 8$, $(0, 8)$

Let $y = 0$, $4x + 0 = 8$, $4x = 8$, $x = 2$, $(2, 0)$

The boundary line is **dashed**.

Choose $(0, 0)$ as a test point.

$4x + y < 8$

$4(0) + 0 ? 8$

$0 ? 8$

Since $0 < 8$, the side containing $(0, 0)$ is **shaded**.

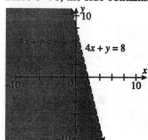

25. $y \geq 2x$

Find the two points.

Let $x = 0$, $y = 2(0)$, $y = 0$, $(0, 0)$

Let $x = 3$, $y = 2(3)$, $y = 6$, $(3, 6)$

The boundary line is solid.

Choose $(5, 0)$ as a test point.

$y \geq 2x$

$0 ? 2(5)$

$0 ? 10$

Since $0 \not\geq 10$, the side *not* containing $(5, 0)$ is shaded.

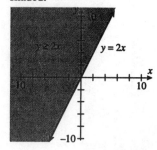

27. $x \geq 0$

$x = 0$ is a vertical line passing through $(0, 0)$.
The boundary line is solid.
Choose $(4, 1)$ as a test point.
$x \geq 0$
$4 ? 0$
Since $4 \geq 0$, the side containing $(4, 1)$ is shaded.

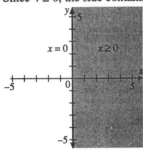

29. $y \leq -3$

$y = -3$ is a horizontal line passing through $(0, -3)$.
The boundary line is solid.
Choose $(1, 0)$ as a test point.
$y \leq -3$
$0 ? -3$
Since $0 \nleq -3$, the side *not* containing $(1, 0)$ is shaded.

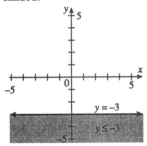

31. $2x - 7y > 0$

Find two points.
Let $x = 0$, $2(0) - 7y = 0$, $-7y = 0$, $y = 0$, $(0, 0)$
Let $x = 2$, $2(2) - 7y = 0$, $4 - 7y = 0$, $-7y = -4$,
$y = \dfrac{4}{7}$, $\left(2, \dfrac{4}{7}\right)$
The boundary line is dashed.
Choose $(0, 4)$ as a test point.
$2x - 7y > 0$
$2(0) - 7(4) ? 0$
$-28 ? 0$
Since $-28 \ngtr 0$, the side *not* containing $(0, 4)$ is shaded.

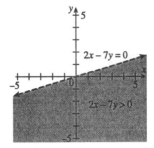

33. $3x - 7y \geq 0$

Find two points.
Let $x = 0$, $3(0) - 7y = 0$, $-7y = 0$, $y = 0$, $(0, 0)$
Let $x = 1$, $3(1) - 7y = 0$, $3 - 7y = 0$, $-7y = -3$,
$y = \dfrac{3}{7}$, $\left(1, \dfrac{3}{7}\right)$
The boundary line is solid.
Choose $(0, 4)$ as a test point.
$3x - 7y \geq 0$
$3(0) - 7(4) ? 0$
$-28 ? 0$
Since $-28 \ngeq 0$, the side *not* containing $(0, 4)$ is shaded.

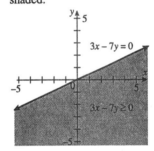

35. $x > y$

Find two points.
Let $x = 0$, $0 = y$, $(0, 0)$
Let $x = 4$, $4 = y$, $(4, 4)$
The boundary line is dashed.
Choose $(5, 0)$ as a test point.
$x > y$
$5 ? 0$
Since $5 > 0$, the side containing $(5, 0)$ is **shaded**.

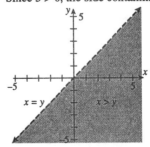

37. $x - y \leq 6$
Find the intercepts.
Let $x = 0$, $0 - y = 6$, $-y = 6$, $y = -6$, $(0, -6)$
Let $y = 0$, $x - 0 = 6$, $x = 6$, $(6, 0)$
The boundary line is solid.
Choose $(0, 0)$ as a test point.
$x - y \leq 6$
$0 - 0 ? 6$
$0 ? 6$
Since $0 \leq 6$, the side containing $(0, 0)$ is shaded.

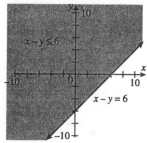

39. $-\dfrac{1}{4}y + \dfrac{1}{3}x > 1$
Find the intercepts.
Let $x = 0$, $-\dfrac{1}{4}y + \dfrac{1}{3}(0) = 1$, $-\dfrac{1}{4}y = 1$, $y = -4$,
$(0, -4)$
Let $y = 0$, $-\dfrac{1}{4}(0) + \dfrac{1}{3}x = 1$, $\dfrac{1}{3}x = 1$, $x = 3$, $(3, 0)$
The boundary line is dashed.
Choose $(0, 0)$ as a test point.
$-\dfrac{1}{4}y + \dfrac{1}{3}x > 1$
$-\dfrac{1}{4}(0) + \dfrac{1}{3}(0) ? 1$
$0 ? 1$
Since $0 \not> 1$, the side *not* containing $(0, 0)$ is shaded.

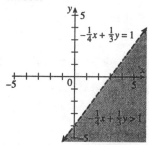

41. $-x < 0.4y$
Find two points.
Let $y = 0$, $-x = 0.4(0)$, $-x = 0$, $x = 0$, $(0, 0)$
Let $y = 10$, $-x = 0.4(10)$, $-x = 4$, $x = -4$, $(-4, 10)$
The boundary line is dashed.
Choose $(5, 0)$ as a test point.

$-x < 0.4y$
$-5 ? 0.4(0)$
$-5 ? 0$
Since $-5 < 0$, the side containing $(5, 0)$ is shaded.

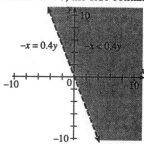

43. $x + y \geq 13$
Find the intercepts.
Let $x = 0$, $0 + y = 13$, $y = 13$, $(0, 13)$
Let $y = 0$, $x + 0 = 13$, $x = 13$, $(13, 0)$
The boundary line is solid.
Choose $(0, 0)$ as a test point.
$x + y \geq 13$
$0 + 0 ? 13$
$0 ? 13$
Since $0 \not\geq 13$ the side *not* containing $(0, 0)$ is
shaded.

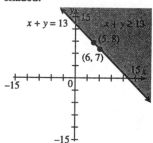

45. e

47. c

49. f

51. Answers may vary.

53. $2^3 = 2 \cdot 2 \cdot 2 = 8$

55. $(-2)^5 = (-2)(-2)(-2)(-2)(-2) = -32$

57. $3 \cdot 4^2 = 3 \cdot 4 \cdot 4 = 48$

59. $x^2 = (-5)^2 = (-5)(-5) = 25$

61. $2x^3 = 2(-1)^3 = 2(-1)(-1)(-1) = -2$

Chapter 3 - Review

1.

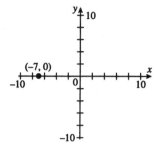

2.

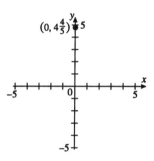

3.

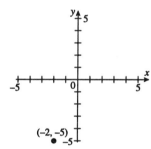

4.

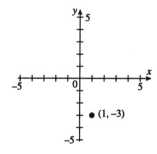

5.

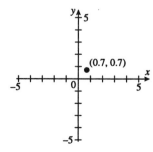

6.

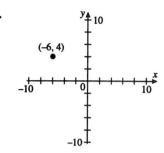

7.

(0, 56)	$7x - 8y = 56$
	$7(0) - 8(56)\ ?\ 56$
	$-448\ ?\ 56$
no	$-448 \neq 56$

(8, 0)	$7x - 8y = 56$
	$7(8) - 8(0)\ ?\ 56$
	$56 - 0\ ?\ 56$
	$56\ ?\ 56$
yes	$56 = 56$

8.

Test $(-5, 0)$:
$-2x + 5y = 10$
$-2(-5) + 5(0) = 10$
$10 + 0 = 10$
$10 = 10$

Test $(1, 1)$:
$-2(1) + 5(1) = 10$
$-2 + 5 = 10$
$3 \neq 10$

$(-5, 0)$ is a solution; $(1, 1)$ is not a solution.

9.

(13, 5)	$x = 13$
	$13\ ?\ 13$
yes	$13 = 13$

(13, 13)	$x = 13$
	$13\ ?\ 13$
yes	$13 = 13$

10.

Test $(7, 2)$:
$y = 2$
$Ax + By = 2$
$0(7) + 1(2) = 2$
$2 = 2$

Test $(2, 7)$:
$Ax + By = 2$
$0(2) + 1(7) = 2$
$7 \neq 2$

$(7, 2)$ is a solution; $(2, 7)$ is not a solution.

11. $-2 + y = 6x,$ $(7, \ )$
$-2 + y = 6(7)$
$-2 + y = 42$
$y = 42 + 2$
$y = 44$ $(7, 44)$

12. $y - 3x = 5$
$-8 - 3x = 5$
$$\frac{-3x}{-3} = \frac{13}{3}$$
$$x = -\frac{13}{3}$$
$$\left(-\frac{13}{3}, -8\right)$$

13. $9 = -3x + 4y;$ $(, 0)$
$9 = -3x + 4(0)$
$9 = -3x + 0$
$9 = -3x$
$-3 = x$ $(-3, 0)$

$9 = -3x + 4y;$ $(, 3)$
$9 = -3x + 4(3)$
$9 = -3x + 12$
$9 - 12 = -3x$
$-3 = -3x$
$1 = x$ $(1, 3)$

$9 = -3x + 4y;$ $(9,)$
$9 = -3(9) + 4y$
$9 = -27 + 4y$
$9 + 27 = 4y$
$36 = 4y$
$9 = y$ $(9, 9)$

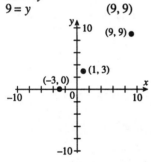

14. $y = 5$

  a. $(7, 5)$

  b. $(-7, 5)$

  c. $(0, 5)$

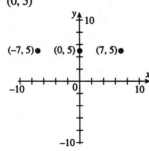

15. $x = 2y;$ $(, 0)$
$x = 2(0)$
$x = 0$ $(0, 0)$

$x = 2y;$ $(, 5)$
$x = 2(5)$
$x = 10$ $(10, 5)$

$x = 2y;$ $(, -5)$
$x = 2(-5)$
$x = -10$ $(-10, -5)$

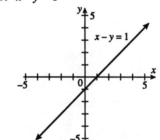

16. a. $y = 5x + 2000$

x	1	100	1000
y	2005	2500	7000

  b. $6430 = 5x + 2000$
$4430 = 5x$
$886 = x$
886 compact disks

17. $x - y = 1$

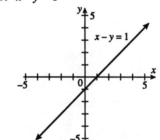

18. $x + y = 6$
Let $x = 0,$ $y = 0$
$0 + y = 6$ $x + 0 = 6$
$y = 6$ $x = 6$
$(0, 6)$ $(6, 0)$

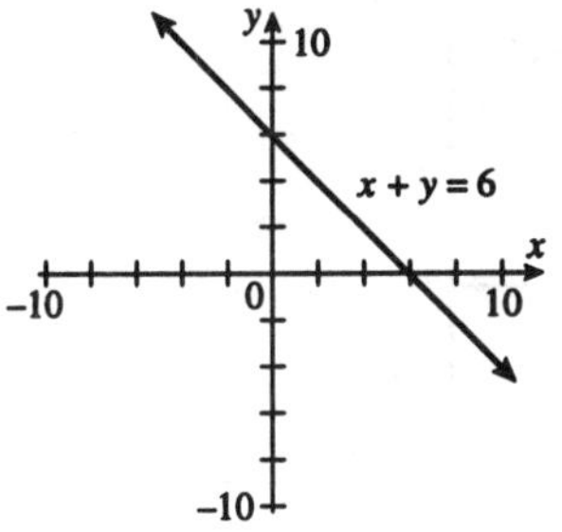

19. $x - 3y = 12$

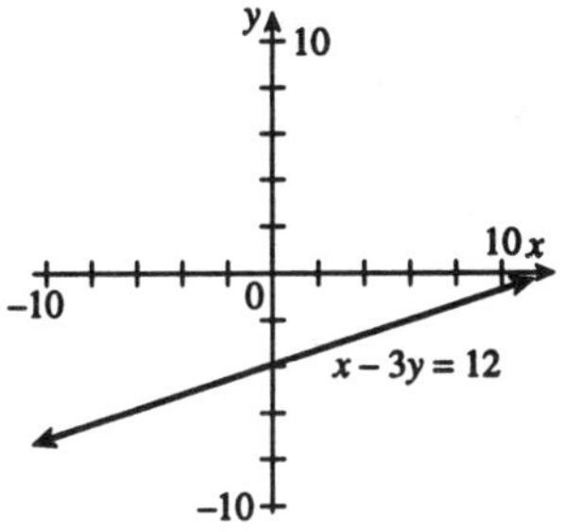

20. $5x - y = -8$

Let $x = 0$,	$y = 0$
$5(0) - y = -8$	$5x - 0 = -8$
$-y = -8$	$5x = -8$
$y = 8$	$y = -\frac{8}{5}$
$(0, 8)$	$\left(-\frac{8}{5}, 0\right)$

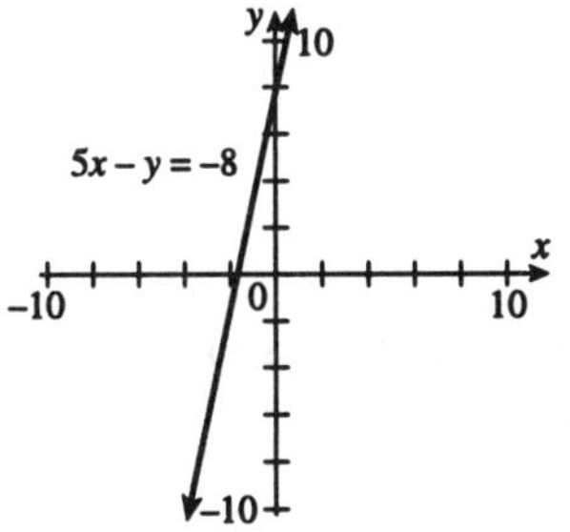

21. $x = 3y$

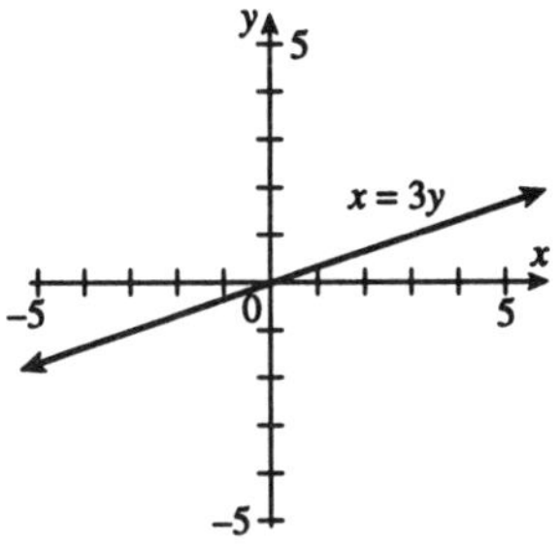

22. $y = -2x$

Let $x = 0$,	$x = 3$
$y = -2(0)$	$y = -2(3)$
$y = 0$	$y = -6$
$(0, 0)$	$(3, -6)$

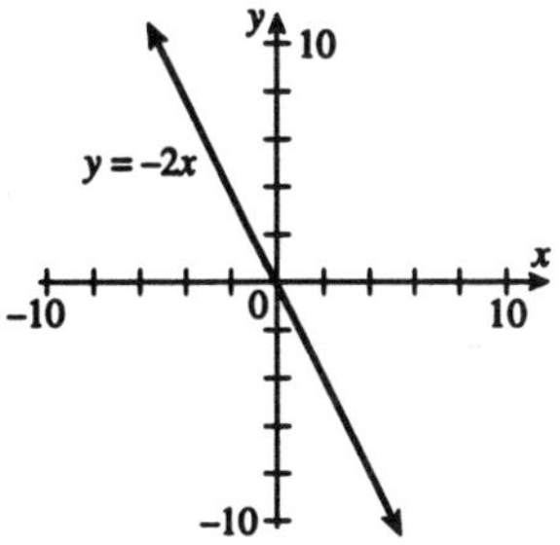

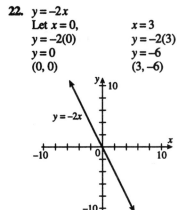

23. $2x - 3y = 6$

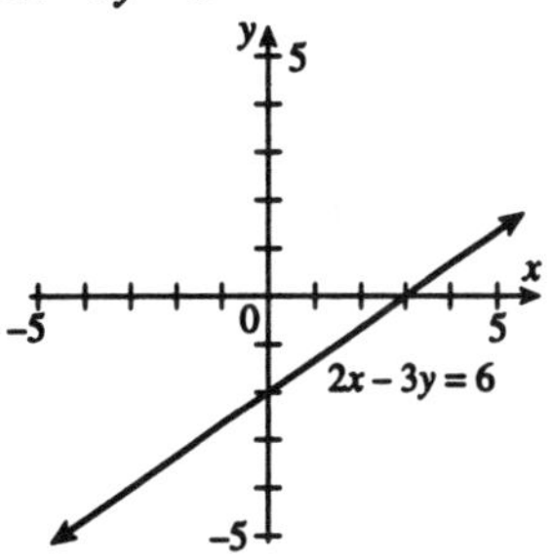

24. $4x - 3y = 12$

Let $x = 0$,	$y = 0$
$4(0) - 3y = 12$	$4x - 3(0) = 12$
$-3y = 12$	$4x = 12$
$y = -4$	$x = 3$
$(0, -4)$	$(3, 0)$

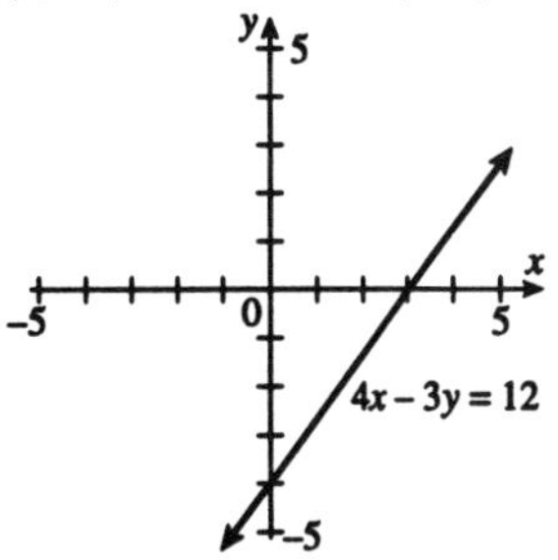

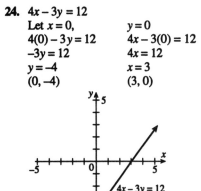

25. $x = 4$; $y = -2$; $(4, 0)$; $(0, -2)$

26. $y = -3$; $(0, -3)$

27. $x = -2$; $x = 2$; $y = 2$; $y = -2$; $(-2, 0)$; $(2, 0)$; $(0, 2)$; $(0, -2)$

28. $x = -1$; $x = 2$; $x = 3$; $y = -2$; $(-1, 0)$; $(2, 0)$; $(3, 0)$; $(0, -2)$

29. $x - 3y = 12$

If $x = 0$, then	If $y = 0$, then
$0 - 3y = 12$	$x - 3(0) = 12$
$-3y = 12$	$x - 0 = 12$
$y = -4$	$x = 12$
$(0, -4)$	$(12, 0)$

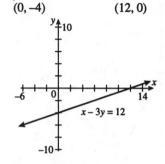

30. $-4x + y = 8$

If $x = 0$, then	If $y = 0$, then
$-4(0) + y = 8$	$-4x + 0 = 8$
$0 + y = 8$	$-4x = 8$
$y = 8$	$x = -2$
$(0, 8)$	$(-2, 0)$

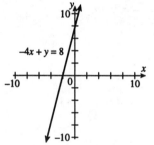

31. $y = -3$
If $x = 0$, then $y = -3$; $(0, -3)$
Since y cannot equal 0 there is no x-intercept.

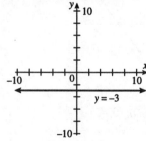

32. $x = 5$
Since x cannot equal 0 there is no y-intercept.
If $y = 0$, then $x = 5$; $(5, 0)$

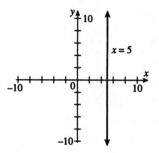

33. $y = -3x$
If $x = 0$, then $y = -3(0) = 0$; $(0, 0)$

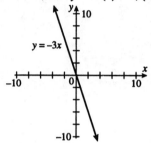

34. $x = 5y$
If $x = 0$, then $0 = 5y$ or $y = 0$; $(0, 0)$

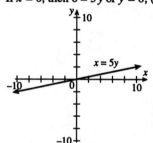

35. $x - 2 = 0$ or $x = 2$
Since x cannot equal 0, there is no y-intercept.
If $y = 0$, then $x = 2$; $(2, 0)$

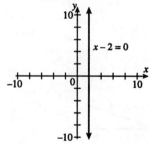

36. $y + 6 = 0$ or $y = -6$
If $x = 0$, then $y = -6$; $(0, -6)$
Since y cannot equal zero there is no x-intercept.

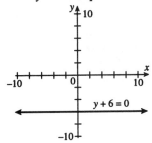

37. $(-1, 2)$ and $(3, -1)$
$$m = \frac{-1 - 2}{3 - (-1)} = \frac{-3}{3 + 1} = -\frac{3}{4}$$

38. $(-2, -2)$ and $(3, -1)$
$$m = \frac{-1 - (-2)}{3 - (-2)} = \frac{-1 + 2}{3 + 2} = \frac{1}{5}$$

39. d

40. b

41. c

42. a

43. e

44. $(2, 5)$ and $(6, 8)$
$$m = \frac{8 - 5}{6 - 2} = \frac{3}{4}$$

45. $(4, 7)$ and $(1, 2)$
$$m = \frac{2 - 7}{1 - 4} = \frac{-5}{-3} = \frac{5}{3}$$

46. $(1, 3)$ and $(-2, -9)$
$$m = \frac{-9 - 3}{-2 - 1} = \frac{-12}{-3} = 4$$

47. $(-4, 1)$ and $(3, -6)$
$$m = \frac{-6 - 1}{3 - (-4)} = \frac{-7}{3 + 4} = \frac{-7}{7} = -1$$

48. $y = 3x + 7$
$m = 3 \qquad\qquad b = 7$

49. $x - 2y = 4$
$-2y = -x + 4$
$$\frac{-2y}{-2} = \frac{-x}{-2} + \frac{4}{-2}$$

$$y = \frac{x}{2} - 2$$
$$y = \frac{1}{2}x - 2$$
$$m = \frac{1}{2}, \ b = -2$$

50. $y = -2$, $m = 0$

51. $x = 0$, undefined slope

52. Line passes through $(0, 3)$ and $(4, 4)$.

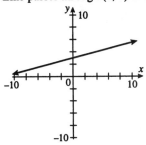

53. Line passes through $(-5, 3)$ and $(-4, 0)$.

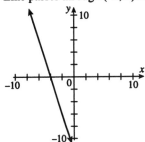

54. $m = \dfrac{-2 - 1}{1 - (-3)} = \dfrac{-3}{1 + 3} = -\dfrac{3}{4}$
$m = \dfrac{1 - 4}{6 - 2} = -\dfrac{3}{4}$
parallel

55. $m = \dfrac{4 - 6}{0 - (-7)} = \dfrac{-2}{0 + 7} = -\dfrac{2}{7}$
$m = \dfrac{5 - (-3)}{1 - (-9)} = \dfrac{5 + 3}{1 + 9} = \dfrac{8}{10} = \dfrac{4}{5}$
neither

56. $m = \dfrac{-7 - 10}{8 - 9} = \dfrac{-17}{-1} = 17$
$m = \dfrac{-8 - (-3)}{2 - (-1)} = \dfrac{-8 + 3}{2 + 1} = -\dfrac{5}{3}$
neither

57. $m = \dfrac{-2-3}{3-(-1)} = \dfrac{-5}{3+1} = -\dfrac{5}{4}$

$m = \dfrac{2-(-2)}{3-(-2)} = \dfrac{2+2}{3+2} = \dfrac{4}{5}$

perpendicular

58. $3x - 4y \le 0$
Test: (0, 1):
$3(0) - 4(1) \le 0$
$0 - 4 \le 0$
$-4 \le 0$ True

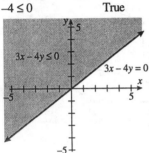

59. $3x - 4y \ge 0$
Find two points.
Let $x = 0, 3(0) - 4y = 0, -4y = 0, y = 0, (0, 0)$
Let $x = 4, 3(4) - 4y = 0, 12 - 4y = 0, -4y = -12,$
$y = 3, (4, 3)$
The boundary line is solid.
Choose (5, 0) for a test point.
$3x - 4y \ge 0$
$3(5) - 4(0) \ge 0$
$15 - 0 \ge 0$
$15 \ge 0$
Since $15 \ge 0$, the side containing (5, 0) is shaded.

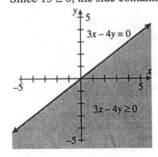

60. $x + 6y < 0$
Test (0, -1):
$0 + 6(-1) < 0$
$0 - 6 < 0$
$-6 < 0$ True

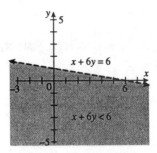

61. $x + y > -2$
Find the intercepts.
Let $x = 0, 0 + y = -2, y = -2, (0, -2)$
Let $y = 0, x + 0 = -2, x = -2, (-2, 0)$
The boundary line is dashed.
Choose (0, 0) as a test point.
$x + y > -2$
$0 + 0 > -2$
$0 > -2$
Since $0 > -2$, the side containing (0, 0) is shaded.

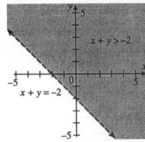

62. $y \ge -7$

63. $y \le -4$
Find two points.
Let $x = 1, y = -4, (1, -4)$
Let $x = 3, y = -4, (3, -4)$
The boundary line is solid.
Choose (0, 0) as a test point.
$y \le -4$
$0 \; ? \; -4$
Since $0 \not\le -4$, the side *not* containing (0, 0) is shaded.

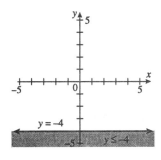

64. $-x \leq y$
Test $(0, 1)$: $-(0) \leq 1$
$0 \leq 1$ True

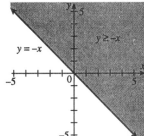

65. $x \geq -y$
Find two points.
Let $y = 0$, $x = -0$, $x = 0$, $(0, 0)$
Let $y = 3$, $x = -3$, $(-3, 3)$
The boundary line is solid.
Choose $(4, 0)$ as a test point.
$x \geq -y$
$4 \geq -0$
$4 \geq 0$
Since $4 \geq 0$ the side containing $(4, 0)$ is shaded.

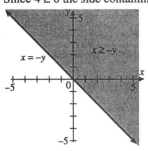

Chapter 3 - Test

1. $x - 2y = 3$ $(1, 1)$
$1 - 2(1)$? 3
$1 - 2$? 3
-1 ? 3
$-1 \neq 3$ No

2. $2x + 3y = 6$ $(0, -2)$
$2(0) + 3(-2)$? 6
$0 - 6$? 6
-6 ? 6
$-6 \neq 6$ No

3. $12y - 7x = 5$ $(1, \ )$
$12y - 7(1) = 5$
$12y - 7 = 5$
$12y = 5 + 7$
$12y = 12$
$y = 1$ $(1, 1)$

4. $y = 17$ $(-4, \ )$
$y = 17$ $(-4, 17)$

5. $(-1, -1)$ and $(4, 1)$
$$m = \frac{1 - (-1)}{4 - (-1)} = \frac{1 + 1}{4 + 1} = \frac{2}{5}$$

6. $m = 0$ for horizontal line

7. $m = \dfrac{2 - (-5)}{-1 - 6} = \dfrac{2 + 5}{-7} = \dfrac{7}{-7} = -1$

8. $m = \dfrac{-1 - (-8)}{-1 - 0} = \dfrac{-1 + 8}{-1} = \dfrac{7}{-1} = -7$

9. $-3x + y = 5$
$y = 3x + 5$
$m = 3$

10. $x = 6$
Undefined slope

11. $m = \dfrac{-3 - 3}{1 - (-1)} = \dfrac{-6}{1 + 1} = \dfrac{-6}{2} = -3$
$m = \dfrac{-7 - (-1)}{4 - 2} = \dfrac{-7 + 1}{2} = \dfrac{-6}{2} = -3$
parallel

12. $m = \dfrac{-2 - (-6)}{-1 - (-6)} = \dfrac{-2 + 6}{-1 + 6} = \dfrac{4}{5}$
$m = \dfrac{-3 - 3}{3 - (-4)} = \dfrac{-6}{3 + 4} = -\dfrac{6}{7}$
neither

13. $2x + y = 8$
Find the intercepts.
Let $x = 0$, $2(0) + y = 8$, $y = 8$, $(0, 8)$
Let $y = 0$, $2x + 0 = 8$, $2x = 8$, $x = 4$ $(4, 0)$

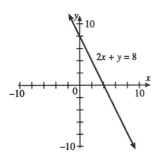

14. $-x + 4y = 5$
Find the intercepts.

Let $x = 0$, $-0 + 4y = 5$, $4y = 5$, $y = \dfrac{5}{4}$

$\left(0, \dfrac{5}{4}\right)$

Let $y = 0$, $-x + 4(0) = 5$, $-x = 5$, $x = -5$, $(-5, 0)$

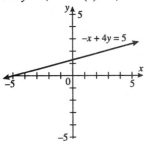

15. $x - y \geq -2$
Find the intercepts.
Let $x = 0$, $0 - y = -2$, $-y = -2$,

$y = \dfrac{-2}{-1} = 2$, $(0, 2)$

let $y = 0$, $x - 0 = -2$, $x = -2$, $(-2, 0)$
The boundary line is solid.
Choose $(0, 0)$ as a test point.
$x - y \geq -2$
$0 - 0 \; ? \; -2$
$0 \; ? \; -2$
Since $0 \geq -2$, the side containing $(0, 0)$ is shaded.

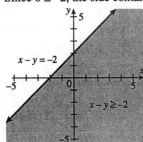

16. $y \geq -4x$
Find two points.
Let $x = 0$, $y = -4(0)$, $y = 0$, $(0, 0)$
Let $x = 1$, $y = -4(1)$, $y = -4$ $(1, -4)$
The boundary line is solid.
Choose $(5, 0)$ as a test point.
$y \geq -4x$
$0 \; ? \; -4(5)$
$0 \; ? \; -20$
Since $0 \geq -20$, the side containing $(5, 0)$ is shaded.

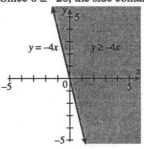

17. $5x - 7y = 10$
Find the intercepts.
Let $x = 0$, $5(0) - 7y = 10$, $-7y = 10$,

$y = -\dfrac{10}{7}$, $\left(0, -\dfrac{10}{7}\right)$

Let $y = 0$, $5x - 7(0) = 10$, $5x = 10$, $x = 2$, $(2, 0)$

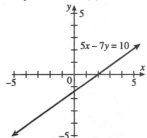

18. $2x - 3y > -6$
Find the intercepts. Let $x = 0$, $2(0) - 3y = -6$,

$-3y = -6$, $y = \dfrac{-6}{-3} = 2$, $(0, 2)$

Let $y = 0$, $2x - 3(0) = -6$, $2x = -6$, $x = -3$, $(-3, 0)$
The boundary line is dashed.
Choose $(0, 0)$ for a test point.
$2x - 3y > -6$
$2(0) - 3(0) \; ? \; -6$
$0 + 0 \; ? \; -6$
$0 \; ? \; -6$
Since $0 > -6$, the side containing $(0, 0)$ is **shaded**.

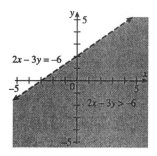

19. $6x + y > -1$
Find the intercepts.
Let $x = 0$, $6(0) + y = -1$, $y = -1$, $(0, -1)$
Let $y = 0$, $6x + 0 = -1$, $6x = -1$, $x = -\dfrac{1}{6}$

$\left(-\dfrac{1}{6},\ 0 \right)$

The boundary line is dashed.
Choose $(0, 0)$ as a test point.
$6x + y > -1$
$6(0) + 0\ ?\ -1$
$0 + 0\ ?\ -1$
$0\ ?\ -1$
Since $0 > -1$ the side containing $(0, 0)$ is shaded.

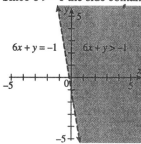

20. $y = -1$
Find two points.
Let $x = 0$, $y = -1$, $(0, -1)$
Let $x = 3$, $y = -1$, $(3, -1)$

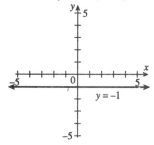

21. $x - 3 = 0$
$x = 3$
Find two points.
Let $y = 0$, $x = 3$ $(3, 0)$
Let $y = 2$, $x = 3$ $(3, 2)$

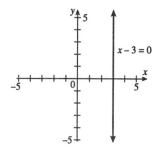

22. $5x - 3y = 15$
Find the intercepts.
Let $x = 0$, $5(0) - 3y = 15$, $-3y = 15$, $y = -5$, $(0, -5)$
Let $y = 0$, $5x - 3(0) = 15$, $5x = 15$, $x = 3$, $(3, 0)$

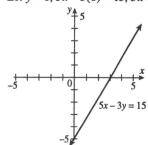

23. $y = 2x + 1$

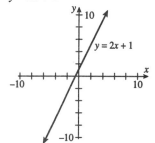

24. $x + 4y < -4$

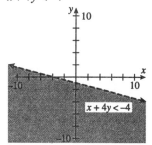

25. $x + x + 2y + 2y = 42$
$2x + 4y = 42$
$x + 2y = 21$
$x + 2(8) = 21$
$x + 16 = 21$
$x = 5$ meters

Chapter 3 - Cumulative Review

1. a. $-1 < 0$

   b. $7 = \dfrac{14}{2}$

   c. $-5 > -6$

2. a. $\dfrac{42}{49} = \dfrac{6 \cdot 7}{7 \cdot 7} = \dfrac{6}{7}$

   b. $\dfrac{11}{27}$ is in lowest terms.

   c. $\dfrac{88}{20} = \dfrac{22 \cdot 4}{5 \cdot 4} = \dfrac{22}{5}$

3. $\dfrac{8 + 2 \cdot 3}{2^2 - 1} = \dfrac{8 + 6}{4 - 1} = \dfrac{14}{3}$

4. a. $x + 3$

   b. $3x$

   c. $2x$

   d. $10 - x$

   e. $5x + 7$

5. a. $\dfrac{x - y}{12 + x} = \dfrac{2 - (-5)}{12 + 2} = \dfrac{2 + 5}{14} = \dfrac{7}{14} = \dfrac{1}{2}$

   b. $x^2 - y = 2^2 - (-5) = 4 + 5 = 9$

6. a. $\dfrac{-24}{-4} = 6$

   b. $\dfrac{-36}{3} = -12$

   c. $\dfrac{2}{3} + \left(-\dfrac{5}{4}\right) = \dfrac{2}{3} \cdot \left(-\dfrac{4}{5}\right) = -\dfrac{8}{15}$

7. a. $5(x + 2) = 5x + 10$

   b. $-2(y + 0.3z - 1) = -2y - 0.6z + 2$

   c. $-(x + y - 2z + 6) = -x - y + 2z - 6$

8. $-5(2a - 1) - (-11a + 6) = 7$
   $-10a + 5 + 11a - 6 = 7$
   $a - 1 = 7$
   $a = 8$

9. $\dfrac{y}{7} = 20$

   $7 \cdot \dfrac{y}{7} = 20 \cdot 7$

   $y = 140$

10. $0.25x + 0.10(x - 3) = 0.05(x + 18)$
   $0.25x + 0.10x - 0.30 = 0.05x + 0.9$
   $0.35x - 0.30 = 0.05x + 0.9$
   $0.30x - 0.30 = 0.9$
   $0.30x = 1.2$
   $x = 4$

11. $2(x + 4) = 4x - 12$
   $2x + 8 = 4x - 12$
   $20 = 2x$
   $10 = x$

12. $x + x + 30 + 30 = 140$
   $2x + 60 = 140$
   $2x = 80$
   $x = 40$
   40 feet

13. $0.72(200) = 144$

14. $-4x + 7 \geq -9$
   $-4x \geq -16$
   $x \leq 4$

15. a. quadrant I

   b. quadrant III

   c. quadrant IV

   d. quadrant II

   e. no quadrant; origin

   f. no quadrant; y-axis

   g. no quadrant; x-axis

h. no quadrant; *y*-axis

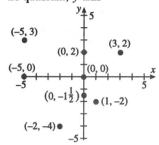

16. $2x + y = 5$
If $x = 0$, then $y = 5$
If $y = 0$, then $x = \dfrac{5}{2}$
If $x = 1$, then $y = 3$

x	y
0	5
$\dfrac{5}{2}$	0
1	3

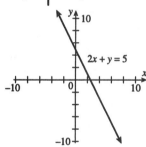

17. $y = 3x$

x	y
−1	−3
0	0
−3	−9

18.

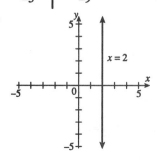

19. Through $(-1, 5)$ and $(2, -3)$
$$m = \frac{-3-5}{2-(-1)} = \frac{-8}{2+1} = -\frac{8}{3}$$

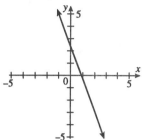

20. Line passes through $(-1, 5)$ and $(0, 3)$

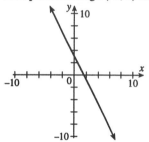

21. $m = 0$

22. $x + y < 7$
Test: $(0, 0)$: $0 + 0 < 7$
$0 < 7$ True

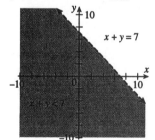

Chapter 4

Section 4.1 Mental Math

1. base: 3, exponent: 2

3. base: -3, exponent: 6

5. base: 4, exponent: 2

7. base: 5, exponent: 1
   base: 3, exponent: 4

9. base: 5, exponent: 1
   base: x, exponent: 2

Exercise Set 4.1

1. $7^2 = 7 \cdot 7 = 49$

3. $(-5)^1 = -5$

5. $-2^4 = -(2 \cdot 2 \cdot 2 \cdot 2) = -(16) = -16$

7. $(-2)^4 = (-2)(-2)(-2)(-2) = 16$

9. $\left(\dfrac{1}{3}\right)^3 = \left(\dfrac{1}{3}\right)\left(\dfrac{1}{3}\right)\left(\dfrac{1}{3}\right) = \dfrac{1}{27}$

11. $7 \cdot 2^4 = 7 \cdot 2 \cdot 2 \cdot 2 \cdot 2 = 112$

13. Answers may vary.

15. $x^2 = (-2)^2 = (-2)(-2) = 4$

17. $5x^3 = 5 \cdot 3^3 = 5 \cdot 3 \cdot 3 \cdot 3 = 135$

19. $2xy^2 = 2 \cdot 3 \cdot 5^2 = 2 \cdot 3 \cdot 5 \cdot 5 = 150$

21. $\dfrac{2z^4}{5} = \dfrac{2(-2)^4}{5} = \dfrac{2(-2)(-2)(-2)(-2)}{5} = \dfrac{32}{5}$

23. $V = x^3$
    $V = 7^3 = 7 \cdot 7 \cdot 7 = 343$ cubic meters

25. Volume

27. $x^2 \cdot x^5 = x^{2+5} = x^7$

29. $(-3)^3 \cdot (-3)^9 = (-3)^{3+9} = (-3)^{12}$

31. $(5y^4)(3y) = 5 \cdot 3 \cdot y^4 \cdot y = 15y^{4+1} = 15y^5$

33. $(4z^{10})(-6z^7)(z^3)$
    $= 4(-6)z^{10} \cdot z^7 \cdot z^3$
    $= -24z^{10+7+3}$
    $= -24z^{20}$

35. $A = lw$
    $A = (5x^3 \text{ ft.})(4x^2 \text{ ft.})$
    $A = 20x^5$ sq. ft.

37. $(pq)^7 = p^{1 \cdot 7}q^{1 \cdot 7} = p^7q^7$

39. $\left(\dfrac{m}{n}\right)^9 = \dfrac{m^{1 \cdot 9}}{n^{1 \cdot 9}} = \dfrac{m^9}{n^9}$

41. $(x^2y^3)^5 = x^{2 \cdot 5}y^{3 \cdot 5} = x^{10}y^{15}$

43. $\left(\dfrac{-2xz}{y^5}\right)^2 = \dfrac{(-2)^2 x^{1 \cdot 2}z^{1 \cdot 2}}{y^{5 \cdot 2}} = \dfrac{4x^2z^2}{y^{10}}$

45. $A = s^2 = (8z^5)^2 = 8^2 \cdot z^{5 \cdot 2} = 64z^{10}$ sq. decimeters

47. $V = lwh$
    $V = (3y^4)(3y^4)(3y^4)$
    $V = 27y^{4+4+4}$
    $V = 27y^{12}$ cubic ft.

49. $\dfrac{x^3}{x} = x^{3-1} = x^2$

51. $\dfrac{(-2)^5}{(-2)^3} = (-2)^{5-3} = (-2)^2 = 4$

53. $\dfrac{p^7 q^{20}}{pq^{15}} = p^{7-1}q^{20-15} = p^6q^5$

55. $\dfrac{7x^2y^6}{14x^2y^3} = \dfrac{7}{14}x^{2-2}y^{6-3} = \dfrac{1}{2}x^0y^3 = \dfrac{1}{2}(1)y^3 = \dfrac{y^3}{2}$

57. $(2x)^0 = 1$

59. $-2x^0 = -2(1) = -2$

61. $5^0 + y^0 = 1 + 1 = 2$

63. Answers may vary.

65. $\left(\dfrac{-3a^2}{b^3}\right)^3 = \dfrac{(-3)^3 a^{2\cdot3}}{b^{3\cdot3}} = \dfrac{-27a^6}{b^9}$

67. $\dfrac{(x^5)^7 \cdot x^8}{x^4}$

$= \dfrac{x^{5\cdot7} x^8}{x^4}$

$= \dfrac{x^{35} x^8}{x^4}$

$= x^{35+8-4} = x^{39}$

69. $\dfrac{(z^3)^6}{(5z)^4} = \dfrac{z^{3\cdot6}}{5^4 \cdot z^4} = \dfrac{z^{18}}{625z^4} = \dfrac{z^{18-4}}{625} = \dfrac{z^{14}}{625}$

71. $\dfrac{(6mn)^5}{mn^2}$

$= \dfrac{6^5 \cdot m^5 \cdot n^5}{mn^2}$

$= 7776m^{5-1}n^{5-2} = 7776m^4 n^3$

73. $-5^2 = -(5)(5) = -25$

75. $\left(\dfrac{1}{4}\right)^3 = \left(\dfrac{1}{4}\right)\left(\dfrac{1}{4}\right)\left(\dfrac{1}{4}\right) = \dfrac{1}{64}$

77. $(9xy)^2 = 9^2 x^2 y^2 = 81x^2 y^2$

79. $(6b)^0 = 1$

81. $2^3 + 2^5$
$= 2\cdot2\cdot2 + 2\cdot2\cdot2\cdot2\cdot2$
$= 8 + 32 = 40$

83. $b^4 b^2 = b^{4+2} = b^6$

85. $a^2 a^3 a^4 = a^{2+3+4} = a^9$

87. $(2x^3)(-8x^4) = (2)(-8)x^{3+4} = -16x^7$

89. $(4a)^3 = 4^{1\cdot3} a^{1\cdot3} = 4^3 a^3 = 64a^3$

91. $(-6xyz^3)^2$
$= (-6)^2 x^{1\cdot2} y^{1\cdot2} z^{3\cdot2} = 36x^2 y^2 z^6$

93. $\left(\dfrac{3y^5}{6x^4}\right)^3$

$= \left(\dfrac{y^5}{2x^4}\right)^3$ Reduce fraction

$= \dfrac{y^{5\cdot3}}{2^{1\cdot3} x^{4\cdot3}}$

$= \dfrac{y^{15}}{2^3 x^{12}}$

$= \dfrac{y^{15}}{8x^{12}}$

95. $\dfrac{x^5}{x^4} = x^{5-4} = x$

97. $\dfrac{2x^3 y^2 z}{xyz}$

$= 2x^{3-1} y^{2-1} z^{1-1}$

$= 2x^2 y^1 z^0$

$= 2x^2 y(1) = 2x^2 y$

99. $\dfrac{(3x^2 y^5)^5}{x^3 y}$

$= \dfrac{3^5 \cdot x^{2\cdot5} \cdot y^{5\cdot5}}{x^3 y}$

$= \dfrac{243x^{10} y^{25}}{x^3 y}$

$= 243x^{10-3} y^{25-1} = 243x^7 y^{24}$

101. $3x - 5x + 7 = -2x + 7$

103. $y - 10 + y = 2y - 10$

105. $7x + 2 - 8x - 6 = -x - 4$

107. $2(x - 5) + 3(5 - x)$
$= 2x - 10 + 15 - 3x = -x + 5$

109. $x^{5a} x^{4a} = x^{5a+4a} = x^{9a}$

111. $(a^b)^5 = a^{b\cdot5} = a^{5b}$

113. $\dfrac{x^{9a}}{x^{4a}} = x^{9a-4a} = x^{5a}$

115. $(x^a y^b z^c)^{5a}$
$= x^{a \cdot 5a} y^{b \cdot 5a} z^{c \cdot 5a} = x^{5a^2} y^{5ab} z^{5ac}$

Section 4.2 Mental Math

1. $-9y - 5y = -14y$

3. $4y^3 + 3y^3 = 7y^3$

5. $x + 6x = 7x$

Exercise Set 4.2

1. $x + 2$
degree is 1; binomial

3. $9m^3 - 5m^2 + 4m - 8$
degree is 3; none of these

5. $12x^4 y - x^2 y^2 - 12x^2 y^4$
degree is 6; trinomial

7. $3zx - 5x^2$
degree is 2; binomial

9. Answers may vary.

11. Answers may vary.

13. $x + 6$

   a. $x = 0$
      $0 + 6 = 6$

   b. $x = -1$
      $-1 + 6 = +5$

15. $x^2 - 5x - 2$

   a. $x = 0$
      $0^2 - 5(0) - 2 = 0 + 0 - 2 = -2$

   b. $x = -1$
      $(-1)^2 - 5(-1) - 2 = 1 + 5 - 2 = 4$

17. $x^3 - 15$

   a. $x = 0$
      $0^3 - 15 = -15$

   b. $x = -1$
      $(-1)^3 - 15 = -1 - 15 = -16$

19. $-16t^2 + 1821 = -16(10.8)^2 + 1821 = -45.24$ ft
The object has reached the ground.

21. $14x^2 + 9x^2 = 23x^2$

23. $15x^2 - 3x^2 - y = 12x^2 - y$

25. $8s - 5s + 4s = 3s + 4s = 7s$

27. $0.1y^2 - 1.2y^2 + 6.7 - 1.9 = -1.1y^2 + 4.8$

29. $(3x + 7) + (9x + 5)$
$= 3x + 7 + 9x + 5 = 12x + 12$

31. $(-7x + 5) + (-3x^2 + 7x + 5)$
$= -7x + 5 - 3x^2 + 7x + 5$
$= -3x^2 + 0x + 10$
$= -3x^2 + 10$

33. $(2x + 5) - (3x - 9)$
$= 2x + 5 - 3x + 9$
$= -x + 14$

35. $3x - (5x - 9)$
$= 3x - 5x + 9 = -2x + 9$

37. $(-5x^2 + 3) + (2x^2 + 1)$
$= -5x^2 + 3 + 2x^2 + 1$
$= -3x^2 + 4$

39. $(2x^2 + 3x - 9) - (-4x + 7)$
$= 2x^2 + 3x - 9 + 4x - 7$
$= 2x^2 + 7x - 16$

41. $P = a + b + c$
$P = (-x^2 + 3x) + (2x^2 + 5) + (4x - 1)$
$P = -x^2 + 3x + 2x^2 + 5 + 4x - 1$
$P = (x^2 + 7x + 4)$ ft.

$(-x^2 + 3x)$ feet $(2x^2 + 5)$ feet $(4x - 1)$ feet

43. $(4y^2 + 4y + 1) - (y^2 - 10)$
$= 4y^2 + 4y + 1 - y^2 + 10$
$= 3y^2 + 4y + 11$

45.
$$3t^2 + 4$$
$$+\ \underline{5t^2 - 8}$$
$$8t^2 - 4$$

47.
$$4z^2 - 8z + 3$$
$$\underline{-(6z^2 + 8z - 3)}$$

$$4z^2 - 8z + 3$$
$$\underline{-\ 6z^2 - 8z + 3}$$
$$-2z^2 - 16z + 6$$

49.
$$5x^3 - 4x^2 + 6x - 2$$
$$\underline{-(3x^3 - 2x^2 - x - 4)}$$

$$5x^3 - 4x^2 + 6x - 2$$
$$\underline{-\ 3x^3 + 2x^2 + x + 4}$$
$$2x^3 - 2x^2 + 7x + 2$$

51.
$$10a^3 - 8a^2 + 9$$
$$\underline{+5a^3 + 9a^2 + 7}$$
$$15a^3 + a^2 + 16$$

53. $81x^2 + 10 - (19x^2 + 5)$
$$= 81x^2 + 10 - 19x^2 - 5$$
$$= 62x^2 + 5$$

55. $[(8x+1)+(6x+3)] - (2x+2)$
$$= [8x + 1 + 6x + 3] - (2x + 2)$$
$$= [14x + 4] - (2x + 2)$$
$$= 14x + 4 - 2x - 2$$
$$= 12x + 2$$

57. $(9xy^2 + 7x - 18) - (8x + 9)$
$$= 9xy^2 + 7x - 18 - 8x - 9$$
$$= 9xy^2 - x - 27$$

59. $-15x - (-4x) = -15x + 4x = -11x$

61. $2x - 5 + 5x - 8 = 7x - 13$

63. $(-3y^2 - 4y) + (2y^2 + y - 1)$
$$= -3y^2 - 4y + 2y^2 + y - 1$$
$$= -y^2 - 3y - 1$$

65. $(-7y^2 + 5) - (-8y^2 + 12)$
$$= -7y^2 + 5 + 8y^2 - 12 = y^2 - 7$$

67. $(5x + 8) - (-2x^2 - 6x + 8)$
$$= 5x + 8 + 2x^2 + 6x - 8 = 2x^2 + 11x$$

69. $(-8x^4 + 7x) + (-8x^4 + x + 9)$
$$= -8x^4 + 7x - 8x^4 + x + 9$$
$$= -16x^4 + 8x + 9$$

71. $(3x^2 + 5x - 8) + (5x^2 + 9x + 12) - (x^2 - 14)$
$$= 3x^2 + 5x - 8 + 5x^2 + 9x + 12 - x^2 + 14$$
$$= 7x^2 + 14x + 18$$

73. $7x - 3 - (4x) = 7x - 3 - 4x = 3x - 3$

75. $(7x^2 + 3x + 9) - (5x + 7)$
$$= 7x^2 + 3x + 9 - 5x - 7$$
$$= 7x^2 - 2x + 2$$

77. $[(8y^2 + 7) + (6y + 9)] - (4y^2 - 6y - 3)$
$$= [8y^2 + 7 + 6y + 9] - (4y^2 - 6y - 3)$$
$$= [8y^2 + 6y + 16] - (4y^2 - 6y - 3)$$
$$= 8y^2 + 6y + 16 - 4y^2 + 6y + 3$$
$$= 4y^2 + 12y + 19$$

79. $[(-x^2 - 2x) + (5x^2 + x + 9)] - (-2x^2 + 4x - 12)$
$$= [-x^2 - 2x + 5x^2 + x + 9] - (-2x^2 + 4x - 12)$$
$$= [4x^2 - x + 9] - (-2x^2 + 4x - 12)$$
$$= 4x^2 - x + 9 + 2x^2 - 4x + 12$$
$$= 6x^2 - 5x + 21$$

81. $[(1.2x^2 - 3x + 9.1) - (7.8x^2 - 3.1 + 8)] + (1.2x - 6)$
$$= 1.2x^2 - 3x + 9.1 - 7.8x^2 + 3.1 - 8 + 1.2x - 6$$
$$= -6.6x^2 - 1.8x - 1.8$$

83. $-16t^2 + 200t$

    a. $t = 1$ second
$$-16(1)^2 + 200(1) = 184 \text{ ft}$$

    b. $t = 5$ seconds
$$-16(5)^2 + 200(5) = 600 \text{ ft}$$

c. $t = 7.6$ seconds

$-16(7.6)^2 + 200(7.6) = 595.84$ ft

d. $t = 10.3$ seconds

$-16(10.3)^2 + 200(10.3) = 362.56$ ft

85. $-16t^2 + 200t = 0$

$t(-16t + 200) = 0$

$-16t + 200 = 0$

$t = \dfrac{-200}{-16} = 12.5$ seconds

87. $3x(2x) = 3 \cdot 2 \cdot x^{1+1} = 6x^2$

89. $(12x^3)(-x^5) = (12)(-1)x^{3+5} = -12x^8$

91. $10x^2(20xy^2) = 10 \cdot 20 \cdot x^{2+1}y^2 = 200x^3y^2$

93. $2x - y = 6$

If $x = 0$, then

$2(0) - y = 6$

$-y = 6$

$y = -6$

If $x = 3$, then

$2(3) - y = 6$

$6 - y = 6$

$-y = 0$

$y = 0$

If $x = 1$, then

$2(1) - y = 6$

$2 - y = 6$

$-y = 4$

$y = -4$

x	y
0	-6
3	0
1	-4

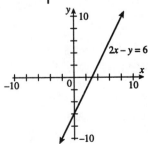

95. $x = -4$

Vertical line with x-intercept at -4

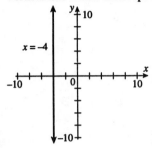

Section 4.3 Mental Math

1. $5x(2y) = 5 \cdot 2xy = 10xy$

3. $x^2x^5 = x^{2+5} = x^7$

5. $6x(3x^2) = 6 \cdot 3 \cdot x^{1+2} = 18x^3$

Exercise Set 4.3

1. $2a(2a - 4)$

$= 2a(2a) + 2a(-4)$

$= 4a^2 - 8a$

3. $7x(x^2 + 2x - 1)$

$= 7x(x^2) + 7x(2x) + 7x(-1)$

$= 7x^3 + 14x^2 - 7x$

5. $3x^2(2x^2 - x)$

$= 3x^2(2x^2) + 3x^2(-x)$

$= 6x^4 - 3x^3$

7. $x^2 + 3x$

9. $(a + 7)(a - 2)$

$= a(a) + a(-2) + 7(a) + 7(-2)$

$= a^2 - 2a + 7a - 14$

$= a^2 + 5a - 14$

11. $(2y - 4)^2$

$= (2y - 4)(2y - 4)$

$= 2y(2y) + 2y(-4) - 4(2y) - 4(-4)$

$= 4y^2 - 8y - 8y + 16$

$= 4y^2 - 16y + 16$

13. $(5x-9y)(6x-5y)$
$= 5x(6x)+5x(-5y)-9y(6x)-9y(-5y)$
$= 30x^2-25xy-54xy+45y^2$
$= 30x^2-79xy+45y^2$

15. $(2x^2-5)^2$
$= (2x^2-5)(2x^2-5)$
$= 2x^2(2x^2)+2x^2(-5)-5(2x^2)-5(-5)$
$= 4x^4-10x^2-10x^2+25$
$= 4x^4-20x^2+25$

17. $x^2+3x+2x+2\cdot3 = x^2+5x+6$

19. $(x-2)(x^2-3x+7)$
$= x(x^2)+x(-3x)+x(7)-2(x^2)-2(-3x)-2(7)$
$= x^3-3x^2+7x-2x^2+6x-14$
$= x^3-5x^2+13x-14$

21. $(x+5)(x^3-3x+4)$
$= x(x^3)+x(-3x)+x(4)+5(x^3)+5(-3x)+5(4)$
$= x^4-3x^2+4x+5x^3-15x+20$
$= x^4+5x^3-3x^2-11x+20$

23. $(2a-3)(5a^2-6a+4)$
$= 2a(5a^2)+2a(-6a)+2a(4)-3(5a^2)-3(-6a)-3(4)$
$= 10a^3-12a^2+8a-15a^2+18a-12$
$= 10a^3-27a^2+26a-12$

25. $(x+2)^3$
$= (x+2)(x+2)(x+2)$
$= (x+2)[x(x)+x(2)+2(x)+2(2)]$
$= (x+2)[x^2+2x+2x+4]$
$= (x+2)[x^2+4x+4]$
$= x(x^2)+x(4x)+x(4)+2(x^2)+2(4x)+2(4)$
$= x^3+4x^2+4x+2x^2+8x+8$
$= x^3+6x^2+12x+8$

27. $(2y-3)^3$
$= (2y-3)(2y-3)(2y-3)$
$= (2y-3)[2y(2y)+2y(-3)-3(2y)-3(-3)]$
$= (2y-3)[4y^2-6y-6y+9]$
$= (2y-3)[4y^2-12y+9]$
$= 2y(4y^2)+2y(-12y)+2y(9)-3(4y^2)$
$\quad -3(-12y)-3(9)$

$= 8y^3-24y^2+18y-12y^2+36y-27$
$= 8y^3-36y^2+54y-27$

29.
$$\begin{array}{r} 2x^2+4x-1 \\ \underline{x+3} \\ 6x^2+12x-3 \\ \underline{2x^3+4x^2-x} \\ 2x^3+10x^2+11x-3 \end{array}$$

31.
$$\begin{array}{r} x^2+5x-7 \\ \underline{x^2-7x-9} \\ -9x^2-45x+63 \\ -7x^3-35x^2+49x \\ \underline{x^4+5x^3-7x^2} \\ x^4-2x^3-51x^2+4x+63 \end{array}$$

33. a. $2^2+3^2 = 4+9 = 13$
$(2+3)^2 = 5^2 = 25$

b. $8^2+10^2 = 64+100 = 164$
$(8+10)^2 = (18)^2 = 324$
No; answers may vary.

35. $2a(a+4)$
$= 2a(a)+2a(4)$
$= 2a^2+8a$

37. $3x(2x^2-3x+4)$
$= 3x(2x^2)+3x(-3x)+3x(4)$
$= 6x^3-9x^2+12x$

39. $(5x+9y)(3x+2y)$
$= 5x(3x)+5x(2y)+9y(3x)+9y(2y)$
$= 15x^2+10xy+27xy+18y^2$
$= 15x^2+37xy+18y^2$

41. $(x+2)(x^2+5x+6)$
$= x(x^2)+x(5x)+x(6)+2(x^2)+2(5x)+2(6)$
$= x^3+5x^2+6x+2x^2+10x+12$
$= x^3+7x^2+16x+12$

43. $(7x+4)^2$
$= (7x+4)(7x+4)$
$= 7x(7x)+7x(4)+4(7x)+4(4)$
$= 49x^2+28x+28x+16$
$= 49x^2+56x+16$

45. $-2a^2(3a^2-2a+3)$
$= -2a^2(3a^2)-2a^2(-2a)-2a^2(3)$
$= -6a^4+4a^3-6a^2$

47. $(x+3)(x^2+7x+12)$
$= x(x^2)+x(7x)+x(12)+3(x^2)+3(7x)+3(12)$
$= x^3+7x^2+12x+3x^2+21x+36$
$= x^3+10x^2+33x+36$

49. $(a+1)^3$
$= (a+1)(a+1)(a+1)$
$= (a+1)[a(a)+a(1)+1(a)+1(1)]$
$= (a+1)[a^2+a+a+1]$
$= (a+1)[a^2+2a+1]$
$= a(a^2)+a(2a)+a(1)+1(a^2)+1(2a)+1(1)$
$= a^3+2a^2+a+a^2+2a+1$
$= a^3+3a^2+3a+1$

51. $(x+y)(x+y)$
$= x(x)+x(y)+y(x)+y(y)$
$= x^2+xy+xy+y^2$
$= x^2+2xy+y^2$

53. $(x-7)(x-6)$
$= x(x)+x(-6)-7(x)-7(-6)$
$= x^2-6x-7x+42$
$= x^2-13x+42$

55. $3a(a^2+2)$
$= 3a(a^2)+3a(2)$
$= 3a^3+6a$

57. $-4y(y^2+3y-11)$
$= -4y(y^2)-4y(3y)-4y(-11)$
$= -4y^3-12y^2+44y$

59. $(5x+1)(5x-1)$
$= 5x(5x)+5x(-1)+1(5x)+1(-1)$
$= 25x^2-5x+5x-1$
$= 25x^2-1$

61. $(5x+4)(x^2-x+4)$
$= 5x(x^2)+5x(-x)+5x(4)+4(x^2)+4(-x)+4(4)$
$= 5x^3-5x^2+20x+4x^2-4x+16$
$= 5x^3-x^2+16x+16$

63. $(2x-5)^3$
$= (2x-5)(2x-5)(2x-5)$
$= (2x-5)[2x(2x)+2x(-5)-5(2x)-5(-5)]$
$= (2x-5)[4x^2-10x-10x+25]$
$= (2x-5)[4x^2-20x+25]$
$= 2x(4x^2)+2x(-20x)+2x(25)-5(4x^2)$
$\quad -5(-20x)-5(25)$
$= 8x^3-40x^2+50x-20x^2+100x-125$
$= 8x^3-60x^2+150x-125$

65. $(4x+5)(8x^2+2x-4)$
$= 4x(8x^2)+4x(2x)+4x(-4)+5(8x^2)+5(2x)+5(-4)$
$= 32x^3+8x^2-16x+40x^2+10x-20$
$= 32x^3+48x^2-6x-20$

67. $(7xy-y)^2$
$= (7xy-y)(7xy-y)$
$= 7xy(7xy)-7xy(-y)-y(7xy)-y(-y)$
$= 49x^2y^2-7xy^2-7xy^2+y^2$
$= 49x^2y^2-14xy^2+y^2$

69.
$$\begin{array}{r} 5y^2-y+3 \\ y^2-3y-2 \\ \hline -10y^2+2y-6 \\ -15y^3+3y^2-9y \\ 5y^4-y^3+3y^2 \\ \hline 5y^4-16y^3-4y^2-7y-6 \end{array}$$

71.
$$\begin{array}{r} 3x^2+2x-4 \\ 2x^2-4x+3 \\ \hline 9x^2+6x-12 \\ -12x^3-8x^2+16x \\ 6x^4+4x^3-8x^2 \\ \hline 6x^4-8x^3-7x^2+22x-12 \end{array}$$

73. $A = lw$
$A = (2x + 5)(2x - 5)$
$A = 2x(2x) + 2x(-5) + 5(2x) + 5(-5)$
$A = 4x^2 - 10x + 10x - 25$
$A = (4x^2 - 25)$ sq. yds.

75. $A = \dfrac{1}{2}bh$

$A = \dfrac{1}{2}(4x)(3x - 2)$

$A = (2x)(3x - 2)$

$A = 6x^2 - 4x$ sq. inches

77. a. $(a + b)(a - b)$
$= a \cdot a - ab + ab - b \cdot b = a^2 - b^2$

b. $(2x + 3y)(2x - 3y)$
$= 2x(2x) - 2x(3y) + 2x(3y) - 3y(3y)$
$= 4x^2 - 9y^2$

c. $(4x + 7)(4x - 7)$
$= 4x(4x) - 4x(7) + 4x(7) - 7(7)$
$= 16x^2 - 49$
Answers may vary.

79. $(4p)^2 = 4p(4p) = 4 \cdot 4p^{1+1} = 16p^2$

81. $(-7m^2)^2$
$= (-7m^2)(-7m^2)$
$= (-7)(-7)m^{2+2} = 49m^4$

83. $3500

85. $6500 - 6000 = 500

87. There is a loss in value each year.

Exercise Set 4.4

1. $(x + 3)(x + 4)$
　　F　O　I　L
$= x(x) + x(4) + 3(x) + 3(4)$
$= x^2 + 4x + 3x + 12$
$= x^2 + 7x + 12$

3. $(x - 5)(x + 10)$
　　F　O　I　L
$= x(x) + x(10) - 5(x) - 5(10)$
$= x^2 + 10x - 5x - 50$
$= x^2 + 5x - 50$

5. $(5x - 6)(x + 2)$
　　F　O　I　L
$= 5x(x) + 5x(2) - 6(x) - 6(2)$
$= 5x^2 + 10x - 6x - 12$
$= 5x^2 + 4x - 12$

7. $(y - 6)(4y - 1)$
　　F　O　I　L
$= y(4y) + y(-1) - 6(4y) - 6(-1)$
$= 4y^2 - y - 24y + 6$
$= 4y^2 - 25y + 6$

9. $(2x + 5)(3x - 1)$
　　F　O　I　L
$= 2x(3x) + 2x(-1) + 5(3x) + 5(-1)$
$= 6x^2 - 2x + 15x - 5$
$= 6x^2 + 13x - 5$

11. $(x - 2)^2$
$= x^2 + 2(x)(-2) + (2)^2$
$= x^2 - 4x + 4$

13. $(2x - 1)^2$
$= (2x)^2 + 2(2x)(-1) + (1)^2$
$= 4x^2 - 4x + 1$

15. $(3a - 5)^2$
$= (3a)^2 + 2(3a)(-5) + (5)^2$
$= 9a^2 - 30a + 25$

17. $(5x + 9)^2$
$= (5x)^2 + 2(5x)(9) + (9)^2$
$= 25x^2 + 90x + 81$

19. Answers may vary.

21. $(a - 7)(a + 7)$
$= (a)^2 - (7)^2$
$= a^2 - 49$

23. $(3x - 1)(3x + 1)$
$= (3x)^2 - (1)^2$
$= 9x^2 - 1$

25. $\left(3x - \dfrac{1}{2}\right)\left(3x + \dfrac{1}{2}\right)$

$= (3x)^2 - \left(\dfrac{1}{2}\right)^2$

$= 9x^2 - \dfrac{1}{4}$

27. $(9x + y)(9x - y)$

$= (9x)^2 - (y)^2$

$= 81x^2 - y^2$

29. $(2x - 3)(2x + 3) - x^2$

$= 4x^2 - 9 - x^2$

$= (3x^2 - 9)$ sq. units

31. $(a + 5)(a + 4)$

$\quad$ F $\quad$ O $\quad$ I $\quad$ L

$= a(a) + a(4) + 5(a) + 5(4)$

$= a^2 + 4a + 5a + 20$

$= a^2 + 9a + 20$

33. $(a + 7)^2$

$= (a)^2 + 2(a)(7) + (7)^2$

$= a^2 + 14a + 49$

35. $(4a + 1)(3a - 1)$

$\quad$ F $\quad$ O $\quad$ I $\quad$ L

$= 4a(3a) + 4a(-1) + 1(3a) + 1(-1)$

$= 12a^2 - 4a + 3a - 1$

$= 12a^2 - a - 1$

37. $(x + 2)(x - 2)$

$= (x)^2 - (2)^2$

$= x^2 - 4$

39. $(3a + 1)^2$

$= (3a)^2 + 2(3a)(1) + (1)^2$

$= 9a^2 + 6a + 1$

41. $(x + y)(4x - y)$

$\quad$ F $\quad$ O $\quad$ I $\quad$ L

$= x(4x) + x(-y) + y(4x) + y(-y)$

$= 4x^2 - xy + 4xy - y^2$

$= 4x^2 + 3xy - y^2$

43. $(2a - 3)^2$

$= (2a)^2 + 2(2a)(-3) + (3)^2$

$= 4a^2 - 12a + 9$

45. $(5x - 6z)(5x + 6z)$

$= (5x)^2 - (6z)^2$

$= 25x^2 - 36z^2$

47. $(x - 3)(x - 5)$

$\quad$ F $\quad$ O $\quad$ I $\quad$ L

$= x(x) + x(-5) - 3(x) - 3(-5)$

$= x^2 - 5x - 3x + 15$

$= x^2 - 8x + 15$

49. $\left(x - \dfrac{1}{3}\right)\left(x + \dfrac{1}{3}\right)$

$= (x)^2 - \left(\dfrac{1}{3}\right)^2$

$= x^2 - \dfrac{1}{9}$

51. $(a + 11)(a - 3)$

$\quad$ F $\quad$ O $\quad$ I $\quad$ L

$= a(a) + a(-3) + 11(a) + 11(-3)$

$= a^2 - 3a + 11a - 33$

$= a^2 + 8a - 33$

53. $(x - 2)^2$

$= (x)^2 + 2(x)(-2) + (2)^2$

$= x^2 - 4x + 4$

55. $(3b + 7)(2b - 5)$

$\quad$ F $\quad$ O $\quad$ I $\quad$ L

$= 3b(2b) + 3b(-5) + 7(2b) + 7(-5)$

$= 6b^2 - 15b + 14b - 35$

$= 6b^2 - b - 35$

57. $(7p - 8)(7p + 8)$

$= (7p)^2 - (8)^2$

$= 49p^2 - 64$

59. $\left(\dfrac{1}{3}a^2 - 7\right)\left(\dfrac{1}{3}a^2 + 7\right)$

$= \left(\dfrac{1}{3}a^2\right)^2 - (7)^2$

$= \dfrac{1}{9}a^4 - 49$

61. $(2r-3s)(2r+3s)$
$= (2r)^2 - (3s)^2$
$= 4r^2 - 9s^2$

63. $(3x-7y)^2$
$= (3x)^2 + 2(3x)(-7y) + (7y)^2$
$= 9x^2 - 42xy + 49y^2$

65. $(4x+5)(4x-5)$
$= (4x)^2 - (5)^2$
$= 16x^2 - 25$

67. $(x+4)(x+4)$
$\quad$ F $\quad$ O $\quad$ I $\quad$ L
$= x(x) + x(4) + 4(x) + 4(4)$
$= x^2 + 4x + 4x + 16$
$= x^2 + 8x + 16$

69. $\left(a - \dfrac{1}{2}y\right)\left(a + \dfrac{1}{2}y\right)$
$= (a)^2 - \left(\dfrac{1}{2}y\right)^2$
$= a^2 - \dfrac{1}{4}y^2$

71. $\left(\dfrac{1}{5}x - y\right)\left(\dfrac{1}{5}x + y\right)$
$= \left(\dfrac{1}{5}x\right)^2 - (y)^2$
$= \dfrac{1}{25}x^2 - y^2$

73. $A = lw$
$A = (2x+1)(2x+1)$
$A = 2x(2x) + 2x(1) + 1(2x) + 1(1)$
$A = 4x^2 + 2x + 2x + 1$
$A = (4x^2 + 4x + 1)$ sq. ft.

75. $(5x-3)^2 - (x+1)^2$
$= 25x^2 - 30x + 9 - (x^2 + 2x + 1)$
$= 25x^2 - 30x + 9 - x^2 - 2x - 1$
$= (24x^2 - 32x + 8)$ sq. meters

77. $\dfrac{50b^{10}}{70b^5} = \dfrac{5b^{10-5}}{7} = \dfrac{5b^5}{7}$

79. $\dfrac{8a^{17}b^5}{-4a^7b^{10}}$
$= \dfrac{8}{-4}a^{17-7}b^{5-10}$
$= -2a^{10}b^{-5}$
$= \dfrac{-2a^{10}}{b^5}$

81. $\dfrac{2x^4y^{12}}{3x^4y^4}$
$= \dfrac{2}{3}x^{4-4}y^{12-4}$
$= \dfrac{2}{3}x^0y^8 = \dfrac{2y^8}{3}$

83. $(-1, 1)$ and $(2, 2)$
$m = \dfrac{2-1}{2-(-1)} = \dfrac{1}{2+1} = \dfrac{1}{3}$

85. $(-1, -2)$ and $(1, 0)$
$m = \dfrac{0-(-2)}{1-(-1)} = \dfrac{2}{1+1} = \dfrac{2}{2} = 1$

87. $[(x+y) - 3][(x+y) + 3]$
$= (x+y)^2 - (3)^2$
$= (x)^2 + 2(x)(y) + (y)^2 - 9$
$= x^2 + 2xy + y^2 - 9$

89. $[(a-3) + b][(a-3) - b]$
$= (a-3)^2 - (b)^2$
$= (a)^2 + 2(a)(-3) + (3)^2 - b^2$
$= a^2 - 6a + 9 - b^2$

91. $[(2x+1) - y][(2x+1) + y]$
$= (2x+1)^2 - (y)^2$
$= (2x)^2 + 2(2x)(1) + (1)^2 - y^2$
$= 4x^2 + 4x + 1 - y^2$

Section 4.5 Mental Math

1. $5x^{-2} = \dfrac{5}{x^2}$

3. $\dfrac{1}{y^{-6}} = y^6$

5. $\dfrac{4}{y^{-3}} = 4y^3$

Exercise Set 4.5

1. $4^{-3} = \dfrac{1}{4^3} = \dfrac{1}{64}$

3. $7x^{-3} = 7\left(\dfrac{1}{x^3}\right) = \dfrac{7}{x^3}$

5. $\left(-\dfrac{1}{4}\right)^{-3} = \dfrac{(-1)^{-3}}{4^{-3}} = \dfrac{4^3}{(-1)^3} = \dfrac{64}{-1} = -64$

7. $3^{-1} + 2^{-1}$
$= \dfrac{1}{3} + \dfrac{1}{2}$
$= \dfrac{1}{3} \cdot \dfrac{2}{2} + \dfrac{1}{2} \cdot \dfrac{3}{3}$
$= \dfrac{2}{6} + \dfrac{3}{6}$
$= \dfrac{5}{6}$

9. $\dfrac{1}{p^{-3}} = p^3$

11. $\dfrac{p^{-5}}{q^{-4}} = \dfrac{q^4}{p^5}$

13. $\dfrac{x^{-2}}{x} = x^{-2-1} = x^{-3} = \dfrac{1}{x^3}$

15. $\dfrac{z^{-4}}{z^{-7}} = z^{-4-(-7)} = z^{-4+7} = z^3$

17. Answers may vary.

19. $(a^{-5}b^2)^{-6} = a^{-5\cdot-6}b^{2\cdot-6} = a^{30}b^{-12} = \dfrac{a^{30}}{b^{12}}$

21. $\left(\dfrac{x^{-2}y^4}{x^3y^7}\right)^2$
$= (x^{-2-3}y^{4-7})^2$

$= (x^{-5}y^{-3})^2$
$= x^{-10}y^{-6} = \dfrac{1}{x^{10}y^6}$

23. $\dfrac{4^2 z^{-3}}{4^3 z^{-5}}$
$= 4^{2-3}z^{-3-(-5)}$
$= 4^{-1}z^{-3+(+5)}$
$= 4^{-1}z^2 = \dfrac{z^2}{4}$

25. Answers may vary.

27. $(-3)^{-2} = \dfrac{1}{(-3)^2} = \dfrac{1}{9}$

29. $\dfrac{-1}{p^{-4}} = -1p^4 = -p^4$

31. $-2^0 - 3^0 = -(1) - (1) = -1 - 1 = -2$

33. $\dfrac{r}{r^{-3}r^{-2}} = r \cdot r^3 \cdot r^2 = r^6$

35. $(x^5 y^3)^{-3}$
$= x^{5\cdot-3}y^{3\cdot-3}$
$= x^{-15}y^{-9} = \dfrac{1}{x^{15}y^9}$

37. $2^0 + 3^{-1} = 1 + \dfrac{1}{3} = \dfrac{3}{3} + \dfrac{1}{3} = \dfrac{4}{3}$

39. $\dfrac{2^{-3}x^{-4}}{2^2 x} = \dfrac{1}{2^2 2^3 xx^4} = \dfrac{1}{2^5 x^5} = \dfrac{1}{32x^5}$

41. $\dfrac{7ab^{-4}}{7^{-1}a^{-3}b^2}$
$= \dfrac{7 \cdot 7aa^3}{b^4 b^2}$
$= \dfrac{7^2 a^4}{b^6}$
$= \dfrac{49a^4}{b^6}$

43. $\left(\dfrac{a^{-5}b}{ab^3}\right)^{-4}$

$= (a^{-5-1}b^{1-3})^{-4}$

$= (a^{-6}b^{-2})^{-4}$

$= a^{-6\cdot-4}b^{-2\cdot-4} = a^{24}b^8$

45. $\dfrac{(xy^3)^5}{(xy)^{-4}}$

$= \dfrac{x^{1\cdot5}y^{3\cdot5}}{x^{1\cdot-4}y^{1\cdot-4}}$

$= \dfrac{x^5y^{15}}{x^{-4}y^{-4}}$

$= x^{5-(-4)}y^{15-(-4)}$

$= x^{5+4}y^{15+4}$

$= x^9y^{19}$

47. $\dfrac{(-2xy^{-3})^{-3}}{(xy^{-1})^{-1}}$

$= \dfrac{(-2)^{1\cdot-3}x^{1\cdot-3}y^{-3\cdot-3}}{x^{1\cdot-1}y^{-1\cdot-1}}$

$= \dfrac{(-2)^{-3}x^{-3}y^9}{x^{-1}y^1}$

$= (-2)^{-3}x^{-3-(-1)}y^{9-1}$

$= (-2)^{-3}x^{-3+1}y^8$

$= (-2)^{-3}x^{-2}y^8$

$= \dfrac{y^8}{(-2)^3x^2}$

$= \dfrac{y^8}{-8x^2}$

$= -\dfrac{y^8}{8x^2}$

49. $\left(\dfrac{3x^{-2}}{z}\right)^3$

$= \dfrac{3^3 \cdot x^{-2\cdot3}}{z^3}$

$= \dfrac{27x^{-6}}{z^3} = \dfrac{27}{x^6z^3}$ cu. in.

51. $\dfrac{14x^5}{7x^{-2}} = 2x^{5+2} = 2x^7$

53. $78,000 = 7.8 \times 10^4$

55. $0.00000167 = 1.67 \times 10^{-6}$

57. $0.00635 = 6.35 \times 10^{-3}$

59. $1,160,000 = 1.16 \times 10^6$

61. $20,000,000 = 2 \times 10^7$

63. $93,000,000 = 9.3 \times 10^7$

65. $7.86 \times 10^8 = 786,000,000$

67. $8.673 \times 10^{-10} = 0.0000000008673$

69. $3.3 \times 10^{-2} = 0.033$

71. $2.032 \times 10^4 = 20,320$

73. $6.25 \times 10^{18} = 6,250,000,000,000,000,000$

75. $9.460 \times 10^{12} = 9,460,000,000,000$

77. $(1.2 \times 10^{-3})(3 \times 10^{-2})$

$= 1.2 \times 3 \times 10^{-3+(-2)}$

$= 3.6 \times 10^{-5} = 0.000036$

79. $(4 \times 10^{-10})(7 \times 10^{-9})$

$= 4 \times 7 \times 10^{-10+(-9)}$

$= 28 \times 10^{-19}$

$= 0.0000000000000000028$

81. $\dfrac{8 \times 10^{-1}}{16 \times 10^5}$

$= \dfrac{8}{16} \times 10^{-1-5}$

$= 0.5 \times 10^{-6}$

$= 0.0000005$

83. $\dfrac{1.4 \times 10^{-2}}{7 \times 10^{-8}}$

$= 0.2 \times 10^{-2-(-8)}$

$= 0.2 \times 10^{-2+8}$

$= 0.2 \times 10^6 = 200,000$

85. $(4.2 \times 10^6)(3600)$

$= (4.2 \times 10^6)(3.6 \times 10^3)$

$= 4.2 \times 3.6 \times 10^{6+3}$

$= 15.12 \times 10^9$

$= 1.512 \times 10^{10}$ cu. ft

87. $F = \dfrac{(6.24 \times 10)(4 \times 10^4)}{2}$

$F = \dfrac{6.24 \times 4}{2} \times 10^{1+4}$

$F = 12.48 \times 10^5$

$F = 1.248 \times 10^6$

89. $(2.63 \times 10^{12})(-1.5 \times 10^{-10})$

$= (2.63)(-1.5)(10^{12-10})$

$= -3.945 \times 10^2 = -394.5$

91. $d = rt$

$238,857 = (1.86 \times 10^5)t$

$\dfrac{238,857}{1.86 \times 10^5} = t$

$1.3 \text{ sec} = t$

93. $\dfrac{5x^7}{3x^4} = \dfrac{5x^{7-4}}{3} = \dfrac{5x^3}{3}$

95. $\dfrac{15z^4 y^3}{21zy} = \dfrac{5z^{4-1} y^{3-1}}{7} = \dfrac{5z^3 y^2}{7}$

97. $\dfrac{1}{y}(5y^2 - 6y + 5)$

$= \dfrac{5y^2}{y} - \dfrac{6y}{y} + \dfrac{5}{y}$

$= 5y - 6 + \dfrac{5}{y}$

99. $2x^2 \left(10x - 6 + \dfrac{1}{x} \right)$

$= 2 \cdot 10x^{2+1} - 6 \cdot 2x^2 + \dfrac{2x^2}{x}$

$= 20x^3 - 12x^2 + 2x$

101. $a^{-4m} a^{5m} = a^{-4m+5m} = a^m$

103. $(3y^{2z})^3 = 3^3 y^{2 \cdot 3z} = 27y^{6z}$

105. $\dfrac{y^{4a}}{y^{-a}} = y^{4a-(-a)} = y^{4a+a} = y^{5a}$

107. $(z^{3a+2})^{-2} = \dfrac{1}{(z^{3a+2})^2} = \dfrac{1}{z^{(3a+2)(2)}} = \dfrac{1}{z^{6a+4}}$

Section 4.6 Mental Math

1. $\dfrac{a^6}{a^4} = a^{6-4} = a^2$

3. $\dfrac{a^3}{a} = a^{3-1} = a^2$

5. $\dfrac{k^5}{k^2} = k^{5-2} = k^3$

Exercise Set 4.6

1. $\dfrac{8k^4}{2k} = 4k^3$

3. $\dfrac{-6m^4}{-2m^3} = 3m$

5. $\dfrac{-24a^6 b}{6ab^2} = \dfrac{-4a^5}{b}$

7. $\dfrac{6x^2 y^3}{-7xy^5} = -\dfrac{6x}{7y^2}$

9. $\dfrac{15p^3 + 18p^2}{3p} = \dfrac{15p^3}{3p} + \dfrac{18p^2}{3p} = 5p^2 + 6p$

11. $\dfrac{-9x^4 + 18x^5}{6x^5}$

$= \dfrac{-9x^4}{6x^5} + \dfrac{18x^5}{6x^5}$

$= -\dfrac{3}{2x} + 3$

13. $\dfrac{-9x^5 + 3x^4 - 12}{3x^3}$

$= \dfrac{-9x^5}{3x^3} + \dfrac{3x^4}{3x^3} - \dfrac{12}{3x^3}$

$= -3x^2 + x - \dfrac{4}{x^3}$

15. $\dfrac{4x^4 - 6x^3 + 7}{-4x^4}$

$= \dfrac{4x^4}{-4x^4} + \dfrac{-6x^3}{-4x^4} + \dfrac{7}{-4x^4}$

$= -1 + \dfrac{3}{2x} - \dfrac{7}{4x^4}$

17. $\dfrac{25x^5 - 15x^3 + 5}{5x^2}$

$= \dfrac{25x^5}{5x^2} - \dfrac{15x^3}{5x^2} + \dfrac{5}{5x^2}$

$= 5x^3 - 3x + \dfrac{1}{x^2}$

19. Perimeter = 4 times a side

$12x^3 + 4x - 16 = 4 \cdot s$

$4s = 12x^3 + 4x - 16$

$s = \dfrac{12x^3 + 4x - 16}{4}$

$s = \dfrac{12x^3}{4} + \dfrac{4x}{4} - \dfrac{16}{4}$

$s = (3x^3 + x - 4)$ ft.

21.
$$\begin{array}{r} x+1 \\ x+3 \overline{\smash{\big)}\, x^2 + 4x + 3} \\ \underline{x^2 + 3x} \\ x + 3 \\ \underline{x + 3} \\ 0 \end{array}$$

$x + 1$

23.
$$\begin{array}{r} 2x+3 \\ x+5 \overline{\smash{\big)}\, 2x^2 + 13x + 15} \\ \underline{2x^2 + 10x} \\ 3x + 15 \\ \underline{3x + 15} \\ 0 \end{array}$$

$2x + 3$

25.
$$\begin{array}{r} 2x+1 \\ x-4 \overline{\smash{\big)}\, 2x^2 - 7x + 3} \\ \underline{2x^2 - 8x} \\ x + 3 \\ \underline{x - 4} \\ 7 \end{array}$$

$2x + 1 + \dfrac{7}{x - 4}$

27.
$$\begin{array}{r} 4x+9 \\ 2x-3 \overline{\smash{\big)}\, 8x^2 + 6x - 27} \\ \underline{8x^2 - 12x} \\ 18x - 27 \\ \underline{18x - 27} \\ 0 \end{array}$$

$4x + 9$

29.
$$\begin{array}{r} 3a^2 - 3a + 1 \\ 3a+2 \overline{\smash{\big)}\, 9a^3 - 3a^2 - 3a + 4} \\ \underline{9a^3 + 6a^2} \\ -9a^2 - 3a \\ \underline{-9a^2 - 6a} \\ 3a + 4 \\ \underline{3a + 2} \\ 2 \end{array}$$

$3a^2 - 3a + 1 + \dfrac{2}{3a + 2}$

31.
$$\begin{array}{r} 2b^2 + b + 2 \\ b+4 \overline{\smash{\big)}\, 2b^3 + 9b^2 + 6b - 4} \\ \underline{2b^3 + 8b^2} \\ b^2 + 6b \\ \underline{b^2 + 4b} \\ 2b - 4 \\ \underline{2b + 8} \\ -12 \end{array}$$

$2b^2 + b + 2 - \dfrac{12}{b + 4}$

33. Answers may vary.

35.
$$\require{enclose}
\begin{array}{r}
2x+5 \\
5x+3 \enclose{longdiv}{10x^2+31x+15} \\
\underline{-(10x^2+6x)} \\
25x+15 \\
\underline{-(25x+15)} \\
0
\end{array}$$

$= (2x + 5)$ meters

37. $\dfrac{20x^2+5x+9}{5x^3}$

$= \dfrac{20x^2}{5x^3} + \dfrac{5x}{5x^3} + \dfrac{9}{5x^3}$

$= \dfrac{4}{x} + \dfrac{1}{x^2} + \dfrac{9}{5x^3}$

39.
$$\begin{array}{r}
5x-2 \\
x+6 \enclose{longdiv}{5x^2+28x-10} \\
\underline{5x^2+30x} \\
-2x-10 \\
\underline{-2x-12} \\
2
\end{array}$$

$5x-2+\dfrac{2}{x+6}$

41. $\dfrac{10x^3-24x^2-10x}{10x}$

$= \dfrac{10x^3}{10x} - \dfrac{24x^2}{10x} - \dfrac{10x}{10x}$

$= x^2 - \dfrac{12x}{5} - 1$

43.
$$\begin{array}{r}
6x-1 \\
x+3 \enclose{longdiv}{6x^2+17x-4} \\
\underline{6x^2+18x} \\
-x-4 \\
\underline{-x-3} \\
-1
\end{array}$$

$6x-1-\dfrac{1}{x+3}$

45. $\dfrac{12x^4+3x^2}{3x^2} = \dfrac{12x^4}{3x^2} + \dfrac{3x^2}{3x^2} = 4x^2+1$

47.
$$\begin{array}{r}
2x^2+6x-5 \\
x-2 \enclose{longdiv}{2x^3+2x^2-17x+8} \\
\underline{2x^2-4x^2} \\
6x^2-17x \\
\underline{6x^2-12x} \\
-5x+8 \\
\underline{-5x+10} \\
-2
\end{array}$$

$2x^2+6x-5-\dfrac{2}{x-2}$

49.
$$\begin{array}{r}
6x-1 \\
5x-2 \enclose{longdiv}{30x^2-17x+2} \\
\underline{30x^2-12x} \\
-5x+2 \\
\underline{-5x+2} \\
0
\end{array}$$

$6x-1$

51. $\dfrac{3x^4-9x^3+12}{-3x}$

$= \dfrac{3x^4}{-3x} + \dfrac{-9x^3}{-3x} + \dfrac{12}{-3x}$

$= -x^3 + 3x^2 - \dfrac{4}{x}$

53.
$$\begin{array}{r}
4x+3 \\
2x+1 \enclose{longdiv}{8x^2+10x+1} \\
\underline{8x^2+4x} \\
6x+1 \\
\underline{6x+3} \\
-2
\end{array}$$

$4x+3-\dfrac{2}{2x+1}$

55.
$$\begin{array}{r}
2x+9 \\
2x-9 \enclose{longdiv}{4x^2+0x-81} \\
\underline{4x^2-18x} \\
18x-81 \\
\underline{18x-81} \\
0
\end{array}$$

$2x+9$

57.
$$2x+3 \overline{\smash{\big)}\,4x^3+12x^2+x-12}$$
quotient: $2x^2+3x-4$

$$\underline{4x^3+6x^2}$$
$$6x^2+x$$
$$\underline{6x^2+9x}$$
$$-8x-12$$
$$\underline{-8x-12}$$
$$0$$

$$2x^2+3x-4$$

59.
$$x-3 \overline{\smash{\big)}\,x^3+0x^2+0x-27}$$
quotient: x^2+3x+9

$$\underline{x^3-3x^2}$$
$$3x^2+0x$$
$$\underline{3x^2-9x}$$
$$9x-27$$
$$\underline{9x-27}$$
$$0$$

$$x^2+3x+9$$

61.
$$x+1 \overline{\smash{\big)}\,x^3+0x^2+0x+1}$$
quotient: x^2-x+1

$$\underline{x^3+x^2}$$
$$-x^2+0x$$
$$\underline{-x^2-x}$$
$$x+1$$
$$\underline{x+1}$$
$$0$$

$$x^2-x+1$$

63.
$$x+2 \overline{\smash{\big)}\,-3x^2+0x+1}$$
quotient: $-3x+6$

$$\underline{-3x^2-6x}$$
$$6x+1$$
$$\underline{6x+12}$$
$$-11$$

$$-3x+6-\dfrac{11}{x+2}$$

65.
$$2b-1 \overline{\smash{\big)}\,4b^2-4b-5}$$
quotient: $2b-1$

$$\underline{4b^2-2b}$$
$$-2b-5$$
$$\underline{-2b+1}$$
$$-6$$

$$2b-1-\dfrac{6}{2b-1}$$

67. $2a(a^2+1)=2a^3+2a$

69. $2x(x^2+7x-5)=2x^3+14x^2-10x$

71. $-3xy(xy^2+7x^2y+8)=-3x^2y^3-21x^3y^2-24xy$

73. $9ab(ab^2c+4bc-8)=9a^2b^3c+36ab^2c-72ab$

75. Thriller

77. Born in the U.S.A.; Eagles Greatest Hits

Chapter 4 - Review

1. 3^2; base = 3, exponent = 2

2. $(-5)^4$; base = –5, exponent = 4

3. -5^4; base = 5, exponent = 4

4. $8^3=8\cdot8\cdot8=512$

5. $(-6)^2=(-6)(-6)=36$

6. $-6^2=-36$

7. $-4^3-4^0=-(64)-(1)=-64-1=-65$

8. $(3b)^0=1$

9. $\dfrac{8b}{8b}=1$

10. $(5b^3)(b^5)(a^6)=5a^6b^8$

11. $2^3\cdot x^0=8\cdot1=8$

12. $[(-3)^2]^3=[(9)]^3=9\cdot9\cdot9=729$

13. $(2x^3)(-5x^2)=(2)(-5)(x^3)(x^2)=10x^5$

14. $\left(\dfrac{mn}{q}\right)^2 \cdot \left(\dfrac{mn}{q}\right) = \dfrac{m^2 n^2}{q^2} \cdot \dfrac{mn}{q} = \dfrac{m^3 n^3}{q^3}$

15. $\left(\dfrac{3ab^2}{6ab}\right)^4 = \left(\dfrac{b}{2}\right)^4 = \dfrac{b^4}{2^4} = \dfrac{b^4}{16}$

16. $\dfrac{x^9}{x^4} = x^{9-4} = x^5$

17. $\dfrac{2x^7 y^8}{8xy^2} = \dfrac{x^6 y^6}{4}$

18. $\dfrac{12xy^6}{3x^4 y^{10}} = \dfrac{4}{x^3 y^4}$

19. $5a^7 (2a^4)^3 = 5a^7 (2^3 a^{12}) = 5a^7 (8a^{12}) = 40a^{19}$

20. $(2x)^2 (9x) = (4x^2)(9x) = 36x^3$

21. $\dfrac{(-4)^2 (3^3)}{(4^5)(3^2)}$

$= \dfrac{(-1 \cdot 4)^2 (3^3)}{(4^5)(3^2)}$

$= \dfrac{(-1)^2 (4^2)(3^3)}{(4^5)(3^2)}$

$= \dfrac{(-1)^2 (3)}{(4^3)}$

$= \dfrac{3}{64}$

22. $\dfrac{(-7)^2 (3^5)}{(-7)^3 (3^4)} = \dfrac{3}{-7} = -\dfrac{3}{7}$

23. $\dfrac{(2x)^0 (-4)^2}{16x} = \dfrac{1 \cdot 16}{16x} = \dfrac{1}{x}$

24. $\dfrac{(8xy)(3xy)}{18x^2 y^2} = \dfrac{24x^2 y^2}{18x^2 y^2} = \dfrac{4}{3}$

25. $m^0 + p^0 + 3q^0$
$= 1 + 1 + 3(1)$
$= 1 + 1 + 3 = 5$

26. $(-5a)^0 + 7^0 + 8^0 = 1 + 1 + 1 = 3$

27. $(3xy^2 + 8x + 9)^0 = 1$

28. $8x^0 + 9^0 = 8(1) + 1 = 9$

29. $6(a^2 b^3)^3 = 6(a^6 b^9) = 6a^6 b^9$

30. $\dfrac{(x^3 z)^a}{x^2 z^2} = \dfrac{x^{3a} z^a}{x^2 z^2} = x^{3a-2} z^{a-2}$

31. $-5x^4 y^3$
The degree is $4 + 3 = 7$.

32. $10x^3 y^2 z$
The degree is 6.

33. $35a^5 bc^2$
The degree is $5 + 1 + 2 = 8$.

34. $95xyz$
The degree is 3.

35. $y^5 + 7x - 8x^4$
The degree is 5.

36. $9y^2 + 30y + 25$
The degree is 2.

37. $-14x^2 yb - 28x^2 y^3 b - 42x^2 y^2$
The degree of $-14x^2 yb$ is 4.
The degree of $-28x^2 y^3 b$ is 6.
The degree of $-42x^2 y^2$ is 4.
Therefore, the degree of the polynomial is 6.

38. $6x^2 y^2 z^2 + 5x^2 y^3 - 12xyz$
The degree is 6.

39.

x	1	3	5.1	10
$2x^2 + 20x$	22	78	154.02	400

40. $6a^2 b^2 + 4ab + 9a^2 b^2 = 15a^2 b^2 + 4ab$

41. $21x^2 y^3 + 3xy + x^2 y^3 + 6 = 22x^2 y^3 + 3xy + 6$

42. $4a^2 b - 3b^2 - 8q^2 - 10a^2 b + 7q^2$
$= -6a^2 b - 3b^2 - q^2$

43. $2s^{14} + 3s^{13} + 12s^{12} - s^{10}$
cannot be combined

44. $(3k^2 + 2k + 6) + (5k^2 + k)$
$= 3k^2 + 2k + 6 + 5k^2 + k$
$= 8k^2 + 3k + 6$

45. $(2s^5 + 3s^4 + 4s^3 + 5s^2) - (4s^2 + 7s + 6)$
$= 2s^5 + 3s^4 + 4s^3 + 5s^2 - 4s^2 - 7s - 6$
$= 2s^5 + 3s^4 + 4s^3 + s^2 - 7s - 6$

46. $(2m^7 + 3x^4 + 7m^6) - (8m^7 + 4m^2 + 6x^4)$
$= 2m^7 + 3x^4 + 7m^6 - 8m^7 - 4m^2 - 6x^4$
$= -6m^7 - 3x^4 + 7m^6 - 4m^2$

47. $(x^2 + 7x + 9) + (x^2 - 4) - (4x^2 + 8x - 7)$
$= x^2 + 7x + 9 + x^2 + 4 - 4x^2 - 8x + 7$
$= -2x^2 - x + 20$

48. $9x(x^2 y) = 9x^3 y$

49. $-7(8xz^2) = -56xz^2$

50. $(6xa^2)(xya^3) = 6x^2 a^5 y$

51. $(4xy)(-3xa^2 y^3) = -12a^2 x^2 y^4$

52. $6(x + 5) = 6x + 6(5) = 6x + 30$

53. $9(x - 7) = 9x - 63$

54. $4(2a + 7) = 4(2a) + 4(7) = 8a + 28$

55. $9(6a - 3) = 54a - 27$

56. $-7x(x^2 + 5)$
$= -7x(x^2) - 7x(5)$
$= -7x^3 - 35x$

57. $-8y(4y^2 - 6) = -32y^3 + 48y$

58. $-2(x^3 - 9x^2 + x)$
$= -2(x^3) - 2(-9x^2) - 2(x)$
$= -2x^3 + 18x^2 - 2x$

59. $-3a(a^2 b + ab + b^2)$
$= -3a^3 b - 3a^2 b - 3ab^2$

60. $(3a^3 - 4a + 1)(-2a)$
$= (3a^3)(-2a) + (-4a)(-2a) + (1)(-2a)$
$= -6a^4 + 8a^2 - 2a$

61. $(6b^3 - 4b + 2)(7b) = 42b^4 - 28b^2 + 14b$

62. $(2x + 2)(x - 7)$
$\quad$ F $\quad$ O $\quad$ I $\quad$ L
$= 2x(x) + 2x(-7) + 2(x) + 2(-7)$
$= 2x^2 - 14x + 2x - 14$
$= 2x^2 - 12x - 14$

63. $(2x - 5)(3x + 2)$
$= 6x^2 + 4x - 15x - 10$
$= 6x^2 - 11x - 10$

64. $(4a - 1)(a + 7)$
$\quad$ F $\quad$ O $\quad$ I $\quad$ L
$= 4a(a) + 4a(7) - 1(a) - 1(7)$
$= 4a^2 + 28a - a - 7$
$= 4a^2 + 27a - 7$

65. $(6a - 1)(7a + 3)$
$= 42a^2 + 18a - 7a - 3$
$= 42a^2 + 11a - 3$

66. $(x + 7)(x^3 + 4x - 5)$
$= x(x^3) + x(4x) + x(-5) + 7(x^3) + 7(4x) + 7(-5)$
$= x^4 + 4x^2 - 5x + 7x^3 + 28x - 35$
$= x^4 + 7x^3 + 4x^2 + 23x - 35$

67. $(x + 2)(x^5 + x + 1)$
$= x^6 + x^2 + x + 2x^5 + 2x + 2$
$= x^6 + 2x^5 + x^2 + 3x + 2$

68.
$$\begin{array}{r} x^2 + 2x + 4 \\ x^2 + 2x - 4 \\ \hline -4x^2 - 8x - 16 \\ 2x^3 + 4x^2 + 8x \\ x^4 + 2x^3 + 4x^2 \\ \hline x^4 + 4x^3 + 4x^2 - 16 \end{array}$$

69. $(x^3 + 4x + 4)(x^3 + 4x - 4)$
$= x^6 + 4x^4 - 4x^3 + 4x^4 + 4x^3$
$\quad + 16x^2 - 16x + 16x - 16$
$= x^6 + 8x^4 + 16x^2 - 16$

70. $(x+7)^3$
$= (x+7)(x+7)(x+7)$
$= (x+7)[x^2 + 2(x)(7) + 7^2]$
$= (x+7)[x^2 + 14x + 49]$
$= x(x^2) + x(14x) + x(49) + 7(x^2) + 7(14x) + 7(49)$
$= x^3 + 14x^2 + 49x + 7x^2 + 98x + 343$
$= x^3 + 21x^2 + 147x + 343$

71. $(2x-5)^3 = (2x-5)(2x-5)(2x-5)$
$= (4x^2 - 20x + 25)(2x-5)$
$= 8x^3 - 40x^2 + 50x - 20x^2 + 100x - 125$
$= 8x^3 - 60x^2 + 150x - 125$

72. $(x+7)^2$
$= x^2 + 2(7)(x) + 7^2$
$= x^2 + 14x + 49$

73. $(x-5)^2 = x^2 - 10x + 25$

74. $(3x-7)^2$
$= (3x)^2 + 2(3x)(-7) + (7)^2$
$= 9x^2 - 42x + 49$

75. $(4x+2)^2 = 16x^2 + 16x + 4$

76. $(5x-9)^2$
$= (5x)^2 + 2(5x)(-9) + (9)^2$
$= 25x^2 - 90x + 81$

77. $(5x+1)(5x-1) = 25x^2 - 1$

78. $(7x+4)(7x-4) = (7x)^2 - (4)^2 = 49x^2 - 16$

79. $(a+2b)(a-2b) = a^2 - 4b^2$

80. $(2x-6)(2x+6) = (2x)^2 - (6)^2 = 4x^2 - 36$

81. $(4a^2 - 2b)(4a^2 + 2b) = 16a^4 - 4b^2$

82. $7^{-2} = \dfrac{1}{7^2} = \dfrac{1}{49}$

83. $-7^{-2} = -\dfrac{1}{7^2} = -\dfrac{1}{49}$

84. $2x^{-4} = 2\left(\dfrac{1}{x^4}\right) = \dfrac{2}{x^4}$

85. $(2x)^{-4} = \dfrac{1}{(2x)^4} = \dfrac{1}{16x^4}$

86. $\left(\dfrac{1}{5}\right)^{-3} = \dfrac{1^{-3}}{5^{-3}} = \dfrac{5^3}{1^3} = \dfrac{125}{1} = 125$

87. $\left(-\dfrac{2}{3}\right)^{-2} = \dfrac{1}{\left(-\dfrac{2}{3}\right)^2} = \dfrac{1}{\frac{4}{9}} = \dfrac{9}{4}$

88. $2^0 + 2^{-4} = 1 + \dfrac{1}{2^4} = 1 + \dfrac{1}{16} = \dfrac{16}{16} + \dfrac{1}{16} = \dfrac{17}{16}$

89. $6^{-1} - 7^{-1} = \dfrac{1}{6} - \dfrac{1}{7} = \dfrac{7}{42} - \dfrac{6}{42} = \dfrac{1}{42}$

90. $\dfrac{1}{(2q)^{-3}} = 1(2q)^3 = 1(2^3 q^3) = 1(8q^3) = 8q^3$

91. $\dfrac{-1}{(qr)^{-3}} = \dfrac{-1}{\frac{1}{(qr)^3}} = \dfrac{-1}{\frac{1}{q^3 r^3}} = -q^3 r^3$

92. $\dfrac{r^{-3}}{s^{-4}} = \dfrac{s^4}{r^3}$

93. $\dfrac{rs^{-3}}{r^{-4}} = r^{1-(-4)} s^{-3} = \dfrac{r^5}{s^3}$

94. $\dfrac{-6}{8x^{-3}r^4} = \dfrac{-6x^3}{8r^4} = -\dfrac{3x^3}{4r^4}$

95. $\dfrac{-4s}{16s^{-3}} = -\dfrac{s^4}{4}$

96. $(2x^{-5})^{-3} = 2^{-3} x^{15} = \dfrac{x^{15}}{2^3} = \dfrac{x^{15}}{8}$

97. $(3y^{-6})^{-1} = 3^{-1} y^6 = \dfrac{y^6}{3}$

98. $(3a^{-1}b^{-1}c^{-2})^{-2}$

$= 3^{-2}a^2b^2c^4$

$= \dfrac{a^2b^2c^4}{3^2}$

$= \dfrac{a^2b^2c^4}{9}$

99. $(4x^{-2}y^{-3}z)^{-3} = 4^{-3}x^6y^9z^{-3} = \dfrac{x^6y^9}{64z^3}$

100. $\dfrac{5^{-2}x^8}{5^{-3}x^{11}}$

$= 5^{-2-(-3)}x^{8-11}$

$= 5^{-2+3}x^{8-11}$

$= 5^1x^{-3} = \dfrac{5}{x^3}$

101. $\dfrac{7^5y^{-2}}{7^7y^{-10}} = \dfrac{y^8}{7^2} = \dfrac{y^8}{49}$

102. $\left(\dfrac{bc^{-2}}{bc^{-3}}\right)^4$

$= (b^{1-1}c^{-2-(-3)})^4$

$= (b^0c^{-2+3})^4$

$= (1 \cdot c^1)^4$

$= 1^4c^4 = c^4$

103. $\left(\dfrac{x^{-3}y^{-4}}{x^{-2}y^{-5}}\right)^{-3} = \dfrac{x^9y^{12}}{x^6y^{15}} = \dfrac{x^3}{y^3}$

104. $\dfrac{x^{-4}y^{-6}}{x^2y^7} = x^{-4-2}y^{-6-7} = x^{-6}y^{-13} = \dfrac{1}{x^6y^{13}}$

105. $\dfrac{a^5b^{-5}}{a^{-5}b^5}$

$= a^{5-(-5)}b^{-5-5}$

$= a^{10}b^{-10} = \dfrac{a^{10}}{b^{10}}$

106. $-2^0 + 2^{-4}$

$= -(1) + \dfrac{1}{2^4}$

$= -1 + \dfrac{1}{16}$

$= -\dfrac{16}{16} + \dfrac{1}{16}$

$= -\dfrac{15}{16}$

107. $-3^{-2} - 3^{-3}$

$= -\dfrac{1}{3^2} - \dfrac{1}{3^3}$

$= -\dfrac{1}{9} - \dfrac{1}{27}$

$= -\dfrac{3}{27} - \dfrac{1}{27}$

$= -\dfrac{4}{27}$

108. $a^{6m}a^{5m} = a^{11m}$

109. $\dfrac{(x^{5+h})^3}{x^5} = \dfrac{x^{15+3h}}{x^5} = x^{15+3h-5} = x^{10+3h}$

110. $(3xy^{2z})^3 = 3^3x^3y^{2z(3)} = 27x^3y^{6z}$

111. $a^{m+2} \cdot a^{m+3} = a^{m+2+m+3} = a^{2m+5}$

112. $0.00027 = 2.7 \times 10^{-4}$

113. $0.8868 = 8.868 \times 10^{-1}$

114. $80,800,000 = 8.08 \times 10^7$

115. $-868,000 = -8.68 \times 10^5$

116. $29,760,000 = 2.976 \times 10^7$

117. $4,000 = 4.0 \times 10^3$

118. $8.67 \times 10^5 = 867,000$

119. $3.86 \times 10^{-3} = 0.00386$

120. $8.6 \times 10^{-4} = 0.00086$

121. $8.936 \times 10^5 = 893,600$

122. $1 \times 10^{20} = 100,000,000,000,000,000,000$

123. $3 \times 10^{-25} = 0.0000000000000000000000003$

124. $(8 \times 10^4)(2 \times 10^{-7})$

$= 8 \times 2 \times 10^{4+(-7)}$

$= 16 \times 10^{-3} = 0.016$

125. $\dfrac{8 \times 10^4}{2 \times 10^{-7}}$

$= \left(\dfrac{8}{2}\right)\left(\dfrac{10^4}{10^{-7}}\right)$

$= (4)[10^{4-(-7)}]$

$= (4)(10^{11})$

$= 4(100000000000)$

$= 400,000,000,000$

126. $\dfrac{4xy^2}{3xz^2y^3}$

$= \dfrac{4x^{1-1}y^{2-3}}{3z^2}$

$= \dfrac{4x^0 y^{-1}}{3z^2}$

$= \dfrac{4(1)}{3z^2(y)}$

$= \dfrac{4}{3z^2 y}$

127. $\dfrac{4xy^3}{32xy^2 z} = \dfrac{y}{8z}$

128. $\dfrac{x^2 + 21x + 49}{7x^2}$

$= \dfrac{x^2}{7x^2} + \dfrac{21x}{7x^2} + \dfrac{49}{7x^2}$

$= \dfrac{1}{7} + \dfrac{3}{x} + \dfrac{7}{x^2}$

129. $\dfrac{5a^3 b - 15ab + 20ab}{-5ab}$

$= \dfrac{5a^3 b}{-5ab} - \dfrac{15ab}{-5ab} + \dfrac{20ab}{-5ab}$

$= -a^2 + 3 - 4$

130.

$$\begin{array}{r} a+1 \\ a-2 \overline{\smash{)}\, a^2 - a + 4} \\ \underline{a^2 - 2a} \\ a + 4 \\ \underline{a - 2} \\ 6 \end{array}$$

$a + 1 + \dfrac{6}{a-2}$

131.

$$\begin{array}{r} 4x \\ x+5 \overline{\smash{)}\, 4x^2 + 20x + 7} \\ \underline{-(4x^2 + 20x)} \\ 0x + 7 \end{array}$$

$4x + \dfrac{7}{x+5}$

132.

$$\begin{array}{r} a^2 + 3a + 8 \\ a-2 \overline{\smash{)}\, a^3 + a^2 + 2a + 6} \\ \underline{a^3 - 2a^2} \\ 3a^2 + 2a \\ \underline{3a^2 + 6a} \\ 8a + 6 \\ \underline{8a - 16} \\ 22 \end{array}$$

$a^2 + 3a + 8 + \dfrac{22}{a-2}$

133.

$$\begin{array}{r} 3b^2 - 4b \\ 3b-2 \overline{\smash{)}\, 9b^3 - 18b^2 + 8b - 1} \\ \underline{-(9b^3 - 6b^2)} \\ -12b^2 + 8b \\ \underline{-(-12b^2 + 8b)} \\ -1 \end{array}$$

$3b^2 - 4b + \dfrac{-1}{3b-2}$

134.

$$2x-1 \overline{) \begin{array}{l} 2x^3 - x^2 + 2 \\ 4x^4 - 4x^3 + x^2 + 4x - 3 \end{array}}$$

$$\underline{4x^4 - 2x^3}$$
$$-2x^3 + x^2$$
$$\underline{-2x^3 + x^2}$$
$$0 + 4x - 3$$
$$\underline{4x - 2}$$
$$-1$$

$$2x^3 - x^2 + 2 - \frac{1}{2x-1}$$

135.

$$x-6 \overline{) \begin{array}{l} -x^2 - 16x - 117 \\ -x^3 - 10x^2 - 21x + 18 \end{array}}$$

$$\underline{-(-x^3 + 6x^2)}$$
$$-16x^2 - 21x$$
$$\underline{-(-16x^2 + 96x)}$$
$$-117x + 18$$
$$\underline{-(-117x + 702)}$$
$$-684$$

$$-x^2 - 16x - 117 + \frac{-684}{x-6}$$

Chapter 4 - Test

1. $2^5 = 2 \cdot 2 \cdot 2 \cdot 2 \cdot 2 = 32$

2. $(-3)^4 = (-3)(-3)(-3)(-3) = 81$

3. $-3^4 = -(3 \cdot 3 \cdot 3 \cdot 3) = -81$

4. $4^{-3} = \frac{1}{4^3} = \frac{1}{64}$

5. $\left(\frac{5x^6 y^3}{35x^7 y} \right)^2$

$$= \left(\frac{1x^{6-7} y^{3-1}}{7} \right)^2$$

$$= \left(\frac{x^{-1} y^2}{7} \right)^2$$

$$= \left(\frac{y^2}{7x} \right)^2$$

$$\frac{y^4}{7^2 x^2} = \frac{y^4}{49x^2}$$

6. $\frac{7(xy)^4}{(xy)^2}$

$$= \frac{7x^4 y^4}{x^2 y^2}$$

$$= 7x^{4-2} y^{4-2} = 7x^2 y^2$$

7. $4(x^2 y^3)^{-3} = 4x^{-6} y^{-9} = \frac{4}{x^6 y^9}$

8. $\left(\frac{x^2 y^3}{x^3 y^{-4}} \right)^{-2}$

$$= (x^{2-3} y^{3-(-4)})^{-2}$$

$$= (x^{-1} y^{3+4})^{-2}$$

$$= (x^{-1} y^7)^{-2}$$

$$= x^2 y^{-14} = \frac{x^2}{y^{14}}$$

9. $\frac{6^2 x^{-4} y^{-1}}{6^3 x^{-3} y^7}$

$$= 6^{2-3} x^{-4-(-3)} y^{-1-7}$$

$$= 6^{-1} x^{-4+3} y^{-8}$$

$$= 6^{-1} x^{-1} y^{-8} = \frac{1}{6xy^8}$$

10. $563,000 = 5.63 \times 10^5$

11. $0.0000863 = 8.63 \times 10^{-5}$

12. $1.5 \times 10^{-3} = 0.0015$

13. $6.23 \times 10^4 = 62,300$

14. $(1.2 \times 10^5)(3 \times 10^{-7})$
$$= 1.2 \times 3 \times 10^{5+(-7)}$$
$$= 3.6 \times 10^{-2} = 0.036$$

15. $4xy^2 + 7xyz + 9x^3 yz$
The degree of $4xy^2$ is 3.
The degree of $7xyz$ is 3.
The degree of $9x^3 yz$ is 5.
Therefore, the degree of the polynomial is 5.

16. $6xyz + 9x^2 y - 3xyz + 9x^2 y = 3xyz + 18x^2 y$

17. $(8x^3 + 7x^2 + 4x - 7) + (8x^3 - 7x - 6)$
$= 8x^3 + 7x^2 + 4x - 7 + 8x^3 - 7x - 6$
$= 16x^3 + 7x^2 - 3x - 13$

18. $5x^3 + x^2 + 5x - 2$
$\underline{-(8x^3 - 4x^2 + x - 7)}$
$5x^3 + x^2 + 5x - 2$
$\underline{-8x^3 + 4x^2 - x + 7}$
$-3x^3 + 5x^2 + 4x + 5$

19. $[(8x^2 + 7x + 5) + (x^3 - 8)] - (4x + 2)$
$= [8x^2 + 7x + 5 + x^3 - 8] - (4x + 2)$
$= [x^3 + 8x^2 + 7x - 3] - (4x + 2)$
$= x^3 + 8x^2 + 7x - 3 - 4x - 2$
$= x^3 + 8x^2 + 3x - 5$

20. $x^2 + 5x + 2$
$\underline{3x + 7}$
$7x^2 + 35x + 14$
$\underline{3x^3 + 15x^2 + 6x}$
$3x^3 + 22x^2 + 41x + 14$

21. $x^3 - x^2 + x + 1$
$\underline{2x^2 - 3x + 7}$
$7x^3 - 7x^2 + 7x + 7$
$-3x^4 + 3x^3 - 3x^2 - 3x$
$\underline{2x^5 - 2x^4 + 2x^3 + 2x^2}$
$2x^5 - 5x^4 + 12x^3 - 8x^2 + 4x + 7$

22. $(x + 7)(3x - 5)$
$\quad$ F $\quad$ O $\quad$ I $\quad$ L
$= x(3x) + x(-5) + 7(3x) + 7(-5)$
$= 3x^2 - 5x + 21x - 35$
$= 3x^2 + 16x - 35$

23. $(3x - 7)(3x + 7) = (3x)^2 - (7)^2 = 9x^2 - 49$

24. $(4x - 2)^2$
$= (4x)^2 + 2(4x)(-2) + (2)^2$
$= 16x^2 - 16x + 4$

25. $(8x + 3)^2$
$= (8x)^2 + 2(8x)(3) + (3)^2$
$= 64x^2 + 48x + 9$

26. $(x^2 - 9b)(x^2 + 9b)$
$= (x^2)^2 - (9b)^2$
$= x^4 - 81b^2$

27.

t	0 sec	1 sec	3 sec	5 sec
$-16t^2 + 1001$	1001 ft	985 ft	857 ft	601 ft

28. $\dfrac{8xy^2}{4x^3y^3z}$
$= \dfrac{8x^{1-3}y^{2-3}}{4z}$
$= \dfrac{2x^{-2}y^{-1}}{z} = \dfrac{2}{x^2yz}$

29. $\dfrac{4x^2 + 2xy - 7x}{8xy}$
$= \dfrac{4x^2}{8xy} + \dfrac{2xy}{8xy} - \dfrac{7x}{8xy}$
$= \dfrac{x}{2y} + \dfrac{1}{4} - \dfrac{7}{8y}$

30.
$$\begin{array}{r} x+2 \\ x+5 \overline{\smash{)}\, x^2 + 7x + 10} \\ \underline{x^2 + 5x} \\ 2x + 10 \\ \underline{2x + 10} \\ 0 \end{array}$$

$x + 2$

31.
$$\begin{array}{r} 9x^2 - 6x + 4 \\ 3x+2 \overline{\smash{)}\, 27x^3 + 0x^2 + 0x - 8} \\ \underline{27x^3 + 18x^2} \\ -18x^2 + 0x \\ \underline{-18x^2 - 12x} \\ 12x - 8 \\ \underline{12x + 8} \\ -16 \end{array}$$

$9x^2 - 6x + 4 - \dfrac{16}{3x + 2}$

Chapter 4 - Cumulative Review

1. a. $9 \le 11$

    b. $8 > 1$

    c. $3 \ne 4$

2. a. $40 = 2 \cdot 2 \cdot 2 \cdot 5$

    b. $63 = 3 \cdot 3 \cdot 7$

3. a. $\frac{2}{7} + \frac{4}{7} = \frac{2+4}{7} = \frac{6}{7}$

    b. $\frac{3}{10} + \frac{2}{10} = \frac{3+2}{10} = \frac{5}{10} = \frac{1}{2}$

    c. $\frac{9}{7} - \frac{2}{7} = \frac{9-2}{7} = \frac{7}{7} = 1$

    d. $\frac{5}{3} - \frac{1}{3} = \frac{5-1}{3} = \frac{4}{3}$

4. $3x - 10 = -2x$
    $3(2) - 10 \overset{?}{=} -2(2)$
    $6 - 10 \overset{?}{=} -4$
    $-4 = -4$
    Solution

5. $-4 - 8 = -12$

6. a. $5x - y = 5(-2) - (-4) = -10 + 4 = -6$

    b. $x^3 - y^2 = (-2)^3 - (-4)^2 = -8 - 16 = -24$

    c. $\frac{3x}{2y} = \frac{3(-2)}{2(-4)} = \frac{-6}{-8} = \frac{3}{4}$

7. a. $2x + 3x + 5 + 2 = 5x + 7$

    b. $-5a - 3 + a + 2 = -4a - 1$

    c. $4y - 3y^2$ cannot be simplified

    d. $2.3x + 5x - 6 = 7.3x - 6$

8. $-3x = 33$
    $\frac{-3x}{-3} = \frac{33}{-3}$
    $x = -11$

9. $3(x - 4) = 3x - 12$
    $3x - 12 = 3x - 12$
    $3x = 3x$
    $x = x$
    All real numbers

10. $V = lwh$
    $\frac{V}{wh} = \frac{lwh}{wh}$
    $\frac{V}{wh} = l$

11. Difference is $7.03 - 5.80 = 1.23$
    $\frac{1.23}{5.80} = 0.212$
    21.2% increase

12. Let $x =$ the amount invested at 7%
    $0.07x + 0.09(20{,}000 - x) = 1550$
    $0.07x + (0.09)(20{,}000) - 0.09x = 1550$
    $1800 - 0.02x = 1550$
    $1800 - 1550 = 0.02x$
    $250 = 0.02x$
    $12{,}500 = x$
    $7{,}500 = 20{,}000 - x$
    \$12,500 at 7%
    \$7,500 at 9%

13. $-5x + 3y = 15$

    If $x = 0$, then
    $-5(0) + 3y = 15$
    $3y = 15$
    $y = 5$

    If $x = -3$, then
    $-5(-3) + 3y = 15$
    $15 + 3y = 15$
    $3y = 0$
    $y = 0$

    If $x = -5$, then
    $-5(-5) + 3y = 15$
    $25 + 3y = 15$
    $3y = -10$
    $y = -\frac{10}{3}$

x	y
0	5
-3	0
-5	$-\frac{10}{3}$

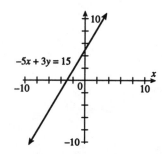

14. $y = x - 1$

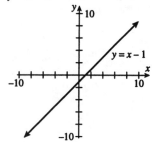

15. $m = \dfrac{4 - (-2)}{2 - (-1)} = \dfrac{4 + 2}{2 + 1} = \dfrac{6}{3} = 2$

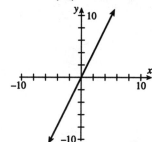

16. a. $(x^2)^5 = x^{2 \cdot 5} = x^{10}$

    b. $(y^8)^2 = y^{8 \cdot 2} = y^{16}$

    c. $[(-5)^3]^4 = (-5)^{3 \cdot 4} = (-5)^{12}$

17. a. $-3x + 7x = 4x$

    b. $11x^2 + 5 + 2x^2 - 7 = 13x^2 - 2$

18. $(2x - y)^2$
$= (2x - y)(2x - y)$
$= 2x(2x) - 2xy - 2xy + (-y)(-y)$
$= 4x^2 - 2xy - 2xy + y^2$
$= 4x^2 - 4xy + y^2$

19. a. $(t + 2)^2 = t^2 + 4t + 4$

    b. $(p - q)^2 = p^2 - 2pq + q^2$

    c. $(2x + 3y)^2 = 4x^2 + 12xy + 9y^2$

    d. $(5r - 7s)^2 = 25r^2 - 70rs + 49s^2$

20. a. $(2x^3)(5x)^{-2} = \dfrac{2x^3}{(5x)^2} = \dfrac{2x^3}{5^2 x^2} = \dfrac{2x^{3-2}}{25} = \dfrac{2x}{25}$

    b. $\left(\dfrac{3a^2}{b}\right)^{-3} = \left(\dfrac{b}{3a^2}\right)^3 = \dfrac{b^3}{3^3 a^{2 \cdot 3}} = \dfrac{b^3}{27a^6}$

    c. $\dfrac{4^{-1} x^{-3} y}{4^{-3} x^2 y^{-6}} = \dfrac{4^{3-1} y^{1+6}}{x^{3+2}} = \dfrac{4^2 y^7}{x^5} = \dfrac{16 y^7}{x^5}$

    d. $(y^{-3} z^{-6})^{-6} = y^{(-3)(-6)} z^{(-6)(-6)} = y^{18} z^{36}$

    e. $\left(\dfrac{-2x^3 y}{xy^{-1}}\right)^3$

    $= \dfrac{(-2)^3 x^{3 \cdot 3} y^3}{x^3 y^{-3}}$

    $= \dfrac{-8x^9 y^3}{x^3 y^{-3}}$

    $= -8x^{9-3} y^{3+3} = -8x^6 y^6$

21. $\dfrac{8x^2 y^2 - 16xy + 2x}{4xy}$

    $= \dfrac{8x^2 y^2}{4xy} - \dfrac{16xy}{4xy} + \dfrac{2x}{4xy}$

    $= 2xy - 4 + \dfrac{1}{2y}$

Chapter 5

Section 5.1 Mental Math

1. $14 = 2 \cdot 7$

3. $10 = 2 \cdot 5$

5. $6 = 2 \cdot 3$
$15 = 3 \cdot 5$
$GCF = 3$

7. 3
$18 = 2 \cdot 3 \cdot 3$
$GCF = 3$

Exercise Set 5.1

1. $32 = 2 \cdot 2 \cdot 2 \cdot 2 \cdot 2$
$36 = 2 \cdot 2 \cdot 3 \cdot 3$
$GCF = 2 \cdot 2 = 4$

3. $12 = 2 \cdot 2 \cdot 3 = 2^2 \cdot 3$
$18 = 2 \cdot 3 \cdot 3 = 2 \cdot 3^2$
$36 = 2 \cdot 2 \cdot 3 \cdot 3 = 2^2 \cdot 3^2$
$GCF = 2 \cdot 3 = 6$

5. $y^2, \; y^4, \; y^7$
$GCF = y^2$

7. $x^{10}y^2, \; xy^2, \; x^3y^3$
$GCF = xy^2$

9. $8x = 2 \cdot 2 \cdot 2 \cdot x$
$4 = 2 \cdot 2$
$GCF = 2 \cdot 2 = 4$

11. $12y^4 = 2 \cdot 2 \cdot 3y^4$
$20y^3 = 2 \cdot 2 \cdot 5y^3$
$GCF = 2 \cdot 2 \cdot y^3 = 4y^3$

13. $12x^3 = 2 \cdot 2 \cdot 3x^3$
$6x^4 = 2 \cdot 3x^4$
$3x^5 = 3x^5$
$GCF = 3x^3$

15. $18x^2y = 2 \cdot 3 \cdot 3 \cdot x^2y$
$9x^3y^3 = 3 \cdot 3 \cdot x^3y^3$
$36x^3y = 2 \cdot 2 \cdot 3 \cdot 3 \cdot x^3y$
$GCF = 3 \cdot 3x^2y = 9x^2y$

17. $3a + 6; \; GCF = 3$
$3a + 3 \cdot 2 = 3(a + 2)$

19. $30x - 15; \; GCF = 15$
$15 \cdot 2x - 15 \cdot 1 = 15(2x - 1)$

21. $24cd^3 - 18c^2d; \; GCF = 6cd$
$6cd \cdot 4d^2 - 6cd \cdot 3c = 6cd(4d^2 - 3c)$

23. $-24a^4x + 18a^3x; \; GCF = 6a^3x$
$-6a^3x(4a) - 6a^3x(-3) = -6a^3x(4a - 3)$

25. $12x^3 + 16x^2 - 8x; \; GCF = 4x$
$4x(3x^2) + 4x(4x) + 4x(-2)$
$= 4x(3x^2 + 4x - 2)$

27. $5x^3y - 15x^2y + 10xy; \; GCF = 5xy$
$5xy(x^2) + 5xy(-3x) + 5xy(2)$
$= 5xy(x^2 - 3x + 2)$

29. Answers may vary.

31. $y(x + 2) + 3(x + 2); \; GCF = (x + 2)$
$(x + 2)(y + 3)$

33. $x(y - 3) - 4(y - 3); \; GCF = (y - 3)$
$(y - 3)(x - 4)$

35. $2x(x + y) - (x + y); \; GCF = (x + y)$
$(x + y)(2x - 1)$

37. $5x + 15 + xy + 3y$
$= 5(x) + 5(3) + x(y) + 3(y)$
$= 5(x + 3) + y(x + 3)$
$= (x + 3)(5 + y)$

39. $2y - 8 + xy - 4x$
$= 2(y - 4) + x(y - 4)$
$= (y - 4)(2 + x)$

41. $3xy - 6x + 8y - 16$
$= 3x(y - 2) + 8(y - 2)$
$= (y - 2)(3x + 8)$

43. $y^3 + 3y^2 + y + 3$

$= y^2(y+3) + 1(y+3)$

$= (y+3)(y^2+1)$

45. $12x(x^2) - 2x$

$= 12x^3 - 2x$

$= 2x(6x^2 - 1)$

47. $20x(10) + \pi \cdot 5^2$

$= 200x + 25\pi$

$= 25(8x + \pi)$

49. $3x - 6$; GCF $= 3$

$3(x-2)$

51. $-8x - 18$; GCF $= -2$

$-2(4x+9)$

53. $32xy - 18x^2$; GCF $= 2x$

$2x(16y - 9x)$

55. $4x - 8y + 4$; GCF $= 4$

$4(x - 2y + 1)$

57. $8(x+2) - y(x+2)$; GCF $= (x+2)$

$(x+2)(8-y)$

59. $-40x^8y^6 - 16x^9y^5$; GCF $= -8x^8y^5$

$-8x^8y^5(5y + 2x)$

61. $5x + 10$; GCF $= 5$

$5(x+2)$

63. $-3x + 12$; GCF $= -3$

$-3(x-4)$

65. $18x^3y^3 - 12x^3y^2 + 6x^5y^2$; GCF $= 6x^3y^2$

$6x^3y^2(3y - 2 + x^2)$

67. $-2a^3 - 6a^2b$; GCF $= -2a^2$

$-2a^2(a + 3b)$

69. $y^2(x-2) + (x-2)$; GCF $= (x-2)$

$(x-2)(y^2+1)$

71. $5xy + 15x + 6y + 18$

$= 5x(y+3) + 6(y+3)$

$= (y+3)(5x+6)$

73. $4x^2 - 8xy - 3x + 6y$

$= 4x(x - 2y) - 3(x - 2y)$

$= (x - 2y)(4x - 3)$

75. $126x^3yz + 210y^4z^3$; GCF $= 42yz$

$42yz(3x^3 + 5y^3z^2)$

77. $4y^2 - 12y + 4yz - 12z$; GCF $= 4$

$4[y^2 - 3y + yz - 3z]$

$= 4[y(y-3) + z(y-3)]$

$= 4[(y-3)(y+z)]$

$= 4(y-3)(y+z)$

79. $3y - 5x + 15 - xy$

$= 3y + 15 - 5x - xy$

$= 3(y+5) - x(5+y)$

$= 3(y+5) - x(y+5)$

$= (y+5)(3-x)$ or $(3-x)(y+5)$

81. $36x + 15y + 30 + 18xy$; GCF $= 3$

$3[12x + 5y + 10 + 6xy]$

$= 3[12x + 10 + 5y + 6xy]$

$= 3[2(6x+5) + y(5+6x)]$

$= 3[2(6x+5) + y(6x+5)]$

$= 3[(6x+5)(2+y)]$

$= 3(6x+5)(2+y)$

83. $12x^2y - 42x^2 - 4y + 14$; GCF $= 2$

$2[6x^2y - 21x^2 - 2y + 7]$

$= 2[3x^2(2y-7) - 1(2y-7)]$

$= 2[(2y-7)(3x^2-1)]$

$= 2(2y-7)(3x^2-1)$

85. Answers may vary.

87. factored

89. not factored

91. $\dfrac{4n^4 - 24n}{4n} = \dfrac{4n^4}{4n} - \dfrac{24n}{4n} = n^3 - 6$

93. $(x+2)(x+5) = x^2 + 2x + 5x + 10 = x^2 + 7x + 10$

95. $(a-7)(a-8) = a^2 - 7a - 8a + 56 = a^2 - 15a + 56$

97. $(b+1)(b-4) = b^2 + b - 4b - 4 = b^2 - 3b - 4$

99. $(y-9)(y+2) = y^2 - 9y + 2y - 18 = y^2 - 7y - 18$

Section 5.2 Mental Math

1. $x^2 + 9x + 20 = (x+4)(x+5)$

3. $x^2 - 7x + 12 = (x-4)(x-3)$

5. $x^2 + 4x + 4 = (x+2)(x+2)$

Exercise Set 5.2

1. $x^2 + 7x + 6 = (x+6)(x+1)$

3. $x^2 + 9x + 20 = (x+5)(x+4)$

5. $x^2 - 8x + 15 = (x-5)(x-3)$

7. $x^2 - 10x + 9 = (x-9)(x-1)$

9. $x^2 - 15x + 5$
   not factorable

11. $x^2 - 3x - 18 = (x-6)(x+3)$

13. $x^2 + 5x + 2$
    not factorable

15. $x^2 + 8xy + 15y^2 = (x+3y)(x+5y)$

17. $x^2 - 2xy + y^2 = (x-y)(x-y) = (x-y)^2$

19. $x^2 - 3xy - 4y^2 = (x-4y)(x+y)$

21. $2z^2 + 20z + 32$; GCF $= 2$
    $2(z^2 + 10z + 16) = 2(z+8)(z+2)$

23. $2x^3 - 18x^2 + 40x$; GCF $= 2x$
    $2x(x^2 - 9x + 20) = 2x(x-5)(x-4)$

25. $7x^2 + 14xy - 21y^2$; GCF $= 7$
    $7(x^2 + 2xy - 3y^2) = 7(x+3y)(x-y)$

27. $x^2 + 15x + 36 = (x+12)(x+3)$

29. $x^2 - x - 2 = (x-2)(x+1)$

31. $r^2 - 16r + 48 = (r-12)(r-4)$

33. $x^2 - 4x - 21 = (x-7)(x+3)$

35. $x^2 + 7xy + 10y^2 = (x+2y)(x+5y)$

37. $r^2 - 3r + 6$
    not factorable

39. $2t^2 + 24t + 64$; GCF $= 2$
    $2(t^2 + 12t + 32) = 2(t+4)(t+8)$

41. $x^3 - 2x^2 - 24x$; GCF $= x$
    $x(x^2 - 2x - 24) = x(x-6)(x+4)$

43. $x^2 - 16x + 63 = (x-9)(x-7)$

45. $x^2 + xy - 2y^2 = (x+2y)(x-y)$

47. $3x^2 + 9x - 30$; GCF $= 3$
    $3(x^2 + 3x - 10) = 3(x+5)(x-2)$

49. $3x^2 - 60x + 108$; GCF $= 3$
    $3(x^2 - 20x + 36) = 3(x-18)(x-2)$

51. $x^2 - 18x - 144 = (x-24)(x+6)$

53. $6x^3 + 54x^2 + 120x$; GCF $= 6x$
    $6x(x^2 + 9x + 20) = 6x(x+5)(x+4)$

55. $2t^5 - 14t^4 + 24t^3$; GCF $= 2t^3$
    $2t^3(t^2 - 7t + 12) = 2t^3(t-4)(t-3)$

57. $5x^3y - 25x^2y^2 - 120xy^3$; GCF $= 5xy$
    $5xy(x^2 - 5xy - 24y^2) = 5xy(x-8y)(x+3y)$

59. $4x^2y + 4xy - 12y$; GCF $= 4y$
    $4y(x^2 + x - 3)$

61. $2a^2b - 20ab^2 + 42b^3$; GCF $= 2b$
    $2b(a^2 - 10ab + 21b^2) = 2b(a-7b)(a-3b)$

63. $b = 8$;
    $x^2 + 8x + 15 = (x+3)(x+5)$
    $b = 16$;
    $x^2 + 16x + 15 = (x+15)(x+1)$

65. $b = 6$;
$$m^2 + 6m - 27 = (m + 9)(m - 3)$$
$b = 26$;
$$m^2 + 26m - 27 = (m + 27)(m - 1)$$

67. $c = 5$;
$$x^2 + 6x + 5 = (x + 5)(x + 1)$$
$c = 8$;
$$x^2 + 6x + 8 = (x + 4)(x + 2)$$
$c = 9$;
$$x^2 + 6x + 9 = (x + 3)(x + 3)$$

69. $c = 3$;
$$y^2 - 4y + 3 = (y - 3)(y - 1)$$
$c = 4$;
$$y^2 - 4y + 4 = (y - 2)(y - 2)$$

71. Answers may vary.

73. $(2x + 1)(x + 5)$
$$= 2x^2 + 10x + x + 5$$
$$= 2x^2 + 11x + 5$$

75. $(5y - 4)(3y - 1)$
$$= 15y^2 - 5y - 12y + 4$$
$$= 15y^2 - 17y + 4$$

77. $(a + 3)(9a - 4)$
$$= 9a^2 - 4a + 27a - 12$$
$$= 9a^2 + 23a - 12$$

79.

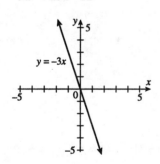

81.

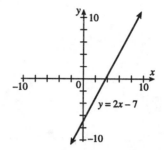

83. $2x^2y + 30xy + 100y$; GCF $= 2y$
$$2y(x^2 + 15x + 50) = 2y(x + 5)(x + 10)$$

85. $-12x^2y^3 - 24xy^3 - 36y^3$; GCF $= -12y^3$
$$-12y^3(x^2 + 2x + 3)$$

87. $y^2(x + 1) - 2y(x + 1) - 15(x + 1)$; GCF $= (x + 1)$
$$(x + 1)(y^2 - 2y - 15) = (x + 1)(y - 5)(y + 3)$$

Section 5.3 Mental Math

1. $x^2 + 14x + 49 = (x + 7)(x + 7) = (x + 7)^2$; Yes

3. No

5. $9y^2 + 6y + 1 = (3y + 1)(3y + 1) = (3y + 1)^2$; Yes

Exercise Set 5.3

1. $2x^2 + 13x + 15 = (2x + 3)(x + 5)$

3. $2x^2 - 9x - 5 = (2x + 1)(x - 5)$

5. $2y^2 - y - 6 = (2y + 3)(y - 2)$

7. $16a^2 - 24a + 9 = (4a - 3)(4a - 3) = (4a - 3)^2$

9. $36r^2 - 5r - 24 = (9r - 8)(4r + 3)$

11. $10x^2 + 17x + 3 = (5x + 1)(2x + 3)$

13. $21x^2 - 48x - 45$; GCF $= 3$
$$3(7x^2 - 16x - 15) = 3(7x + 5)(x - 3)$$

15. $12x^2 - 14x - 6$; GCF $= 2$
$$2(6x^2 - 7x - 3) = 2(2x - 3)(3x + 1)$$

17. $4x^3 - 9x^2 - 9x$; GCF $= x$

$x(4x^2 - 9x - 9) = x(4x + 3)(x - 3)$

19. $x^2 + 22x + 121 = (x + 11)(x + 11) = (x + 11)^2$

21. $x^2 - 16x + 64 = (x - 8)(x - 8) = (x - 8)^2$

23. $16y^2 - 40y + 25 = (4y - 5)(4y - 5) = (4y - 5)^2$

25. $x^2y^2 - 10xy + 25 = (xy - 5)(xy - 5) = (xy - 5)^2$

27. Answers may vary.

29. $2x^2 - 7x - 99 = (2x + 11)(x - 9)$

31. $4x^2 - 8x - 21 = (2x - 7)(2x + 3)$

33. $30x^2 - 53x + 21 = (6x - 7)(5x - 3)$

35. $24x^2 - 58x + 9 = (4x - 9)(6x - 1)$

37. $9x^2 - 24xy + 16y^2$

$= (3x - 4y)(3x - 4y)$

$= (3x - 4y)^2$

39. $x^2 - 14xy + 49y^2$

$= (x - 7y)(x - 7y)$

$= (x - 7y)^2$

41. $2x^2 + 7x + 5 = (2x + 5)(x + 1)$

43. $3x^2 - 5x + 1$

not factorable

45. $-2y^2 + y + 10 = 10 + y - 2y^2 = (5 - 2y)(2 + y)$

47. $16x^2 + 24xy + 9y^2$

$= (4x + 3y)(4x + 3y)$

$= (4x + 3y)^2$

49. $8x^2y + 34xy - 84y$; GCF $= 2y$

$2y(4x^2 + 17x - 42)$

$= 2y(4x - 7)(x + 6)$

51. $3x^2 + x - 2 = (3x - 2)(x + 1)$

53. $x^2y^2 + 4xy + 4$

$= (xy + 2)(xy + 2)$

$= (xy + 2)^2$

55. $49y^2 + 42xy + 9x^2$

$= (7y + 3x)(7y + 3x)$

$= (7y + 3x)^2$

57. $3x^2 - 42x + 63$; GCF $= 3$

$3(x^2 - 14x + 21)$

59. $42a^2 - 43a + 6 = (7a - 6)(6a - 1)$

61. $18x^2 - 9x - 14 = (6x - 7)(3x + 2)$

63. $25p^2 - 70pq + 49q^2$

$= (5p - 7q)(5p - 7q)$

$= (5p - 7q)^2$

65. $15x^2 - 16x - 15 = (5x + 3)(3x - 5)$

67. $-27t + 7t^2 - 4 = 7t^2 - 27t - 4 = (7t + 1)(t - 4)$

69. $b = 2$; $3x^2 + 2x - 5 = (3x + 5)(x - 1)$

$b = 14$; $3x^2 + 14x - 5 = (3x - 1)(x + 5)$

71. $b = 5$; $2z^2 + 5z - 7 = (2z + 7)(z - 1)$

$b = 13$; $2z^2 + 13z - 7 = (2z - 1)(z + 7)$

73. $c = 2$; $5x^2 + 7x + 2 = (5x + 2)(x + 1)$

75. $c = 4$; $3x^2 - 8x + 4 = (3x - 2)(x - 2)$

$c = 5$; $3x^2 - 8x + 5 = (3x - 5)(x - 1)$

77. $(x - 2)(x + 2) = x^2 - 2x + 2x - 4 = x^2 - 4$

79. $(a + 3)(a^2 - 3a + 9)$

$= a^3 - 3a^2 + 9a + 3a^2 - 9a + 27$

$= a^3 + 27$

81. $(y - 5)(y^2 + 5y + 25)$

$= y^3 + 5y^2 + 25y - 5y^2 - 25y - 125$

$= y^3 - 125$

83. $60,000 and above

85. Answers may vary.

87. $-12x^3y^2 + 3x^2y^2 + 15xy^2$; $GCF = -3xy^2$

$\quad -3xy^2(4x^2 - x - 5) = -3xy^2(4x - 5)(x + 1)$

89. $-30p^3q + 88p^2q^2 + 6pq^3$; $GCF = -2pq$

$\quad -2pq(15p^2 - 44pq - 3q^2)$

$\quad = -2pq(15p + q)(p - 3q)$

91. $4x^2(y-1)^2 + 10x(y-1)^2 + 25(y-1)^2$

$\quad GCF = (y-1)^2$

$\quad (y-1)^2(4x^2 + 10x + 25)$

Section 5.4 Mental Math

1. $1 = 1^2$

3. $81 = 9^2$

5. $9 = 3^2$

7. $1 = 1^3$

9. $8 = 2^3$

Exercise Set 5.4

1. $25y^2 - 9$

$\quad = (5y)^2 - 3^2$

$\quad = (5y + 3)(5y - 3)$

3. $121 - 100x^2$

$\quad = 11^2 - (10x)^2$

$\quad = (11 - 10x)(11 + 10x)$

5. $12x^2 - 27$; $GCF = 3$

$\quad 3(4x^2 - 9)$

$\quad = 3[(2x)^2 - 3^2]$

$\quad = 3(2x + 3)(2x - 3)$

7. $169a^2 - 49b^2$

$\quad = (13a)^2 - (7b)^2$

$\quad = (13a + 7b)(13a - 7b)$

9. $x^2y^2 - 1$

$\quad = (xy)^2 - 1^2$

$\quad = (xy + 1)(xy - 1)$

11. $x + 6$

13. $a^3 + 27$

$\quad = a^3 + 3^3$

$\quad = (a + 3)(a^2 - 3a + 9)$

15. $8a^3 + 1$

$\quad = (2a)^3 + 1^3$

$\quad = (2a + 1)(4a^2 - 2a + 1)$

17. $5k^3 + 40$; $GCF = 5$

$\quad 5(k^3 + 8)$

$\quad = 5(k^3 + 2^3)$

$\quad = 5(k + 2)(k^2 - 2k + 4)$

19. $x^3y^3 - 64$

$\quad = (xy)^3 - 4^3$

$\quad = (xy - 4)(x^2y^2 + 4xy + 16)$

21. $x^3 + 125$

$\quad = x^3 + 5^3$

$\quad = (x + 5)(x^2 - 5x + 25)$

23. $24x^4 - 81xy^3$; $GCF = 3x$

$\quad 3x(8x^3 - 27y^3)$

$\quad = 3x[(2x)^3 - (3y)^3]$

$\quad = 3x(2x - 3y)(4x^2 + 6xy + 9y^2)$

25. $(2x + y)$

27. $x^2 - 4 = x^2 - 2^2 = (x - 2)(x + 2)$

29. $81 - p^2 = 9^2 - p^2 = (9 - p)(9 + p)$

31. $4r^2 - 1 = (2r)^2 - 1^2 = (2r - 1)(2r + 1)$

33. $9x^2 - 16^2 = (3x)^2 - 4^2 = (3x - 4)(3x + 4)$

35. $16r^2 + 1$

not factorable

37. $27 - t^3 = 3^3 - t^3 = (3 - t)(9 + 3t + t^2)$

39. $8r^3 - 64$; GCF $= 8$

$8(r^3 - 8)$

$= 8(r^3 - 2^3)$

$= 8(r-2)(r^2 + 2r + 4)$

41. $t^3 - 343$

$= t^3 - 7^3$

$= (t-7)(t^2 + 7t + 49)$

43. $x^2 - 169y^2$

$= x^2 - (13y)^2$

$= (x - 13y)(x + 13y)$

45. $x^2 y^2 - z^2$

$= (xy)^2 - z^2$

$= (xy - z)(xy + z)$

47. $x^3 y^3 + 1$

$= (xy)^3 + 1^3$

$= (xy + 1)(x^2 y^2 - xy + 1)$

49. $s^3 - 64t^3$

$= s^3 - (4t)^3$

$= (s - 4t)(s^2 + 4st + 16t^2)$

51. $18r^2 - 8$; GCF $= 2$

$2(9r^2 - 4)$

$= 2[(3r)^2 - 2^2]$

$= 2(3r - 2)(3r + 2)$

53. $9xy^2 - 4x$; GCF $= x$

$x(9y^2 - 4)$

$= x[(3y)^2 - 2^2]$

$= x(3y + 2)(3y - 2)$

55. $25y^4 - 100y^2$; GCF $= 25y^2$

$25y^2(y^2 - 4)$

$= 25y^2(y^2 - 2^2)$

$= 25y^2(y + 2)(y - 2)$

57. $x^3 y - 4xy^3$; GCF $= xy$

$xy(x^2 - 4y^2)$

$= xy[x^2 - (2y)^2]$

$= xy(x - 2y)(x + 2y)$

59. $8s^6 t^3 + 100s^3 t^6$; GCF $= 4s^3 t^3$

$4s^3 t^3(2s^3 + 25t^3)$

61. $27x^2 y^3 - xy^2$; GCF $= xy^2$

$xy^2(27xy - 1)$

63. $\dfrac{8x^4 + 4x^3 - 2x + 6}{2x}$

$= \dfrac{8x^4}{2x} + \dfrac{4x^3}{2x} - \dfrac{2x}{2x} + \dfrac{6}{2x}$

$= 4x^3 + 2x^2 - 1 + \dfrac{3}{x}$

65.
$$\begin{array}{r} 2x+1 \\ x-2 \overline{)\, 2x^2 - 3x - 2} \\ \underline{2x^2 - 4x} \\ x - 2 \\ \underline{x - 2} \\ 0 \end{array}$$

$2x + 1$

67.
$$\begin{array}{r} 3x+4 \\ x+3 \overline{)\, 3x^2 + 13x + 10} \\ \underline{3x^2 + 9x} \\ 4x + 10 \\ \underline{4x + 12} \\ -2 \end{array}$$

$3x + 4 - \dfrac{2}{x+3}$

69. $x^4 - 16$

$= (x^2)^2 - 4^2$

$= (x^2 + 4)(x^2 - 4)$

$= (x^2 + 4)(x^2 - 2^2)$

$= (x^2 + 4)(x + 2)(x - 2)$

71. $a^2 - (2 + b)^2$

$= [a - (2 + b)][a + (2 + b)]$

$= (a - 2 - b)(a + 2 + b)$

73. $(x^2-4)^2-(x-2)^2$
$= [(x^2-4)+(x-2)][(x^2-4)-(x-2)]$
$= [x^2-4+x-2][x^2-4-x+2]$
$= (x^2+x-6)(x^2-x-2)$
$= (x+3)(x-2)(x+1)(x-2)$
$= (x+3)(x+1)(x-2)^2$

Exercise Set 5.5

1. $a^2+2ab+b^2$
$= (a+b)(a+b)$
$= (a+b)^2$

3. $a^2+a-12 = (a-3)(a+4)$

5. $a^2-a-6 = (a-3)(a+2)$

7. $x^2+2x+1 = (x+1)(x+1) = (x+1)^2$

9. $x^2+4x+3 = (x+3)(x+1)$

11. $x^2+7x+12 = (x+3)(x+4)$

13. $x^2+3x-4 = (x+4)(x-1)$

15. $x^2+2x-15 = (x+5)(x-3)$

17. $x^2-x-30 = (x-6)(x+5)$

19. $2x^2-98$; GCF $= 2$
$2(x^2-49)$
$= 2(x^2-7^2)$
$= 2(x+7)(x-7)$

21. $x^2+3x+xy+3y$
$= x(x+3)+y(x+3)$
$= (x+3)(x+y)$

23. $x^2+6x-16 = (x+8)(x-2)$

25. $4x^3+20x^2-56x$; GCF $= 4x$
$4x(x^2+5x-14) = 4x(x+7)(x-2)$

27. $12x^2+34x+24$; GCF $= 2$
$2(6x^2+17x+12) = 2(3x+4)(2x+3)$

29. $4a^2-b^2 = (2a)^2-b^2 = (2a+b)(2a-b)$

31. $20-3x-2x^2 = (5-2x)(4+x)$

33. a^2+a-3
not factorable

35. $4x^2-x-5 = (4x-5)(x+1)$

37. $4t^2+36$; GCF $= 4$
$4(t^2+9)$

39. $ax+2x+a+2$
$= x(a+2)+1(a+2)$
$= (a+2)(x+1)$

41. $12a^3-24a^2+4a$; GCF $= 4a$
$4a(3a^2-6a+1)$

43. $x^2-14x-48$
not factorable

45. $25p^2-70pq+49q^2$
$= (5p-7q)(5p-7q)$
$= (5p-7q)^2$

47. $125-8y^3$
$= 5^3-(2y)^3$
$= (5-2y)(25+10y+4y^2)$

49. $-x^2-x+30$
$= 30-x-x^2$
$= (6+x)(5-x)$

51. $14+5x-x^2 = (7-x)(2+x)$

53. $3x^4y+6x^3y-72x^2y$; GCF $= 3x^2y$
$3x^2y(x^2+2x-24) = 3x^2y(x+6)(x-4)$

55. $5x^3y^2-40x^2y^3+35xy^4$; GCF $= 5xy^2$
$5xy^2(x^2-8xy+7y^2) = 5xy^2(x-7y)(x-y)$

57. $12x^3y+243xy$; GCF $= 3xy$
$3xy(4x^2+81)$

59. $(x-y)^2-z^2$
$= [(x-y)+z][(x-y)-z]$
$= (x-y+z)(x-y-z)$

61. $3rs - s + 12r - 4$
$= s(3r - 1) + 4(3r - 1)$
$= (3r - 1)(s + 4)$

63. $4x^2 - 8xy - 3x + 6y$
$= 4x(x - 2y) - 3(x - 2y)$
$= (x - 2y)(4x - 3)$

65. $6x^2 + 18xy + 12y^2$; GCF $= 6$
$6(x^2 + 3xy + 2y^2) = 6(x + 2y)(x + y)$

67. $xy^2 - 4x + 3y^2 - 12$
$= x(y^2 - 4) + 3(y^2 - 4)$
$= (y^2 - 4)(x + 3)$
$= (y^2 - 2^2)(x + 3)$
$= (y + 2)(y - 2)(x + 3)$

69. $5(x + y) + x(x + y)$
$= (x + y)(5 + x)$

71. $14t^2 - 9t + 1 = (7t - 1)(2t - 1)$

73. $3x^2 + 2x - 5 = (3x + 5)(x - 1)$

75. $x^2 + 9xy - 36y^2 = (x + 12y)(x - 3y)$

77. $1 - 8ab - 20a^2b^2 = (1 - 10ab)(1 + 2ab)$

79. $x^4 - 10x^2 + 9$
$= (x^2 - 1)(x^2 - 9)$
$= (x + 1)(x - 1)(x + 3)(x - 3)$

81. $x^4 - 14x^2 - 32$
$= (x^2 - 16)(x^2 + 2)$
$= (x + 4)(x - 4)(x^2 + 2)$

83. $x^2 - 23x + 120 = (x - 15)(x - 8)$

85. $6x^3 - 28x^2 + 16x$; GCF $= 2x$
$2x(3x^2 - 14x + 8) = 2x(3x - 2)(x - 4)$

87. $27x^3 - 125y^3$
$= (3x)^3 - (5y)^3$
$= (3x - 5y)(9x^2 + 15xy + 25y^2)$

89. $x^3y^3 + 8z^3$
$= (xy)^3 + (2z)^3$
$= (xy + 2z)(x^2y^2 - 2xyz + 4z^2)$

91. $2xy - 72x^3y$; GCF $= 2xy$
$2xy(1 - 36x^2) = 2xy(1 - 6x)(1 + 6x)$

93. $x^3 + 6x^2 - 4x - 24$
$= x^2(x + 6) - 4(x + 6)$
$= (x + 6)(x^2 - 4)$
$= (x + 6)(x + 2)(x - 2)$

95. $6a^3 + 10a^2$; GCF $= 2a^2$
$2a^2(3a + 5)$

97. $a^2(a + 2) + 2(a + 2) = (a + 2)(a^2 + 2)$

99. $x^3 - 28 + 7x^2 - 4x$
$= x^3 + 7x^2 - 28 - 4x$
$= x^2(x + 7) - 4(7 + x)$
$= x^2(x + 7) - 4(x + 7)$
$= (x + 7)(x^2 - 4)$
$= (x + 7)(x + 2)(x - 2)$

101. Answers may vary.

103. $x - 6 = 0$
$x - 6 + 6 = 0 + 6$
$x = 6$

105. $2m + 4 = 0$
$2m + 4 - 4 = 0 - 4$
$2m = -4$
$\dfrac{2m}{2} = \dfrac{-4}{2}$
$m = -2$

107. $5z - 1 = 0$
$5z - 1 + 1 = 0 + 1$
$5z = 1$
$\dfrac{5z}{5} = \dfrac{1}{5}$
$z = \dfrac{1}{5}$

109. $V = lwh$
$960 = (12)(x)(10)$
$960 = 120x$
$\dfrac{960}{120} = \dfrac{120x}{120}$
$8 = x$
8 inches

111. $(-2, 0), (4, 0), (0, 2), (0, -2)$

113. $(-1, 0), (3, 0), (0, -3)$

Section 5.6 Mental Math

1. $(a - 3)(a - 7) = 0$
$a - 3 = 0$ or $a - 7 = 0$
$a = 3$ or $a = 7$

3. $(x + 8)(x + 6) = 0$
$x + 8 = 0$ or $x + 6 = 0$
$x = -8$ or $x = -6$

5. $(x + 1)(x - 3) = 0$
$x + 1 = 0$ or $x - 3 = 0$
$x = -1$ or $x = 3$

Exercise Set 5.6

1. $(x - 2)(x + 1) = 0$
$x - 2 = 0$ or $x + 1 = 0$
$x = 2$ or $x = -1$

3. $x(x + 6) = 0$
$x = 0$ or $x + 6 = 0$
$x = 0$ or $x = -6$

5. $(2x + 3)(4x - 5) = 0$
$2x + 3 = 0$ or $4x - 5 = 0$
$2x = -3$ or $4x = 5$
$x = -\dfrac{3}{2}$ or $x = \dfrac{5}{4}$

7. $(2x - 7)(7x + 2) = 0$
$2x - 7 = 0$ or $7x + 2 = 0$
$2x = 7$ or $7x = -2$
$x = \dfrac{7}{2}$ or $x = -\dfrac{2}{7}$

9. $(x - 6)(x + 1) = 0$

11. $x^2 - 13x + 36 = 0$
$(x - 4)(x - 9) = 0$
$x - 4 = 0$ or $x - 9 = 0$
$x = 4$ or $x = 9$

13. $x^2 + 2x - 8 = 0$
$(x + 4)(x - 2) = 0$
$x + 4 = 0$ or $x - 2 = 0$
$x = -4$ or $x = 2$

15. $x^2 - 4x = 32$
$x^2 - 4x - 32 = 0$
$(x - 8)(x + 4) = 0$
$x - 8 = 0$ or $x + 4 = 0$
$x = 8$ or $x = -4$

17. $x(3x - 1) = 14$
$3x^2 - x = 14$
$3x^2 - x - 14 = 0$
$(3x - 7)(x + 2) = 0$
$3x - 7 = 0$ or $x + 2 = 0$
$3x = 7$ or $x = -2$
$x = \dfrac{7}{3}$ or $x = -2$

19. $3x^2 + 19x - 72 = 0$
$(3x - 8)(x + 9) = 0$
$3x - 8 = 0$ or $x + 9 = 0$
$3x = 8$ or $x = -9$
$x = \dfrac{8}{3}$ or $x = -9$

21. Two solutions, 5 and 7.
$x = 5$ or $x = 7$
$x - 5 = 0$ or $x - 7 = 0$
$(x - 5)(x - 7) = 0$
$x^2 - 7x - 5x + 35 = 0$
$x^2 - 12x + 35 = 0$

23. $x^3 - 12x^2 + 32x = 0$
$x(x^2 - 12x + 32) = 0$
$x(x - 8)(x - 4) = 0$
$x = 0$ or $x - 8 = 0$ or $x - 4 = 0$
$x = 0$ or $x = 8$ or $x = 4$

25. $(4x - 3)(16x^2 - 24x + 9) = 0$
$(4x - 3)(4x - 3)(4x - 3) = 0$
Since all factors are the same:
$4x - 3 = 0$
$4x = 3$
$x = \dfrac{3}{4}$

27. $4x^3 - x = 0$

$x(4x^2 - 1) = 0$

$x(2x + 1)(2x - 1) = 0$

$x = 0$	or	$2x + 1 = 0$	or	$2x - 1 = 0$
$x = 0$	or	$2x = -1$	or	$2x = 1$
$x = 0$	or	$x = -\dfrac{1}{2}$	or	$x = \dfrac{1}{2}$

29. $32x^3 - 4x^2 - 6x = 0$

$2x(16x^2 - 2x - 3) = 0$

$2x(2x - 1)(8x + 3) = 0$

$2x = 0$	or	$2x - 1 = 0$	or	$8x + 3 = 0$
$x = \dfrac{0}{2}$	or	$2x = 1$	or	$8x = -3$
$x = 0$	or	$x = \dfrac{1}{2}$	or	$x = -\dfrac{3}{8}$

31. Let $y = 0$ and solve for x.

$y = (3x + 4)(x - 1)$

$0 = (3x + 4)(x - 1)$

$3x + 4 = 0$	or	$x - 1 = 0$
$3x = -4$	or	$x = 1$

$x = -\dfrac{4}{3}$

The x-intercepts are $\left(-\dfrac{4}{3},\ 0\right)$ and (1, 0).

33. Let $y = 0$ and solve for x.

$y = x^2 - 3x - 10$

$0 = x^2 - 3x - 10$

$0 = (x - 5)(x + 2)$

$x - 5 = 0$	or	$x + 2 = 0$
$x = 5$	or	$x = -2$

The x-intercepts are (5, 0) and (–2, 0).

35. Let $y = 0$ and solve for x.

$y = 2x^2 + 11x - 6$

$0 = 2x^2 + 11x - 6$

$0 = (2x - 1)(x + 6)$

$2x - 1 = 0$	or	$x + 6 = 0$
$2x = 1$	or	$x = -6$

$x = \dfrac{1}{2}$

The x-intercepts are $\left(\dfrac{1}{2},\ 0\right)$ and (–6, 0).

37. E; x-intercepts are (–2, 0), (1, 0)

39. B; x-intercepts are (0, 0), (–3, 0)

41. C; $y = 2x^2 - 8 = 2(x - 2)(x + 2)$

x-intercepts are (2, 0) and (–2, 0)

43. $x(x + 7) = 0$

$x = 0$	or	$x + 7 = 0$
$x = 0$	or	$x = -7$

45. $(x + 5)(x - 4) = 0$

$x + 5 = 0$	or	$x - 4 = 0$
$x = -5$	or	$x = 4$

47. $x^2 - x = 30$

$x^2 - x - 30 = 0$

$(x - 6)(x + 5) = 0$

$x - 6 = 0$	or	$x + 5 = 0$
$x = 6$	or	$x = -5$

49. $6y^2 - 22y - 40 = 0$

$2(3y^2 - 11y - 20) = 0$

$2(3y + 4)(y - 5) = 0$

$3y + 4 = 0$	or	$y - 5 = 0$
$3y = -4$	or	$y = 5$
$y = -\dfrac{4}{3}$	or	$y = 5$

51. $(2x + 3)(2x^2 - 5x - 3) = 0$

$(2x + 3)(2x + 1)(x - 3) = 0$

$2x + 3 = 0$	or	$2x + 1 = 0$	or	$x - 3 = 0$
$2x = -3$	or	$2x = -1$	or	$x = 3$
$x = -\dfrac{3}{2}$	or	$x = -\dfrac{1}{2}$	or	$x = 3$

53. $x^2 - 15 = -2x$

$x^2 + 2x - 15 = 0$

$(x + 5)(x - 3) = 0$

$x + 5 = 0$	or	$x - 3 = 0$
$x = -5$	or	$x = 3$

55. $x^2 - 16x = 0$

$x(x - 16) = 0$

$x = 0$	or	$x - 16 = 0$
$x = 0$	or	$x = 16$

57. $-18y^2 - 33y + 216 = 0$

$-3(6y^2 + 11y - 72) = 0$

$-3(3y - 8)(2y + 9) = 0$

$3y - 8 = 0$	or	$2y + 9 = 0$
$3y = 8$	or	$2y = -9$
$y = \dfrac{8}{3}$	or	$y = -\dfrac{9}{2}$

59. $12x^2 - 59x + 55 = 0$
$(4x - 5)(3x - 11) = 0$

$4x - 5 = 0$	or	$3x - 11 = 0$
$4x = 5$	or	$3x = 11$
$x = \dfrac{5}{4}$	or	$x = \dfrac{11}{3}$

61. $18x^2 + 9x - 2 = 0$
$(3x + 2)(6x - 1) = 0$

$3x + 2 = 0$	or	$6x - 1 = 0$
$3x = -2$	or	$6x = 1$
$x = -\dfrac{2}{3}$	or	$x = \dfrac{1}{6}$

63. $x(6x + 7) = 5$
$6x^2 + 7x = 5$
$6x^2 + 7x - 5 = 0$
$(3x + 5)(2x - 1) = 0$

$3x + 5 = 0$	or	$2x - 1 = 0$
$3x = -5$	or	$2x = 1$
$x = -\dfrac{5}{3}$	or	$x = \dfrac{1}{2}$

65. $4(x - 7) = 6$
$4x - 28 = 6$
$4x = 6 + 28$
$4x = 34$
$x = \dfrac{34}{4} = \dfrac{17}{2}$

67. $5x^2 - 6x - 8 = 0$
$(5x + 4)(x - 2) = 0$

$5x + 4 = 0$	or	$x - 2 = 0$
$5x = -4$	or	$x = 2$
$x = -\dfrac{4}{5}$	or	$x = 2$

69. $(y - 2)(y + 3) = 6$
$y^2 + 3y - 2y - 6 = 6$
$y^2 + y - 6 = 6$
$y^2 + y - 6 - 6 = 0$
$y^2 + y - 12 = 0$
$(y + 4)(y - 3) = 0$

$y + 4 = 0$	or	$y - 3 = 0$
$y = -4$	or	$y = 3$

71. $4y^2 - 1 = 0$
$(2y + 1)(2y - 1) = 0$

$2y + 1 = 0$	or	$2y - 1 = 0$
$2y = -1$	or	$2y = 1$
$y = -\dfrac{1}{2}$	or	$y = \dfrac{1}{2}$

73. $t^2 + 13t + 22 = 0$
$(t + 11)(t + 2) = 0$

$t + 11 = 0$	or	$t + 2 = 0$
$t = -11$	or	$t = -2$

75. $5t - 3 = 12$
$5t = 12 + 3$
$5t = 15$
$t = \dfrac{15}{5}$
$t = 3$

77. $x^2 + 6x - 17 = -26$
$x^2 + 6x - 17 + 26 = 0$
$x^2 + 6x + 9 = 0$
$(x + 3)(x + 3) = 0$
$x + 3 = 0$
$x = -3$

79. $12x^2 + 7x - 12 = 0$
$(3x + 4)(4x - 3) = 0$

$3x + 4 = 0$	or	$4x - 3 = 0$
$3x = -4$	or	$4x = 3$
$x = -\dfrac{4}{3}$	or	$x = \dfrac{3}{4}$

81. $10t^3 - 25t - 15t^2 = 0$
$10t^3 - 15t^2 - 25t = 0$
$5t(2t^2 - 3t - 5) = 0$
$5t(2t - 5)(t + 1) = 0$

$5t = 0$	or	$2t - 5 = 0$	or	$t + 1 = 0$
$t = \dfrac{0}{5}$	or	$2t = 5$	or	$t = -1$
$t = 0$	or	$t = \dfrac{5}{2}$	or	$t = -1$

83. a. $y = -16x^2 + 20x + 300$

x	0	1	2	3	4	5	6
y	300	304	276	216	124	0	-156

b. 5 seconds

c. 304 feet

d. Answers may vary.

85. $\dfrac{3}{5}+\dfrac{4}{9}$

$=\dfrac{3}{5}\cdot\dfrac{9}{9}+\dfrac{4}{9}\cdot\dfrac{5}{5}$

$=\dfrac{27}{45}+\dfrac{20}{45}=\dfrac{47}{45}$

87. $\dfrac{7}{10}-\dfrac{5}{12}$

$=\dfrac{7}{10}\cdot\dfrac{6}{6}-\dfrac{5}{12}\cdot\dfrac{5}{5}$

$=\dfrac{42}{60}-\dfrac{25}{60}=\dfrac{17}{60}$

89. $\dfrac{7}{8}+\dfrac{7}{15}=\dfrac{7}{8}\cdot\dfrac{15}{7}=\dfrac{15}{8}$

91. $\dfrac{4}{5}\cdot\dfrac{7}{8}=\dfrac{4}{5}\cdot\dfrac{7}{8}=\dfrac{7}{10}$

93. $(x-3)(3x+4)=(x+2)(x-6)$

$3x^2+4x-9x-12=x^2-6x+2x-12$

$3x^2-5x-12=x^2-4x-12$

$3x^2-x^2-5x+4x-12+12=0$

$2x^2-x=0$

$x(2x-1)=0$

$x=0$	or	$2x-1=0$
$x=0$	or	$2x=1$
$x=0$	or	$x=\dfrac{1}{2}$

95. $(2x-3)(x+8)=(x-6)(x+4)$

$2x^2+16x-3x-24=x^2+4x-6x-24$

$2x^2+13x-24=x^2-2x-24$

$2x^2-x^2+13x+2x-24+24=0$

$x^2+15x=0$

$x(x+15)=0$

$x=0$	or	$x+15=0$
$x=0$	or	$x=-15$

97. $(4x-1)(x-8)=(x+2)(x+4)$

$4x^2-32x-x+8=x^2+4x+2x+8$

$4x^2-33x+8=x^2+6x+8$

$4x^2-x^2-33x-6x+8-8=0$

$3x^2-39x=0$

$3x(x-13)=0$

$3x=0$	or	$x-13=0$
$\dfrac{3x}{3}=\dfrac{0}{3}$	or	$x=13$
$x=0$	or	$x=13$

Exercise Set 5.7

1. 1st number $=x$
2nd number $=36-x$

3. 1st odd integer $=x$
2nd consecutive odd integer $=x+2$

5. width $=x$
length $=3x-4$

7. age now $=x$
woman's age ten years ago $=x-10$

9. 1st integer $=x$
2nd integer $=x+1$
3rd integer $=x+2$

11. 1st side $=2x-2$
2nd side $=x$
3rd side $=x+10$

13. $h=-16t^2+64t+80$

$0=-16t^2+64t+80$

$0=-16(t^2-4t-5)$

$0=-16(t-5)(t+1)$

$t-5=0$ or $t=5$ seconds

15. $A=lw$
length $=2x-7$
width $=x$

$30=x(2x-7)$

$30=2x^2-7x$

$2x^2-7x-30=0$

$(2x+5)(x-6)=0$

$2x+5=0 \qquad x=6$

width $=6$ cm

length $=[2(6)-7]=5$ cm

17. base $= x$
altitude $= 2x + 8$
$$A = \frac{1}{2}ab$$
$$96 = \frac{1}{2}(x)(2x + 8)$$
$$96 = x(x + 4)$$
$$96 = x^2 + 4x$$
$$x^2 + 4x - 96 = 0$$
$$(x + 12)(x - 8) = 0$$
$x = -12$ or $x = 8$
base $= 8$ cm
altitude $= 2(8) + 8 = 24$ cm
base $= 8$ cm
altitude $= 24$ cm

19. original side $= x$
new side $= x + 3$
$$A = s^2$$
$$64 = (x + 3)^2$$
$$64 = x^2 + 6x + 9$$
$$x^2 + 6x - 55 = 0$$
$$(x + 11)(x - 5) = 0$$
$x + 11 = 0$ or $x - 5 = 0$
$x = -11$ or $x = 5$
Since length cannot be negative, $x = 5$ inches.

21. short leg $= x - 10$
long leg $= x - 5$
hypotenuse $= x$
$$a^2 + b^2 = c^2$$
$$(x - 10)^2 + (x - 5)^2 = x^2$$
$$x^2 - 20x + 100 + x^2 - 10x + 25 = x^2$$
$$x^2 - 30x + 125 = 0$$
$$(x - 5)(x - 25) = 0$$
$x - 5 = 0$ or $x - 25 = 0$
$x = 5$ or $x = 25$
When $x = 5$, one of the legs would have a negative
length, so hypotenuse $= 25$ cm;
leg $= x - 10 = 15$ cm; leg $= x - 5 = 20$ cm.

23. short leg $= x$
long leg $= x + 12$
hypotenuse $= 2x - 12$
$$x^2 + (x + 12)^2 = (2x - 12)^2$$
$$x^2 + x^2 + 24x + 144 = 4x^2 - 48x + 144$$
$$0 = 2x^2 - 72x$$
$$0 = 2(x^2 - 36x)$$
$$2(x)(x - 36) = 0$$
$x = 0$ or $x = 36$
Short leg $= 36$ feet

25. $x^2 + x = 132$
$$x^2 + x - 132 = 0$$
$$(x + 12)(x - 11) = 0$$
$x + 12 = 0$ or $x - 11 = 0$
$x = -12$ or $x = 11$
$x = -12, 11$

27. 1st $\# = x$
2nd $\# = 20 - x$
$$x^2 + (20 - x)^2 = 218$$
$$x^2 + 400 - 40x + x^2 = 218$$
$$2x^2 - 40x + 182 = 0$$
$$2(x^2 - 20x + 91) = 0$$
$$2(x - 13)(x - 7) = 0$$
$x - 13 = 0$ or $x - 7 = 0$
$x = 13$ or $x = 7$
13, 7

29. $D = \dfrac{n(n - 3)}{2}$
$$2(5) = 2\left(\frac{n(n - 3)}{2}\right)$$
$$10 = n(n - 3)$$
$$10 = n^2 - 3n$$
$$0 = n^2 - 3n - 10$$
$$0 = (n - 5)(n + 2)$$
$n - 5 = 0$ or $n + 2 = 0$
$n = 5$ or $n = -2$
$\#$ of sides $= 5$

31. $P = 2x + 2w$
$$42 = 2x + 2w$$
$$42 - 2x = 2w$$
$$21 - x = w$$
length $= x$
width $= 21 - x$ so
$$x(21 - x) = 104$$
$$21x - x^2 = 104$$
$$0 = x^2 - 21x + 104$$
$$0 = (x - 13)(x - 8)$$
$x = 13$ or $x = 8$
length $= 13$ miles
width $= 8$ miles

33. 1st integer $= x$

2nd integer $= x + 1$

$(x)^2 + (x+1)^2 = 8x + 9$

$x^2 + x^2 + 2x + 1 = 8x + 9$

$2x^2 - 6x - 8 = 0$

$2(x^2 - 3x - 4) = 0$

$2(x - 4)(x + 1) = 0$

$x - 4 = 0$	$x + 1 = 0$
$x = 4$	$x = -1$
$x = 4$ and	$x + 1 = 5$
or	
$x = -1$ and	$x + 1 = 0$

35. pool width $= x$

pool length $= x + 6$

entire area's width $= x + 4$

entire area's length $= x + 6 + 4 = x + 10$

$x(x + 6) + 272 = (x + 4)(x + 10)$

$x^2 + 6x + 272 = x^2 + 14x + 40$

$272 + 6x = 14x$

$\dfrac{232}{8} = \dfrac{8x}{8}$

$x = 29$

29 m = width of pool

35 m = length of pool

37. 1st # $= x$

2nd # $= x + 2$

$x(x + 2) = 624$

$x^2 + 2x - 624 = 0$

$(x + 26)(x - 24) = 0$

$x + 26 = 0$	or	$x - 24 = 0$
$x = -26$	or	$x = 24$

The two integers are either –26 and –24 or 24 and 26.

39. smaller leg $= x$

longer leg $= x + 4$

hypotenuse $= x + 8$

$x^2 + (x+4)^2 = (x+8)^2$

$x^2 + x^2 + 8x + 16 = x^2 + 16x + 64$

$x^2 - 8x - 48 = 0$

$(x - 12)(x + 4) = 0$

$x - 12 = 0$	or	$x + 4 = 0$
$x = 12$	or	$x = -4$

smaller leg = 12 mm

longer leg = 16 mm

hypotenuse = 20 mm

41. base $= 2x$

altitude $= x$

$100 = \dfrac{1}{2} x(2x)$

$100 = x^2$

$\pm 10 = x$

altitude = 10 km

43. 1st integer $= x$

2nd integer $= x + 1$

$x^2 + (x+1)^2 = 221$

$x^2 + x^2 + 2x + 1 = 221$

$2x^2 + 2x - 220 = 0$

$2(x^2 + x - 110) = 0$

$(x + 11)(x - 10) = 0$

$x + 11 = 0$	or	$x - 10 = 0$
$x = -11$	or	$x = 10$

The integers are –10 and –11.

45. width $= x$

length $= 2x + 2$

$x(2x + 2) = 60$

$2x^2 + 2x - 60 = 0$

$2(x^2 + x - 30) = 0$

$(x + 6)(x - 5) = 0$

$x + 6 = 0$	or	$x - 5 = 0$
$x = -6$	or	$x = 5$

width = 5 yds

length = 12 yds

47. length $= x$

width $= x - 7$

$x(x - 7) = 120$

$x^2 - 7x - 120 = 0$

$(x - 15)(x + 8) = 0$

$x - 15 = 0$	or	$x + 8 = 0$
$x = 15$	or	$x = -8$

length = 15 miles

width = 8 miles

49. $A = P(1 + r)^2$

$144 = 100(1 + r)^2$

$144 = 100(1 + 2r + r^2)$

$144 = 100 + 200r + 100r^2$

$0 = 100r^2 + 200r - 44$

$0 = 50r^2 + 100r - 22$

$0 = 25r^2 + 50r - 11$

$0 = (5r - 1)(5r + 11)$

$5r - 1 = 0$ or $5r + 11 = 0$
$5r = 1$ or $5r = -11$
$r = \dfrac{1}{5}$ or $r = -\dfrac{11}{5}$
$r = \dfrac{1}{5} = 0.20 = 20\%$

51. $C = x^2 - 15x + 50$
$9500 = x^2 - 15x + 50$
$0 = x^2 - 15x - 9450$
$0 = (x - 105)(x + 90)$
$x - 105 = 0$ or $x + 90 = 0$
$x = 105$ or $x = -90$
105 units

53. 1st boat $= x$
2nd boat $= x + 7$
$x^2 + (x + 7)^2 = 17^2$
$x^2 + x^2 + 14x + 49 = 289$
$2x^2 + 14x - 240 = 0$
$2(x^2 + 7x - 120) = 0$
$(x + 15)(x - 8) = 0$
$x + 15 = 0$ or $x - 8 = 0$
$x = -15$ or $x = 8$
8 mph and 15 mph

55. $42 = 2x + 2w$
$42 - 2x = 2w$
$21 - x = w$
length $= x$
width $= 21 - x$
$x(21 - x) = 104$
$21x - x^2 = 104$
$0 = x^2 - 21x + 104$
$0 = (x - 13)(x - 8)$
$x - 13 = 0$ or $x - 8 = 0$
$x = 13$ or $x = 8$
length $= 8$ yds.
width $= 13$ yds.

57. Answers may vary.

59. 467 acres

61. 2.1 million

63. Answers may vary.

65. $\dfrac{20}{35} = \dfrac{2 \cdot 2 \cdot 5}{5 \cdot 7} = \dfrac{4}{7}$

67. $\dfrac{27}{18} = \dfrac{3 \cdot 3 \cdot 3}{2 \cdot 3 \cdot 3} = \dfrac{3}{2}$

69. $\dfrac{14}{42} = \dfrac{2 \cdot 7}{2 \cdot 3 \cdot 7} = \dfrac{1}{3}$

Chapter 5 - Review

1. $6x^2 - 15x = 3x(2x - 5)$

2. $2x^3y - 6x^2y^2 - 8xy^3$
$= 2xy(x^2 - 3xy - 4y^2)$
$= 2xy(x - 4y)(x + y)$

3. $20x^2 + 12x$; GCF $= 4x$
$4x(5x + 3)$

4. $6x^2y^2 - 3xy^3 = 3xy^2(2x - y)$

5. $-8x^3y + 6x^2y^2$; GCF $= -2x^2y$
$-2x^2y(4x - 3y)$

6. $3x(2x + 3) - 5(2x + 3) = (2x + 3)(3x - 5)$

7. $5x(x + 1) - (x + 1)$; GCF $= (x + 1)$
$(x + 1)(5x - 1)$

8. $3x^2 - 3x + 2x - 2$
$= 3x(x - 1) + 2(x - 1)$
$= (x - 1)(3x + 2)$

9. $6x^2 + 10x - 3x - 5$
$= 2x(3x + 5) - 1(3x + 5)$
$= (3x + 5)(2x - 1)$

10. $3a^2 + 9ab + 3b^2 + ab$
$= 3a(a + 3b) + b(a + 3b)$
$= (a + 3b)(3a + b)$

11. $x^2 + 6x + 8 = (x + 4)(x + 2)$

12. $x^2 - 11x + 24 = (x - 8)(x - 3)$

13. $x^2 + x + 2$
not factorable

14. $x^2 - 5x - 6 = (x - 6)(x + 1)$

15. $x^2 + 2x - 8 = (x + 4)(x - 2)$

16. $x^2 + 4xy - 12y^2 = (x + 6y)(x - 2y)$

17. $x^2 + 8xy + 15y^2 = (x + 3y)(x + 5y)$

18. $3x^2y + 6xy^2 + 3y^3$
$$= 3y(x^2 + 2xy + y^2)$$
$$= 3y(x + y)(x + y)$$
$$= 3y(x + y)^2$$

19. $72 - 18x - 2x^2$; GCF $= 2$
$$2(36 - 9x - x^2) = 2(12 + x)(3 - x)$$

20. $32 + 12x - 4x^2 = 4(8 + 3x - x^2)$

21. $2x^2 + 11x - 6 = (2x - 1)(x + 6)$

22. $4x^2 - 7x + 4$
not factorable

23. $4x^2 + 4x - 3 = (2x + 3)(2x - 1)$

24. $6x^2 + 5xy - 4y^2$
$$= 6x^2 + 8xy - 3xy - 4y^2$$
$$= 2x(3x + 4y) - y(3x + 4y)$$
$$= (3x + 4y)(2x - y)$$

25. $6x^2 - 25xy + 4y^2 = (6x - y)(x - 4y)$

26. $18x^2 - 60x + 50$
$$= 2(9x^2 - 30x + 25)$$
$$= 2(3x - 5)(3x - 5)$$
$$= 2(3x - 5)^2$$

27. $2x^2 - 23xy - 39y^2 = (2x + 3y)(x - 13y)$

28. $4x^2 - 28xy + 49y^2$
$$= (2x - 7y)(2x - 7y)$$
$$= (2x - 7y)^2$$

29. $18x^2 - 9xy - 20y^2 = (6x + 5y)(3x - 4y)$

30. $36x^3y + 24x^2y^2 - 45xy^3$
$$= 3xy(12x^2 + 8xy - 15y^2)$$
$$= 3xy(12x^2 + 18xy - 10y^2 - 15y^2)$$
$$= 3xy[6x(2x + 3y) - 5y(2x + 3y)]$$
$$= 3xy(2x + 3y)(6x - 5y)$$

31. $4x^2 - 9 = (2x)^2 - 3^2 = (2x + 3)(2x - 3)$

32. $9t^2 - 25s^2 = (3t - 5s)(3t + 5s)$

33. $16x^2 + y^2$
not factorable

34. $x^3 - 8y^3 = (x - 2y)(x^2 + 2xy + 4y^2)$

35. $8x^3 + 27 = (2x)^3 + 3^3 = (2x + 3)(4x^2 - 6x + 9)$

36. $2x^3 + 8x = 2x(x^2 + 4)$

37. $54 - 2x^3y^3$; GCF $= 2$
$$2(27 - x^3y^3)$$
$$= 2[3^3 - (xy)^3]$$
$$= 2(3 - xy)(9 + 3xy + x^2y^2)$$

38. $9x^2 - 4y^2 = (3x - 2y)(3x + 2y)$

39. $16x^4 - 1$
$$= (4x^2)^2 - 1^2$$
$$= (4x^2 + 1)(4x^2 - 1)$$
$$= (4x^2 + 1)[(2x)^2 - 1^2]$$
$$= (4x^2 + 1)(2x + 1)(2x - 1)$$

40. $x^4 + 16$
not factorable

41. $2x^2 + 5x - 12 = (2x - 3)(x + 4)$

42. $3x^2 - 12 = 3(x^2 - 4) = 3(x - 2)(x + 2)$

43. $x(x - 1) + 3(x - 1)$; GCF $= (x - 1)$
$(x - 1)(x + 3)$

44. $x^2 + xy - 3x - 3y$
$$= x(x + y) - 3(x + y)$$
$$= (x + y)(x - 3)$$

45. $4x^2y - 6xy^2$; GCF $= 2xy$
$2xy(2x - 3y)$

46. $8x^2 - 15x - x^3$
$$= -x(-8x + 15 + x^2)$$
$$= -x(x^2 - 8x + 15)$$
$$= -x(x - 5)(x - 3)$$

47. $125x^3 + 27$
$$= (5x)^3 + 3^3$$
$$= (5x + 3)(25x^2 - 15x + 9)$$

48. $24x^2 - 3x - 18 = 3(8x^2 - x - 6)$

49. $(x+7)^2 - y^2$
$= [(x+7) + y][(x+7) - y]$
$= (x+7+y)(x+7-y)$

50. $x^2(x+3) - 4(x+3)$
$= (x+3)(x^2 - 4)$
$= (x+3)(x-2)(x+2)$

51. $(x+6)(x-2) = 0$
$x+6 = 0$ or $x-2 = 0$
$x = -6$ or $x = 2$

52. $3x(x+1)(7x-2) = 0$
$3x = 0$ or $x+1 = 0$ or $7x-2 = 0$
$x = 0$ or $x = -1$ or $7x = 2$
$x = \dfrac{2}{7}$

53. $4(5x+1)(x+3) = 0$
$5x+1 = 0$ or $x+3 = 0$
$5x = -1$ or $x = -3$
$x = -\dfrac{1}{5}$ or $x = -3$

54. $x^2 + 8x + 7 = 0$
$(x+7)(x+1) = 0$
$x+7 = 0$ or $x+1 = 0$
$x = -7$ or $x = -1$

55. $x^2 - 2x - 24 = 0$
$(x-6)(x+4) = 0$
$x-6 = 0$ or $x+4 = 0$
$x = 6$ or $x = -4$

56. $x^2 + 10x = -25$
$x^2 + 10x + 25 = 0$
$(x+5)(x+5) = 0$
$x+5 = 0$
$x = -5$

57. $x(x-10) = -16$
$x^2 - 10x = -16$
$x^2 - 10x + 16 = 0$
$(x-8)(x-2) = 0$
$x-8 = 0$ or $x-2 = 0$
$x = 8$ or $x = 2$

58. $(3x-1)(9x^2 + 3x + 1) = 0$
$3x - 1 = 0$
$3x = 1$
$x = \dfrac{1}{3}$

59. $56x^2 - 5x - 6 = 0$
$(7x+2)(8x-3) = 0$
$7x+2 = 0$ or $8x-3 = 0$
$7x = -2$ or $8x = 3$
$x = -\dfrac{2}{7}$ or $x = \dfrac{3}{8}$

60. $20x^2 - 7x - 6 = 0$
$(4x-3)(5x+2) = 0$
$4x-3 = 0$ or $5x+2 = 0$
$\dfrac{4x}{4} = \dfrac{3}{4}$ or $\dfrac{5x}{5} = -\dfrac{2}{5}$
$x = \dfrac{3}{4}$ or $x = -\dfrac{2}{5}$

61. $5(3x+2) = 4$
$15x + 10 = 4$
$15x = 4 - 10$
$15x = -6$
$x = -\dfrac{6}{15} = -\dfrac{2}{5}$

62. $6x^2 - 3x + 8 = 0$
no real solution

63. $12 - 5t = -3$
$-5t = -3 - 12$
$-5t = -15$
$t = \dfrac{-15}{-5}$
$t = 3$

64. $5x^3 + 20x^2 + 20x = 0$
$5x(x^2 + 4x + 4) = 0$
$5x(x+2)(x+2) = 0$
$x+2 = 0$ or $5x = 0$
$x = -2$ or $x = 0$

65. $4t^3 - 5t^2 - 21t = 0$
$t(4t^2 - 5t - 21) = 0$
$t(4t+7)(t-3) = 0$
$t = 0$ or $4t+7 = 0$ or $t-3 = 0$
$t = 0$ or $4t = -7$ or $t = 3$
$t = 0$ or $t = -\dfrac{7}{4}$ or $t = 3$

66. Let x = width
Then $2x - 15$ = length
$A = lw$
$500 = (2x - 15)(x)$
$500 = 2x^2 - 15x$
$0 = 2x^2 - 15x - 500$
$0 = (2x + 25)(x - 20)$
$2x + 25 = 0$　　or　　$x - 20 = 0$
$x = -\dfrac{25}{2}$　　or　　$x = 20$
cannot be negative
width = 20 inches
length = 25 inches

67. base = $4x$
altitude = x
$162 = \dfrac{1}{2}(x)(4x)$
$162 = 2x^2$
$81 = x^2$
$x^2 - 81 = 0$
$(x + 9)(x - 9) = 0$
$x - 9 = 0$　　or　　$x + 9 = 0$
$x = 9$　　or　　$x = -9$
base = 36 yards

68. 1st integer = x
2nd integer = $x + 1$
$x(x + 1) = 380$
$x^2 + x - 380 = 0$
$(x + 20)(x - 19) = 0$
$x + 20 = 0$　　or　　$x - 19 = 0$
$x = -20$　　or　　$x = 19$
1st integer = 19
2nd integer = 20

69. a.　$h = -16t^2 + 440t$
$2800 = -16t^2 + 440t$
$0 = -16t^2 + 440t - 2800$
$0 = -8(2t^2 - 55t + 350)$
$0 = -8(2t - 35)(t - 10)$
$2t - 35 = 0$　　or　　$t - 10 = 0$
$t = \dfrac{35}{2} = 17\dfrac{1}{2}$　　or　　$t = 10$
17.5 seconds and 10 seconds
The rocket reaches a height of 2800 ft on its
way up and on its way back down.

b.　$h = -16t^2 + 440t$
$0 = -16t^2 + 440t$
$0 = -16t(t - 27.5)$
$t - 27.5 = 0$
$t = 27.5$ seconds

70. short leg = $x - 8$
long leg = x
hypotenuse = $x + 8$
$(x - 8)^2 + x^2 = (x + 8)^2$
$x^2 + 2(x)(-8) + 8^2 + x^2 = x^2 + 2(x)(8) + 8^2$
$x^2 - 16x + 64 + x^2 = x^2 + 16x + 64$
$2x^2 - 16x + 64 = x^2 + 16x + 64$
$2x^2 - x^2 - 16x - 16x + 64 - 64 = 0$
$x^2 - 32x = 0$
$x(x - 32) = 0$
$x = 0$　　　or　　$x - 32 = 0$
$x = 0$　　　or　　$x = 32$
The long leg is 32 cm.

Chapter 5 - Test

1. $9x^3 + 39x^2 + 12x$; GCF = $3x$
$3x(3x^2 + 13x + 4) = 3x(3x + 1)(x + 4)$

2. $x^2 + x - 10$
not factorable

3. $x^2 + 4$
not factorable

4. $y^2 - 8y - 48 = (y - 12)(y + 4)$

5. $3a^2 + 3ab - 7a - 7b$
$= 3a(a + b) - 7(a + b)$
$= (a + b)(3a - 7)$

6. $3x^2 - 5x + 2 = (3x - 2)(x - 1)$

7. $x^2 + 20x + 90$
not factorable

8. $x^2 + 14xy + 24y^2 = (x + 12y)(x + 2y)$

9. $26x^6 - x^4$; GCF = x^4
$x^4(26x^2 - 1)$

10. $50x^3 + 10x^2 - 35x$; GCF = $5x$
$5x(10x^2 + 2x - 7)$

11. $180 - 5x^2$; GCF $= 5$

$5(36 - x^2) = 5(6 - x)(6 + x)$

12. $64x^3 - 1$

$= (4x)^3 - 1^3$

$= (4x - 1)(16x^2 + 4x + 1)$

13. $6t^2 - t - 5 = (6t + 5)(t - 1)$

14. $xy^2 - 7y^2 - 4x + 28$

$= y^2(x - 7) - 4(x - 7)$

$= (x - 7)(y^2 - 4)$

$= (x - 7)(y + 2)(y - 2)$

15. $x - x^5$; GCF $= x$

$x(1 - x^4)$

$= x[1 - (x^2)^2]$

$= x(1 - x^2)(1 + x^2)$

$= x(1 - x)(1 + x)(1 + x^2)$

16. $-xy^3 - x^3y$; GCF $= -xy$

$-xy(y^2 + x^2)$

17. $x^2 + 5x = 14$

$x^2 + 5x - 14 = 0$

$(x + 7)(x - 2) = 0$

$x + 7 = 0$ or $x - 2 = 0$

$x = -7$ or $x = 2$

18. $(x + 3)^2 = 16$

$x^2 + 2(x)(3) + 3^2 = 16$

$x^2 + 6x + 9 = 16$

$x^2 + 6x + 9 - 16 = 0$

$x^2 + 6x - 7 = 0$

$(x + 7)(x - 1) = 0$

$x + 7 = 0$ or $x - 1 = 0$

$x = -7$ or $x = 1$

19. $3x(2x - 3)(3x + 4) = 0$

$3x = 0$ or $2x - 3 = 0$ or $3x + 4 = 0$

$x = \dfrac{0}{3}$ or $2x = 3$ or $3x = -4$

$x = 0$ or $x = \dfrac{3}{2}$ or $x = -\dfrac{4}{3}$

20. $5t^3 - 45t = 0$

$5t(t^2 - 9) = 0$

$5t(t + 3)(t - 3) = 0$

$5t = 0$ or $t + 3 = 0$ or $t - 3 = 0$

$t = \dfrac{0}{5} = 0$ or $t = -3$ or $t = 3$

21. $3x^2 = -12x$

$3x^2 + 12x = 0$

$3x(x + 4) = 0$

$3x = 0$ or $x + 4 = 0$

$x = \dfrac{0}{3} = 0$ or $x = -4$

22. $t^2 - 2t - 15 = 0$

$(t - 5)(t + 3) = 0$

$t - 5 = 0$ or $t + 3 = 0$

$t = 5$ or $t = -3$

23. $7x^2 = 168 + 35x$

$7x^2 - 35x - 168 = 0$

$7(x^2 - 5x - 24) = 0$

$7(x - 8)(x + 3) = 0$

$x - 8 = 0$ or $x + 3 = 0$

$x = 8$ or $x = -3$

24. $6x^2 = 15x$

$6x^2 - 15x = 0$

$3x(2x - 5) = 0$

$3x = 0$ or $2x - 5 = 0$

$x = \dfrac{0}{3}$ or $2x = 5$

$x = 0$ or $x = \dfrac{5}{2}$

25. width $= x$

length $= x + 5$

$A = lw$

$66 = x(x + 5)$

$66 = x^2 + 5x$

$0 = x^2 + 5x - 66$

$0 = (x + 11)(x - 6)$

$x + 11 = 0$ or $x - 6 = 0$

$x = -11$ or $x = 6$

The dimensions are 6 ft. by 11 ft.

26. altitude $= x$

base $= x + 9$

$A = \dfrac{1}{2}bh$

$68 = \dfrac{1}{2}(x+9)(x)$

$2(68) = 2\left[\dfrac{1}{2}(x+9)(x)\right]$

$136 = (x+9)(x)$

$136 = x^2 + 9x$

$0 = x^2 + 9x - 136$

$0 = (x+17)(x-8)$

$x + 17 = 0 \qquad$ or $\qquad x - 8 = 0$

$x = -17 \qquad\;\;$ or $\qquad x = 8$

The base is 17 feet

27. one number $= x$

other number $= 17 - x$

$x^2 + (17-x)^2 = 145$

$x^2 + 289 + 2(17)(-x) + x^2 = 145$

$x^2 + 289 - 34x + x^2 = 145$

$2x^2 - 34x + 289 = 145$

$2x^2 - 34x + 289 - 145 = 0$

$2x^2 - 34x + 144 = 0$

$2(x^2 - 17x + 72) = 0$

$2(x - 8)(x - 9) = 0$

$x - 8 = 0 \qquad$ or $\qquad x - 9 = 0$

$x = 8 \qquad\;\;$ or $\qquad x = 9$

The numbers are 8 and 9.

28. $h = -16t^2 + 784$

$0 = -16t^2 + 784$

$0 = -16(t^2 - 49)$

$0 = -16(t - 7)(t + 7)$

$t - 7 = 0$

$t = 7$

7 seconds

Chapter 5 - Cumulative Review

1. a. $|4| = 4$

   b. $|-5| = 5$

   c. $|0| = 0$

2. a. $2x - y = 2(3) - 2 = 6 - 2 = 4$

   b. $\dfrac{3x}{2y} = \dfrac{3(3)}{2(2)} = \dfrac{9}{4}$

   c. $\dfrac{x}{y} + \dfrac{y}{2} = \dfrac{3}{2} + \dfrac{2}{2} = \dfrac{3+2}{2} = \dfrac{5}{2}$

   d. $x^2 - y^2 = 3^2 - 2^2 = 9 - 4 = 5$

3. $14{,}494 - (-282) = 14{,}494 + 282 = 14{,}776$ ft

4. $5t - 5 = 6t + 2$

$5t - 5t - 5 = 6t - 5t + 2$

$-5 = t + 2$

$-5 - 2 = t + 2 - 2$

$-7 = t$

5. $4(2x - 3) + 7 = 3x + 5$

$8x - 12 + 7 = 3x + 5$

$8x - 5 = 3x + 5$

$8x - 3x - 5 = 3x - 3x + 5$

$5x - 5 = 5$

$5x - 5 + 5 = 5 + 5$

$5x = 10$

$\dfrac{5x}{5} = \dfrac{10}{5}$

$x = 2$

6. Let $x =$ number of minutes

$99.68 = 50 + 0.36x$

$49.68 = 0.36x$

$138 = x$

138 minutes

7. $-2x \le -4$

$\dfrac{-2x}{-2} \ge \dfrac{-4}{-2}$

$x \ge 2$

8. a. $x - 2y = 6$

$6 - 2(0) \overset{?}{=} 6$

$6 - 0 \overset{?}{=} 6$

$6 = 6$

Solution

   b. $x - 2y = 6$

$0 - 2(3) \overset{?}{=} 6$

$0 - 6 \overset{?}{=} 6$

$-6 \ne 6$

Not a solution

   c. $x - 2y = 6$

$2 - 2(-2) \overset{?}{=} 6$

$2 + 4 \overset{?}{=} 6$

$6 = 6$

Solution

9.

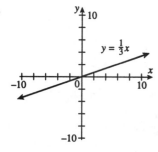

$y = \frac{1}{3}x$

10. a. $x = -3;\ y = 2;$
    $(-3, 0);\ (0, 2)$

b. $x = -4;\ x = -1;\ y = 1;$
    $(-4, 0);\ (-1, 0);\ (0, 1)$

c. $x = 0;\ y = 0;\ (0, 0)$

d. $x = 2;\ (2, 0)$

e. $x = -1;\ x = 3;\ y = -1;\ y = 2;$
    $(-1, 0);\ (3, 0);\ (0, -1);\ (0, 2)$

11. $-2x + 3y = 12$
    $3y = 2x + 12$
    $y = \dfrac{2}{3}x + 4$

    slope $= \dfrac{2}{3}$

12.

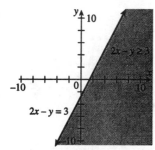

$2x - y = 3$

13. a. $(st)^4 = s^4 t^4$

b. $\left(\dfrac{m}{n}\right)^7 = \dfrac{m^7}{n^7}$

c. $(2a)^3 = 2^3 a^3 = 8a^3$

d. $(-5x^2 y^3 z)^2 = (-5)^2 x^{2\cdot 2} y^{3\cdot 2} z^2 = 25x^4 y^6 z^2$

e. $\left(\dfrac{2x^4}{3y^5}\right)^4 = \dfrac{2^4 x^{4\cdot 4}}{3^4 y^{5\cdot 4}} = \dfrac{16x^{16}}{81y^{20}}$

14. $4x^3 - 6x^2 + 2x + 7$
    $+\quad\ \ 5x^2 - 2x$
    $\overline{\ 4x^3 - x^2 + 7}$

15. $\dfrac{x^2 + 7x + 12}{x + 3} = \dfrac{(x+4)(x+3)}{(x+3)} = x + 4$

16. a. $6t + 18 = 6(t + 3)$

b. $y^5 - y^7 = y^5(1 - y^2) = y^5(1 - y)(1 + y)$

17. $x^2 - 8x + 15 = (x - 5)(x - 3)$

18. $3x^2 + 11x + 6 = (3x + 2)(x + 3)$

19. $9x^2 - 36 = 9(x^2 - 4) = 9(x - 2)(x + 2)$

20. $(x - 5)(2x + 7) = 0$
    $x - 5 = 0$ \quad or \quad $2x + 7 = 0$
    $x = 5$ \quad\ \ or \quad $2x = -7$
    $x = -\dfrac{7}{2}$

21. Let $x =$ length of base
    Then $2x - 2 =$ height
    $A = \dfrac{1}{2}bh$
    $30 = \dfrac{1}{2}x(2x - 2)$
    $30 = x^2 - x$
    $0 = x^2 - x - 30$
    $0 = (x - 6)(x + 5)$
    $x - 6 = 0$
    $x = 6$
    base $= 6$ m
    height $= 10$ m

Chapter 6

Section 6.1 Mental Math

1. $\dfrac{x+5}{x}$; $x = 0$

3. $\dfrac{x^2 + 4x - 2}{x(x-1)}$; $x = 0$, $x = 1$

Exercise Set 6.1

1. $x = 2$

$$\frac{x+5}{x+2} = \frac{2+5}{2+2} = \frac{7}{4}$$

3. $z = -5$

$$\frac{z-8}{z+2} = \frac{-5-8}{-5+2} = \frac{-13}{-3} = \frac{13}{3}$$

5. $x = 2$

$$\frac{x^2 + 8x + 2}{x^2 - x - 6} = \frac{(2)^2 + 8(2) + 2}{(2)^2 - 2 - 6} = \frac{4 + 16 + 2}{4 - 2 - 6}$$

$$= \frac{22}{-4} = -\frac{11}{2}$$

7. $x = 2$

$$\frac{x+5}{x^2 + 4x - 8} = \frac{2+5}{2^2 + 4(2) - 8} = \frac{7}{4 + 8 - 8} = \frac{7}{4}$$

9. $y = -2$

$$\frac{y^3}{y^2 - 1} = \frac{(-2)^3}{(-2)^2 - 1} = \frac{-8}{4 - 1} = \frac{-8}{3}$$

11. a. $R = \dfrac{150x^2}{x^2 + 3} = \dfrac{150(1)^2}{1^2 + 3} = \dfrac{150}{4} = \$37.5$ million

b. $R = \dfrac{150x^2}{x^2 + 3} = \dfrac{150(2)^2}{2^2 + 3} = \dfrac{600}{7} = \$85.7$ million

c. $85.7 - 37.5 = \$48.2$ million

13. $\dfrac{x+3}{x+2}$ undefined if

$$x + 2 = 0$$
$$x = -2$$

15. $\dfrac{4x^2 + 9}{2x - 8}$ undefined if

$$2x - 8 = 0$$

$$2x = 8$$
$$x = \frac{8}{2} = 4$$

17. $\dfrac{9x^3 + 4x}{15x + 30}$ undefined if

$$15x + 30 = 0$$
$$15x = -30$$
$$x = \frac{-30}{15} = -2$$

19. $\dfrac{x^2 - 5x - 2}{x^2 + 4}$ undefined if

$$x^2 + 4 = 0$$
$$x^2 = -4$$
$$x = \pm\sqrt{-4}$$

Not a real number; always defined

21. Answers may vary.

23. $\dfrac{8x^5}{4x^9} = \dfrac{2}{x^4}$

25. $\dfrac{5(x-2)}{(x-2)(x+1)} = \dfrac{5}{x+1}$

27. $\dfrac{-5a - 5b}{a+b} = \dfrac{-5(a+b)}{a+b} = -5$

29. $\dfrac{x+5}{x^2 - 4x - 45} = \dfrac{x+5}{(x-9)(x+5)} = \dfrac{1}{(x-9)}$

31. $\dfrac{5x^2 + 11x + 2}{x+2} = \dfrac{(5x+1)(x+2)}{x+2} = 5x + 1$

33. $\dfrac{x^2 + x - 12}{2x^2 - 5x - 3} = \dfrac{(x+4)(x-3)}{(2x+1)(x-3)} = \dfrac{x+4}{2x+1}$

35. Answers may vary.

37. $\dfrac{x-7}{7-x} = \dfrac{x-7}{-1(x-7)} = -1$

39. $\dfrac{y^2 - 2y}{4 - 2y} = \dfrac{y(y-2)}{-2(y-2)} = -\dfrac{y}{2}$

41.
$$\frac{x^2-4x+4}{4-x^2}=\frac{(x-2)(x-2)}{-1(x^2-4)}$$
$$=\frac{(x-2)(x-2)}{-1(x+2)(x-2)}=-\frac{x-2}{x+2}=\frac{2-x}{x+2}$$

43.
$$\frac{x^2+xy+2x+2y}{x+2}=\frac{x(x+y)+2(x+y)}{x+2}$$
$$=\frac{(x+y)(x+2)}{(x+2)}=x+y$$

45.
$$\frac{5x+15-xy-3y}{2x+6}=\frac{5(x+3)-y(x+3)}{2(x+3)}$$
$$=\frac{(x+3)(5-y)}{2(x+3)}=\frac{5-y}{2}$$

47.
$$\frac{15x^4y^8}{-5x^8y^3}=\frac{-3y^5}{x^4}$$

49.
$$\frac{(x-2)(x+3)}{5(x+3)}=\frac{x-2}{5}$$

51.
$$\frac{-6a-6b}{a+b}=\frac{-6(a+b)}{a+b}=-6$$

53.
$$\frac{2x^2-8}{4x-8}=\frac{2(x^2-4)}{4(x-2)}$$
$$=\frac{2(x+2)(x-2)}{4(x-2)}=\frac{x+2}{2}$$

55.
$$\frac{11x^2-22x^3}{6x-12x^2}=\frac{11x^2(1-2x)}{6x(1-2x)}=\frac{11x^2}{6x}=\frac{11x}{6}$$

57.
$$\frac{x+7}{x^2+5x-14}=\frac{x+7}{(x+7)(x-2)}=\frac{1}{x-2}$$

59.
$$\frac{2x^2+3x-2}{2x-1}=\frac{(2x-1)(x+2)}{2x-1}=x+2$$

61.
$$\frac{x^2-1}{x^2-2x+1}=\frac{(x+1)(x-1)}{(x-1)(x-1)}=\frac{x+1}{x-1}$$

63.
$$\frac{m^2-6m+9}{m^2-9}=\frac{(m-3)(m-3)}{(m+3)(m-3)}=\frac{m-3}{m+3}$$

65.
$$\frac{-2a^2+12a-18}{9-a^2}=\frac{-2(a^2-6a+9)}{-1(a^2-9)}$$
$$=\frac{-2(a-3)(a-3)}{-1(a+3)(a-3)}=\frac{2(a-3)}{a+3}=\frac{2a-6}{a+3}$$

67.
$$\frac{2-x}{x-2}=\frac{-1(x-2)}{x-2}=-1$$

69.
$$\frac{x^2-1}{1-x}=\frac{(x+1)(x-1)}{-1(x-1)}=-(x+1)=-x-1$$

71.
$$\frac{x^2+7x+10}{x^2-3x-10}=\frac{(x+5)(x+2)}{(x-5)(x+2)}=\frac{x+5}{x-5}$$

73.
$$\frac{3x^2+7x+2}{3x^2+13x+4}=\frac{(3x+1)(x+2)}{(3x+1)(x+4)}=\frac{x+2}{x+4}$$

75.
$$\frac{x^2+3x-2x-6}{x^2-2x}=\frac{x(x+3)-2(x+3)}{x(x-2)}$$
$$=\frac{(x+3)(x-2)}{x(x-2)}=\frac{x+3}{x}$$

77.
$$\frac{x^3+8}{x+2}=\frac{(x+2)(x^2-2x+4)}{x+2}=x^2-2x+4$$

79.
$$\frac{x^2+xy+5x+5y}{3x+3y}=\frac{x(x+y)+5(x+y)}{3(x+y)}$$
$$=\frac{(x+y)(x+5)}{3(x+y)}=\frac{x+5}{3}$$

81.
$$\frac{x^3-1}{1-x}=\frac{(x-1)(x^2+x+1)}{-1(x-1)}$$
$$=-1(x^2+x+1)=-x^2-x-1$$

83. $y=\dfrac{x^2-25}{x+5}=\dfrac{(x+5)(x-5)}{(x+5)}=x-5,\ x\neq-5$

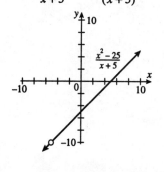

85. $y = \dfrac{x^2 + x - 12}{x + 4} = \dfrac{(x+4)(x-3)}{(x+4)} = x - 3, \ x \neq -4$

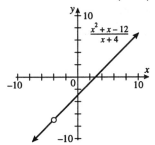

87. $\dfrac{1}{3} \cdot \dfrac{9}{11} = \dfrac{1 \cdot 3 \cdot 3}{3 \cdot 11} = \dfrac{3}{11}$

89. $\dfrac{1}{3} + \dfrac{1}{4} = \dfrac{1}{3} \cdot \dfrac{4}{1} = \dfrac{4}{3}$

91. $\dfrac{5}{6} \cdot \dfrac{10}{11} \cdot \dfrac{2}{3} = \dfrac{5 \cdot 2 \cdot 5 \cdot 2}{2 \cdot 3 \cdot 11 \cdot 3} = \dfrac{50}{99}$

93. $\dfrac{13}{20} + \dfrac{2}{9} = \dfrac{13}{20} \cdot \dfrac{9}{2} = \dfrac{13 \cdot 3 \cdot 3}{2 \cdot 2 \cdot 5 \cdot 2} = \dfrac{117}{40}$

Section 6.2 Mental Math

1. $\dfrac{2}{y} \cdot \dfrac{x}{3} = \dfrac{2 \cdot x}{y \cdot 3} = \dfrac{2x}{3y}$

3. $\dfrac{5}{7} \cdot \dfrac{y^2}{x^2} = \dfrac{5y^2}{7x^2}$

5. $\dfrac{9}{x} \cdot \dfrac{x}{5} = \dfrac{9 \cdot x}{x \cdot 5} = \dfrac{9}{5}$

Exercise Set 6.2

1. $\dfrac{3x}{y^2} \cdot \dfrac{7y}{4x} = \dfrac{21xy}{4xy^2} = \dfrac{21}{4y}$

3. $\dfrac{8x}{2} \cdot \dfrac{x^5}{4x^2} = \dfrac{8x^6}{8x^2} = x^4$

5. $-\dfrac{5a^2 b}{30a^2 b^2} \cdot b^3 = -\dfrac{5a^2 b^4}{30a^2 b^2} = -\dfrac{b^2}{6}$

7. $\dfrac{x}{2x-14} \cdot \dfrac{x^2 - 7x}{5} = \dfrac{x}{2(x-7)} \cdot \dfrac{x(x-7)}{5} = \dfrac{x^2}{10}$

9. $\dfrac{6x+6}{5} \cdot \dfrac{10}{36x+36} = \dfrac{6(x+1)}{5} \cdot \dfrac{10}{36(x+1)} = \dfrac{60}{180} = \dfrac{1}{3}$

11. $\dfrac{m^2 - n^2}{m+n} \cdot \dfrac{m}{m^2 - mn} = \dfrac{(m+n)(m-n)}{m+n} \cdot \dfrac{m}{m(m-n)}$

$= 1$

13. $\dfrac{x^2 - 25}{x^2 - 3x - 10} \cdot \dfrac{x+2}{x} = \dfrac{(x+5)(x-5)}{(x+2)(x-5)} \cdot \dfrac{x+2}{x}$

$= \dfrac{x+5}{x}$

15. $A = lw$

$A = \dfrac{2x}{x^2 - 25} \cdot \dfrac{x+5}{9x^3}$

$A = \dfrac{2x}{(x+5)(x-5)} \cdot \dfrac{(x+5)}{9x^3}$

$A = \dfrac{2}{9x^2(x-5)}$

17. $\dfrac{5x^7}{2x^5} \div \dfrac{10x}{4x^3} = \dfrac{5x^7}{2x^5} \cdot \dfrac{4x^3}{10x} = \dfrac{20x^{10}}{20x^6} = x^4$

19. $\dfrac{8x^2}{y^3} \div \dfrac{4x^2 y^3}{6} = \dfrac{8x^2}{y^3} \cdot \dfrac{6}{4x^2 y^3} = \dfrac{48x^2}{4x^2 y^6} = \dfrac{12}{y^6}$

21. $\dfrac{(x-6)(x+4)}{4x} \div \dfrac{2x-12}{8x^2}$

$= \dfrac{(x-6)(x+4)}{4x} \cdot \dfrac{8x^2}{2x-12}$

$= \dfrac{(x-6)(x+4)}{4x} \cdot \dfrac{8x^2}{2(x-6)}$

$\dfrac{8x^2(x+4)}{8x} = x(x+4)$

23. $\dfrac{3x^2}{x^2 - 1} \div \dfrac{x^5}{(x+1)^2}$

$= \dfrac{3x^2}{(x+1)(x-1)} \cdot \dfrac{(x+1)(x+1)}{x^5}$

$= \dfrac{3x^2(x+1)}{x^5(x-1)} = \dfrac{3(x+1)}{x^3(x-1)}$

25. $\dfrac{m^2-n^2}{m+n} \div \dfrac{m}{m^2+nm} = \dfrac{m^2-n^2}{m+n} \cdot \dfrac{m^2+nm}{m}$

$= \dfrac{(m+n)(m-n)}{m+n} \cdot \dfrac{m(m+n)}{m} = (m-n)(m+n)$

$= m^2-n^2$

27. $\dfrac{x+2}{7-x} \div \dfrac{x^2-5x+6}{x^2-9x+14} = \dfrac{x+2}{7-x} \cdot \dfrac{x^2-9x+14}{x^2-5x+6}$

$= \dfrac{x+2}{-1(x-7)} \cdot \dfrac{(x-7)(x-2)}{(x-3)(x-2)} = -\dfrac{x+2}{x-3}$

29. $\dfrac{x^2+7x+10}{1-x} \div \dfrac{x^2+2x-15}{x-1}$

$= \dfrac{x^2+7x+10}{1-x} \cdot \dfrac{x-1}{x^2+2x-15}$

$= \dfrac{(x+5)(x+2)}{-1(x-1)} \cdot \dfrac{x-1}{(x+5)(x-3)} = -\dfrac{x+2}{x-3}$

31. Answers may vary.

33. $\dfrac{5a^2b}{30a^2b^2} \cdot \dfrac{1}{b^3} = \dfrac{5a^2b}{30a^2b^5} = \dfrac{1}{6b^4}$

35. $\dfrac{12x^3y}{8xy^7} \div \dfrac{7x^5y}{6x} = \dfrac{12x^3y}{8xy^7} \cdot \dfrac{6x}{7x^5y} = \dfrac{72x^4y}{56x^6y^8}$

$= \dfrac{9}{7x^2y^7}$

37. $\dfrac{5x-10}{12} \div \dfrac{4x-8}{8} = \dfrac{5x-10}{12} \cdot \dfrac{8}{4x-8}$

$= \dfrac{5(x-2)}{12} \cdot \dfrac{8}{4(x-2)} = \dfrac{40}{48} = \dfrac{5}{6}$

39. $\dfrac{x^2+5x}{8} \cdot \dfrac{9}{3x+15} = \dfrac{x(x+5)}{8} \cdot \dfrac{9}{3(x+5)} = \dfrac{3x}{8}$

41. $\dfrac{7}{6p^2+q} \div \dfrac{14}{18p^2+3q} = \dfrac{7}{6p^2+q} \cdot \dfrac{18p^2+3q}{14}$

$= \dfrac{7}{6p^2+q} \cdot \dfrac{3(6p^2+q)}{14} = \dfrac{3}{2}$

43. $\dfrac{3x+4y}{x^2+4xy+4y^2} \cdot \dfrac{x+2y}{2}$

$= \dfrac{3x+4y}{(x+2y)(x+2y)} \cdot \dfrac{x+2y}{2} = \dfrac{3x+4y}{2(x+2y)}$

45. $\dfrac{x^2-9}{x^2+8} \div \dfrac{3-x}{2x^2+16} = \dfrac{x^2-9}{x^2+8} \cdot \dfrac{2x^2+16}{3-x}$

$= \dfrac{(x+3)(x-3)}{x^2+8} \cdot \dfrac{2(x^2+8)}{-1(x-3)} = -2(x+3)$

47. $\dfrac{(x+2)^2}{x-2} \div \dfrac{x^2-4}{2x-4} = \dfrac{(x+2)^2}{(x-2)} \cdot \dfrac{2x-4}{x^2-4}$

$= \dfrac{(x+2)(x+2)}{(x-2)} \cdot \dfrac{2(x-2)}{(x+2)(x-2)} = \dfrac{2(x+2)}{x-2}$

49. $\dfrac{a^2+7a+12}{a^2+5a+6} \cdot \dfrac{a^2+8a+15}{a^2+5a+4}$

$= \dfrac{(a+3)(a+4)}{(a+3)(a+2)} \cdot \dfrac{(a+3)(a+5)}{(a+1)(a+4)}$

$= \dfrac{(a+3)(a+5)}{(a+2)(a+1)}$

51. $\dfrac{1}{-x-4} \div \dfrac{x^2-7x}{x^2-3x-28} = \dfrac{1}{-x-4} \cdot \dfrac{x^2-3x-28}{x^2-7x}$

$= \dfrac{1}{-1(x+4)} \cdot \dfrac{(x-7)(x+4)}{x(x-7)} = -\dfrac{1}{x}$

53. $\dfrac{x^2-5x-24}{2x^2-2x-24} \cdot \dfrac{4x^2+4x-24}{x^2-10x+16}$

$= \dfrac{(x-8)(x+3)}{2(x-4)(x+3)} \cdot \dfrac{2(x+3)(x-2)}{(x-8)(x-2)} = \dfrac{2(x+3)}{x-4}$

55. $x-5 \div \dfrac{5-x}{x^2+2} = x-5 \cdot \dfrac{x^2+2}{5-x}$

$= x-5 \cdot \dfrac{x^2+2}{-1(x-5)} = -(x^2+2)$

57. $\dfrac{x^2-y^2}{x^2-2xy+y^2} \cdot \dfrac{y-x}{x+y} = \dfrac{(x+y)(x-y)}{(x-y)(x-y)} \cdot \dfrac{-1(x-y)}{x+y}$

$= -1$

59. $\dfrac{a^2+ac+ba+bc}{a-b} \div \dfrac{a+c}{a+b}$

$= \dfrac{a(a+c)+b(a+c)}{a-b} \cdot \dfrac{a+b}{a+c}$

$= \dfrac{(a+c)(a+b)}{a-b} \cdot \dfrac{a+b}{a+c} = \dfrac{(a+b)^2}{a-b}$

61. $\dfrac{3x^2+8x+5}{x^2+8x+7}\cdot\dfrac{x+7}{x^2+4}$

$=\dfrac{(3x+5)(x+1)}{(x+7)(x+1)}\cdot\dfrac{x+7}{x^2+4}=\dfrac{3x+5}{x^2+4}$

63. $\dfrac{x^2-9}{2x}\div\dfrac{x+3}{8x^4}=\dfrac{x^2-9}{2x}\cdot\dfrac{8x^4}{x+3}$

$=\dfrac{(x+3)(x-3)}{2x}\cdot\dfrac{8x^4}{x+3}=4x^3(x-3)$

65. $\dfrac{x^3+8}{x^2-2x+4}\cdot\dfrac{4}{x^2-4}$

$=\dfrac{(x+2)(x^2-2x+4)}{x^2-2x+4}\cdot\dfrac{4}{(x+2)(x-2)}=\dfrac{4}{x-2}$

67. $\dfrac{a^2-ab}{6a^2+6ab}\div\dfrac{a^3-b^3}{a^2-b^2}=\dfrac{a^2-ab}{6a^2+6ab}\cdot\dfrac{a^2-b^2}{a^3-b^3}$

$=\dfrac{a(a-b)}{6a(a+b)}\cdot\dfrac{(a-b)(a+b)}{(a-b)(a^2+ab+b^2)}$

$=\dfrac{a-b}{6(a^2+ab+b^2)}$

69. $\dfrac{1}{5}+\dfrac{4}{5}=\dfrac{5}{5}=1$

71. $\dfrac{9}{9}-\dfrac{19}{9}=-\dfrac{10}{9}$

73. $\dfrac{6}{5}+\left(\dfrac{1}{5}-\dfrac{8}{5}\right)=\dfrac{6}{5}+\left(-\dfrac{7}{5}\right)=-\dfrac{1}{5}$

75.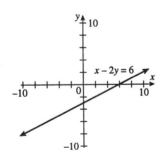

77. $\left(\dfrac{x^2-y^2}{x^2+y^2}\div\dfrac{x^2-y^2}{3x}\right)\cdot\dfrac{x^2+y^2}{6}$

$=\dfrac{x^2-y^2}{x^2+y^2}\cdot\dfrac{3x}{x^2-y^2}\cdot\dfrac{x^2+y^2}{6}$

$=\dfrac{(x+y)(x-y)}{(x^2+y^2)}\cdot\dfrac{3x}{(x+y)(x-y)}\cdot\dfrac{x^2+y^2}{6}=\dfrac{x}{2}$

79. $\left(\dfrac{2a+b}{b^2}\cdot\dfrac{3a^2-2ab}{ab+2b^2}\right)\div\dfrac{a^2-3ab+2b^2}{5ab-10b^2}$

$=\dfrac{2a+b}{b^2}\cdot\dfrac{3a^2-2ab}{ab+2b^2}\cdot\dfrac{5ab-10b^2}{a^2-3ab+2b^2}$

$=\dfrac{2a+b}{b^2}\cdot\dfrac{a(3a-2b)}{b(a+2b)}\cdot\dfrac{5b(a-2b)}{(a-2b)(a-b)}$

$=\dfrac{5a(2a+b)(3a-2b)}{b^2(a+2b)(a-b)}$

Section 6.3 Mental Math

1. $\dfrac{2}{3}+\dfrac{1}{3}=\dfrac{2+1}{3}=\dfrac{3}{3}=1$

3. $\dfrac{3x}{9}+\dfrac{4x}{9}=\dfrac{3x+4x}{9}=\dfrac{7x}{9}$

5. $\dfrac{8}{9}-\dfrac{7}{9}=\dfrac{8-7}{9}=\dfrac{1}{9}$

7. $\dfrac{7}{5}-\dfrac{10y}{5}=\dfrac{7-10y}{5}$

Exercise Set 6.3

1. $\dfrac{a}{13}+\dfrac{9}{13}=\dfrac{a+9}{13}$

3. $\dfrac{9}{3+y}+\dfrac{y+1}{3+y}=\dfrac{9+y+1}{3+y}=\dfrac{10+y}{3+y}$

5. $\dfrac{4m}{3n}+\dfrac{5m}{3n}=\dfrac{4m+5m}{3n}=\dfrac{9m}{3n}=\dfrac{3m}{n}$

7. $\dfrac{2x+1}{x-3}+\dfrac{3x+6}{x-3}=\dfrac{2x+1+3x+6}{x-3}$

$=\dfrac{5x+7}{x-3}$

9. $\dfrac{7}{8} - \dfrac{3}{8} = \dfrac{4}{8} = \dfrac{1}{2}$

11. $\dfrac{4m}{m-6} - \dfrac{24}{m-6} = \dfrac{4m-24}{m-6} = \dfrac{4(m-6)}{m-6} = 4$

13. $\dfrac{2x^2}{x-5} - \dfrac{25+x^2}{x-5} = \dfrac{2x^2 - (25+x^2)}{x-5}$

$= \dfrac{2x^2 - 25 - x^2}{x-5} = \dfrac{x^2 - 25}{x-5} = \dfrac{(x+5)(x-5)}{x-5}$

$= x+5$

15. $\dfrac{-3x^2 - 4}{x-4} - \dfrac{12 - 4x^2}{x-4} = \dfrac{-3x^2 - 4 - (12 - 4x^2)}{x-4}$

$= \dfrac{-3x^2 - 4 - 12 + 4x^2}{x-4} = \dfrac{x^2 - 16}{x-4}$

$= \dfrac{(x+4)(x-4)}{x-4} = x+4$

17. $\dfrac{2x+3}{x+1} - \dfrac{x+2}{x+1} = \dfrac{2x+3 - (x+2)}{x+1}$

$= \dfrac{2x+3 - x - 2}{x+1} = \dfrac{x+1}{x+1} = 1$

19. $\dfrac{3}{x^3} + \dfrac{9}{x^3} = \dfrac{12}{x^3}$

21. $\dfrac{5}{x+4} - \dfrac{10}{x+4} = -\dfrac{5}{x+4}$

23. $\dfrac{x}{x+y} - \dfrac{2}{x+y} = \dfrac{x-2}{x+y}$

25. $\dfrac{8x}{2x+5} + \dfrac{20}{2x+5} = \dfrac{8x+20}{2x+5} = \dfrac{4(2x+5)}{2x+5} = 4$

27. $\dfrac{5x+4}{x-1} - \dfrac{2x+7}{x-1} = \dfrac{5x+4 - (2x+7)}{x-1}$

$= \dfrac{5x+4 - 2x - 7}{x-1} = \dfrac{3x-3}{x-1}$

$= \dfrac{3(x-1)}{x-1} = 3$

29. $\dfrac{a}{a^2 + 2a - 15} - \dfrac{3}{a^2 + 2a - 15} = \dfrac{a-3}{a^2 + 2a - 15}$

$= \dfrac{a-3}{(a+5)(a-3)} = \dfrac{1}{a+5}$

31. $\dfrac{2x+3}{x^2 - x - 30} - \dfrac{x-2}{x^2 - x - 30} = \dfrac{2x+3 - (x-2)}{x^2 - x - 30}$

$= \dfrac{2x+3 - x + 2}{x^2 - x - 30} = \dfrac{x+5}{x^2 - x - 30} = \dfrac{x+5}{(x-6)(x+5)}$

$= \dfrac{1}{x-6}$

33. $P = 4s$

$P = 4\left(\dfrac{5}{x-2}\right) = \dfrac{20}{x-2}$ m

35. Answers may vary.

37. $3 = 3$

$33 = 3 \cdot 11$

$\text{LCD} = 3 \cdot 11 = 33$

39. $2x = 2 \cdot x$

$4x^3 = 2^2 \cdot x^3$

$\text{LCD} = 2^2 \cdot x^3 = 4x^3$

41. $8x = 2^3 \cdot x$

$2x + 4 = 2(x+2)$

$\text{LCD} = 2^3 \cdot x \cdot (x+2) = 8x(x+2)$

43. $3x + 3 = 3(x+1)$

$2x^2 + 4x + 2 = 2(x+1)^2$

$\text{LCD} = 3 \cdot 2 \cdot (x+1)^2 = 6(x+1)^2$

45. $x - 8 = x - 8$

$8 - x = -1(x - 8)$

$\text{LCD} = x - 8$

47. $8x^2(x-1)^2 = 2^3 \cdot x^2 \cdot (x-1)^2$

$10x^3(x-1) = 2 \cdot 5 \cdot x^3 \cdot (x-1)$

$\text{LCD} = 2^3 \cdot 5 \cdot x^3 \cdot (x-1)^2 = 40x^3(x-1)^2$

49. $2x + 1 = (2x+1)$

$2x - 1 = (2x-1)$

$\text{LCD} = (2x+1)(2x-1)$

51. $2x^2 + 7x - 4 = (2x-1)(x+4)$

$2x^2 + 5x - 3 = (2x-1)(x+3)$

$\text{LCD} = (2x-1)(x+4)(x+3)$

53. Answers may vary.

55. $\dfrac{3}{2x} \cdot \dfrac{2x}{2x} = \dfrac{6x}{4x^2}$

57. $\dfrac{6}{3a} \cdot \dfrac{4b^2}{4b^2} = \dfrac{24b^2}{12ab^2}$

59. $\dfrac{9}{x+3} \cdot \dfrac{2}{2} = \dfrac{18}{2(x+3)}$

61. $\dfrac{9a+2}{5a+10} \cdot \dfrac{b}{b} = \dfrac{9ab+2b}{5b(a+2)}$

63. $\dfrac{x}{x^3+6x^2+8x} = \dfrac{x}{x(x+4)(x+2)} \cdot \dfrac{(x+1)}{(x+1)}$

$= \dfrac{x(x+1)}{x(x+4)(x+2)(x+1)}$

65. $\dfrac{5}{2x^2-9x-5} = \dfrac{5}{(2x+1)(x-5)} \cdot \dfrac{3x(x-7)}{3x(x-7)}$

$= \dfrac{15x(x-7)}{3x(2x+1)(x-7)(x-5)}$

67. $\dfrac{9y-1}{15x^2-30} = \dfrac{}{30x^2-60}$

$\dfrac{9y-1}{15(x^2-2)} \cdot \dfrac{2}{2} = \dfrac{18y-2}{30(x^2-2)} = \dfrac{18y-2}{30x^2-60}$

69. $\dfrac{1}{x^2-16} = \dfrac{1}{(x+4)(x-4)} \cdot \dfrac{x(x-4)}{x(x-4)}$

$= \dfrac{x(x-4)}{x(x-4)^2(x+4)}$

71. $\dfrac{5}{2-x} = -\dfrac{5}{x-2}$

73. $-\dfrac{7+x}{2-x} = \dfrac{7+x}{x-2}$

75. $x(x-3)=0$

$x=0$ or $x-3=0$

                  $x=3$

77. $x^2+6x+5=0$

$(x+5)(x+1)=0$

$x+5=0$ or $x+1=0$

$x=-5$ $x=-1$

79. $\dfrac{2}{3}+\dfrac{5}{7} = \dfrac{2}{3}\cdot\dfrac{7}{7}+\dfrac{5}{7}\cdot\dfrac{3}{3} = \dfrac{14}{21}+\dfrac{15}{21} = \dfrac{29}{21}$

81. $\dfrac{2}{6}-\dfrac{3}{4} = \dfrac{2}{6}\cdot\dfrac{2}{2}-\dfrac{3}{4}\cdot\dfrac{3}{3} = \dfrac{4}{12}-\dfrac{9}{12} = -\dfrac{5}{12}$

Exercise Set 6.4

1. $\dfrac{4}{2x}+\dfrac{9}{3x};$ $LCD=6x$

$\dfrac{4}{2x}\cdot\dfrac{3}{3}+\dfrac{9}{3x}\cdot\dfrac{2}{2} = \dfrac{12}{6x}+\dfrac{18}{6x} = \dfrac{30}{6x} = \dfrac{5}{x}$

3. $\dfrac{15a}{b}+\dfrac{6b}{5};$ $LCD=5b$

$\dfrac{15a}{b}\cdot\dfrac{5}{5}+\dfrac{6b}{5}\cdot\dfrac{b}{b} = \dfrac{75a}{5b}+\dfrac{6b^2}{5b} = \dfrac{75a+6b^2}{5b}$

5. $\dfrac{3}{x}+\dfrac{5}{2x^2};$ $LCD=2x^2$

$\dfrac{3}{x}\cdot\dfrac{2x}{2x}+\dfrac{5}{2x^2} = \dfrac{6x}{2x^2}+\dfrac{5}{2x^2} = \dfrac{6x+5}{2x^2}$

7. $\dfrac{6}{x+1}+\dfrac{9}{2x+2} = \dfrac{6}{x+1}+\dfrac{9}{2(x+1)};$ $LCD=2(x+1)$

$\dfrac{6}{(x+1)}\cdot\dfrac{2}{2}+\dfrac{9}{2(x+1)} = \dfrac{12}{2(x+1)}+\dfrac{9}{2(x+1)}$

$= \dfrac{21}{2(x+1)}$

9. $\dfrac{15}{2x-4}+\dfrac{x}{x^2-4} = \dfrac{15}{2(x-2)}+\dfrac{x}{(x+2)(x-2)}$

$LCD=2(x-2)(x+2)$

$\dfrac{15}{2(x-2)}\cdot\dfrac{(x+2)}{(x+2)}+\dfrac{x}{(x+2)(x-2)}\cdot\dfrac{2}{2}$

$= \dfrac{15(x+2)}{2(x-2)(x+2)}+\dfrac{2x}{2(x-2)(x+2)}$

$= \dfrac{15x+30+2x}{2(x-2)(x+2)} = \dfrac{17x+30}{2(x-2)(x+2)}$

11. $\dfrac{3}{4x}+\dfrac{8}{x-2};$ $LCD=4x(x-2)$

$\dfrac{3}{4x}\cdot\dfrac{(x-2)}{(x-2)}+\dfrac{8}{(x-2)}\cdot\dfrac{4x}{4x}$

$= \dfrac{3(x-2)}{4x(x-2)}+\dfrac{32x}{4x(x-2)}$

$= \dfrac{3x-6+32x}{4x(x-2)} = \dfrac{35x-6}{4x(x-2)}$

13. $\dfrac{5}{y^2} - \dfrac{y}{2y+1}$;　　　$LCD = y^2(2y+1)$

$$\dfrac{5}{y^2} \cdot \dfrac{(2y+1)}{(2y+1)} - \dfrac{y}{(2y+1)} \cdot \dfrac{y^2}{y^2}$$

$$= \dfrac{5(2y+1)}{y^2(2y+1)} - \dfrac{y^3}{y^2(2y+1)} = \dfrac{10y+5-y^3}{y^2(2y+1)}$$

15. Answers may vary.

17. $\dfrac{6}{x-3} + \dfrac{8}{3-x} = \dfrac{6}{x-3} - \dfrac{8}{x-3} = \dfrac{6-8}{x-3} = \dfrac{-2}{x-3}$

19. $\dfrac{-8}{x^2-1} - \dfrac{7}{1-x^2} = \dfrac{-8}{x^2-1} + \dfrac{7}{x^2-1} = \dfrac{-1}{x^2-1}$

21. $\dfrac{x}{x^2-4} - \dfrac{2}{4-x^2} = \dfrac{x}{x^2-4} + \dfrac{2}{x^2-4}$

$$= \dfrac{x+2}{x^2-4} = \dfrac{x+2}{(x+2)(x-2)} = \dfrac{1}{x-2}$$

23. $\dfrac{5}{x} + 2$;　　　$LCD = x$

$$\dfrac{5}{x} + 2 \cdot \dfrac{x}{x} = \dfrac{5}{x} + \dfrac{2x}{x} = \dfrac{5+2x}{x}$$

25. $\dfrac{5}{x-2} + 6$;　　$LCD = x-2$

$$\dfrac{5}{x-2} + 6 \cdot \dfrac{(x-2)}{(x-2)} = \dfrac{5}{x-2} + \dfrac{6(x-2)}{x-2}$$

$$= \dfrac{5+6x-12}{x-2} = \dfrac{6x-7}{x-2}$$

27. $\dfrac{y+2}{y+3} - 2$;　　$LCD = y+3$

$$\dfrac{y+2}{y+3} - 2 \cdot \dfrac{(y+3)}{(y+3)} = \dfrac{y+2}{y+3} - \dfrac{2(y+3)}{y+3}$$

$$= \dfrac{y+2-2y-6}{y+3} = \dfrac{-y-4}{y+3} = \dfrac{-1(y+4)}{y+3}$$

$$= -\dfrac{y+4}{y+3}$$

29. $90 - \dfrac{40}{x} = \dfrac{90x-40}{x}$ degrees

31. $\dfrac{5x}{x+2} - \dfrac{3x-4}{x+2} = \dfrac{5x-(3x-4)}{x+2}$

$$= \dfrac{5x-3x+4}{x+2} = \dfrac{2x+4}{x+2} = \dfrac{2(x+2)}{x+2}$$

$$= 2$$

33. $\dfrac{3x^4}{x} - \dfrac{4x^2}{x^2} = 3x^3 - 4$

35. $\dfrac{1}{x+3} - \dfrac{1}{(x+3)^2}$;　　$LCD = (x+3)^2$

$$\dfrac{1}{x+3} \cdot \dfrac{(x+3)}{(x+3)} - \dfrac{1}{(x+3)^2}$$

$$= \dfrac{x+3}{(x+3)^2} - \dfrac{1}{(x+3)^2} = \dfrac{x+3-1}{(x+3)^2} = \dfrac{x+2}{(x+3)^2}$$

37. $\dfrac{4}{5b} + \dfrac{1}{b-1}$;　　$LCD = 5b(b-1)$

$$\dfrac{4}{5b} \cdot \dfrac{(b-1)}{(b-1)} + \dfrac{1}{(b-1)} \cdot \dfrac{5b}{5b}$$

$$= \dfrac{4(b-1)}{5b(b-1)} + \dfrac{5b}{5b(b-1)} = \dfrac{4b-4+5b}{5b(b-1)}$$

$$= \dfrac{9b-4}{5b(b-1)}$$

39. $\dfrac{2}{m} + 1$;　　　$LCD = m$

$$\dfrac{2}{m} + 1 \cdot \dfrac{m}{m} = \dfrac{2}{m} + \dfrac{m}{m} = \dfrac{2+m}{m}$$

41. $\dfrac{6}{1-2x} - \dfrac{4}{2x-1} = \dfrac{6}{1-2x} + \dfrac{4}{1-2x} = \dfrac{10}{1-2x}$

43. $\dfrac{7}{(x+1)(x-1)} + \dfrac{8}{(x+1)^2}$

$$LCD = (x+1)^2(x-1)$$

$$\dfrac{7}{(x+1)(x-1)} \cdot \dfrac{(x+1)}{(x+1)} + \dfrac{8}{(x+1)^2} \cdot \dfrac{(x-1)}{(x-1)}$$

$$= \dfrac{7(x+1)}{(x+1)^2(x-1)} + \dfrac{8(x-1)}{(x+1)^2(x-1)}$$

$$= \dfrac{7x+7+8x-8}{(x+1)^2(x-1)} = \dfrac{15x-1}{(x+1)^2(x-1)}$$

45. $\dfrac{x}{x^2-1}-\dfrac{2}{x^2-2x+1}=\dfrac{x}{(x+1)(x-1)}-\dfrac{2}{(x-1)^2}$

$\text{LCD}=(x+1)(x-1)^2$

$\dfrac{x}{(x+1)(x-1)}\cdot\dfrac{(x-1)}{(x-1)}-\dfrac{2}{(x-1)^2}\cdot\dfrac{(x+1)}{(x+1)}$

$=\dfrac{x(x-1)}{(x+1)(x-1)}-\dfrac{2(x+1)}{(x+1)^2(x+1)}$

$=\dfrac{x^2-x-2x-2}{(x+1)(x-1)^2}=\dfrac{x^2-3x-2}{(x+1)(x-1)^2}$

47. $\dfrac{3a}{2a+6}-\dfrac{a-1}{a+3}=\dfrac{3a}{2(a+3)}-\dfrac{a-1}{a+3}$

$\text{LCD}=2(a+3)$

$\dfrac{3a}{2(a+3)}-\dfrac{(a-1)}{(a+3)}\cdot\dfrac{2}{2}=\dfrac{3a}{2(a+3)}-\dfrac{2(a-1)}{2(a+3)}$

$=\dfrac{3a-2a+2}{2(a+3)}=\dfrac{a+2}{2(a+3)}$

49. $\dfrac{5}{2-x}+\dfrac{x}{2x-4}=-\dfrac{5}{x-2}+\dfrac{x}{2(x-2)}$

$\text{LCD}=2(x-2)$

$-\dfrac{5}{(x-2)}\cdot\dfrac{2}{2}+\dfrac{x}{2(x-2)}=\dfrac{-10}{2(x-2)}+\dfrac{x}{2(x-2)}$

$=\dfrac{-10+x}{2(x-2)}$

51. $\dfrac{-7}{y^2-3y+2}-\dfrac{2}{y-1}=\dfrac{-7}{(y-2)(y-1)}-\dfrac{2}{y-1}$

$\text{LCD}=(y-2)(y-1)$

$\dfrac{-7}{(y-2)(y-1)}-\dfrac{2}{(y-1)}\cdot\dfrac{(y-2)}{(y-2)}$

$=\dfrac{-7-2y+4}{(y-2)(y-1)}=\dfrac{-2y-3}{(y-2)(y-1)}$

53. $\dfrac{13}{x^2-5x+6}-\dfrac{5}{x-3}=\dfrac{13}{(x-3)(x-2)}-\dfrac{5}{x-3}$

$\text{LCD}=(x-3)(x-2)$

$\dfrac{13}{(x-3)(x-2)}-\dfrac{5}{(x-3)}\cdot\dfrac{(x-2)}{(x-2)}$

$=\dfrac{13}{(x-3)(x-2)}-\dfrac{5(x-2)}{(x-3)(x-2)}$

$=\dfrac{13-5x+10}{(x-3)(x-2)}=\dfrac{-5x+23}{(x-3)(x-2)}$

55. $\dfrac{8}{(x+2)(x-2)}+\dfrac{4}{(x+2)(x-3)}$

$\text{LCD}=(x+2)(x-2)(x-3)$

$\dfrac{8}{(x+2)(x-2)}\cdot\dfrac{(x-3)}{(x-3)}+\dfrac{4}{(x+2)(x-3)}\cdot\dfrac{(x-2)}{(x-2)}$

$=\dfrac{8(x-3)}{(x+2)(x-2)(x-3)}+\dfrac{4(x-2)}{(x+2)(x-3)(x-2)}$

$=\dfrac{8x-24+4x-8}{(x+2)(x-2)(x-3)}=\dfrac{12x-32}{(x+2)(x-2)(x-3)}$

57. $\dfrac{5}{9x^2-4}+\dfrac{2}{3x-2}=\dfrac{5}{(3x+2)(3x-2)}+\dfrac{2}{3x-2}$

$\text{LCD}=(3x+2)(3x-2)$

$\dfrac{5}{(3x+2)(3x-2)}+\dfrac{2}{(3x-2)}\cdot\dfrac{(3x+2)}{(3x+2)}$

$=\dfrac{5}{(3x+2)(3x-2)}+\dfrac{2(3x+2)}{(3x-2)(3x+2)}$

$=\dfrac{5+6x+4}{(3x+2)(3x-2)}=\dfrac{6x+9}{(3x+2)(3x-2)}$

59. $\dfrac{x+8}{x^2-5x-6}+\dfrac{x+1}{x^2-4x-5}$

$=\dfrac{x+8}{(x-6)(x+1)}+\dfrac{x+1}{(x-5)(x+1)}$

$\text{LCD}=(x-6)(x+1)(x-5)$

$\dfrac{(x+8)}{(x-6)(x+1)}\cdot\dfrac{(x-5)}{(x-5)}+\dfrac{(x+1)}{(x-5)(x+1)}\cdot\dfrac{(x-6)}{(x-6)}$

$=\dfrac{(x+8)(x-5)}{(x-6)(x+1)(x-5)}+\dfrac{(x+1)(x-6)}{(x-5)(x+1)(x-6)}$

$=\dfrac{x^2-5x+8x-40+x^2-6x+x-6}{(x-6)(x+1)(x-5)}$

$=\dfrac{2x^2-2x-46}{(x-6)(x+1)(x-5)}$

61. $\dfrac{3}{x+4}-\dfrac{1}{x-4}$

$\text{LCD}=(x+4)(x-4)$

$\dfrac{3}{(x+4)}\cdot\dfrac{(x-4)}{(x-4)}-\dfrac{1}{(x-4)}\cdot\dfrac{(x+4)}{(x+4)}$

$=\dfrac{3(x-4)}{(x+4)(x-4)}-\dfrac{1(x+4)}{(x-4)(x+4)}$

$=\dfrac{3x-12-x-4}{(x+4)(x-4)}=\dfrac{2x-16}{(x+4)(x-4)}$

63. $\dfrac{15x}{x+8} \cdot \dfrac{2x+16}{3x} = \dfrac{15x}{x+8} \cdot \dfrac{2(x+8)}{3x} = \dfrac{30x}{3x} = 10$

65. $\dfrac{8x+7}{3x+5} - \dfrac{2x-3}{3x+5} = \dfrac{8x+7-(2x-3)}{3x+5}$

$= \dfrac{8x+7-2x+3}{3x+5} = \dfrac{6x+10}{3x+5} = \dfrac{2(3x+5)}{3x+5} = 2$

67. $\dfrac{5a+10}{18} + \dfrac{a^2-4}{10a} = \dfrac{5a+10}{18} \cdot \dfrac{10a}{a^2-4}$

$= \dfrac{5(a+2)}{18} \cdot \dfrac{10a}{(a+2)(a-2)}$

$= \dfrac{50a}{18(a-2)} = \dfrac{25a}{9(a-2)}$

69. $\dfrac{5}{x^2-3x+2} + \dfrac{1}{x-2} = \dfrac{5}{(x-2)(x-1)} + \dfrac{1}{x-2}$

$\text{LCD} = (x-2)(x-1)$

$\dfrac{5}{(x-2)(x-1)} + \dfrac{1}{(x-2)} \cdot \dfrac{(x-1)}{(x-1)}$

$= \dfrac{5}{(x-2)(x-1)} + \dfrac{x-1}{(x-2)(x-1)}$

$= \dfrac{5+x-1}{(x-2)(x-1)} = \dfrac{x+4}{(x-2)(x-1)}$

71. Answers may vary.

73. $x^3 - 1 = (x-1)(x^2+x+1)$

75. $125z^3 + 8 = (5z)^3 + 2^3 = (5z+2)(25z^2-10z+4)$

77. $xy + 2x + 3y + 6 = x(y+2) + 3(y+2)$
$= (y+2)(x+3)$

79. $(1, 2)$ and $(-1, -2)$

$m = \dfrac{-2-2}{-1-1} = \dfrac{-4}{-2} = 2$

81. $(0, 0)$ and $(3, -1)$

$m = \dfrac{-1-0}{3-0} = -\dfrac{1}{3}$

83. $\dfrac{5}{x^2-4} + \dfrac{2}{x^2-4x+4} - \dfrac{3}{x^2-x-6} = \dfrac{5}{(x+2)(x-2)} + \dfrac{2}{(x-2)^2} - \dfrac{3}{(x-3)(x+2)}$

$\text{LCD} = (x+2)(x-2)^2(x-3)$

$\dfrac{5}{(x+2)(x-2)} \cdot \dfrac{(x-2)(x-3)}{(x-2)(x-3)} + \dfrac{2}{(x-2)^2} \cdot \dfrac{(x+2)(x-3)}{(x+2)(x-3)} - \dfrac{3}{(x-3)(x+2)} \cdot \dfrac{(x-2)^2}{(x-2)^2}$

$= \dfrac{5(x-2)(x-3)}{(x+2)(x-2)^2(x-3)} + \dfrac{2(x+2)(x-3)}{(x-2)^2(x+2)(x-3)} - \dfrac{3(x-2)^2}{(x-3)(x+2)(x-2)^2}$

$= \dfrac{5(x^2-5x+6) + 2(x^2-x-6) - 3(x^2-4x+4)}{(x+2)(x-2)^2(x-3)} = \dfrac{5x^2-25x+30+2x^2-2x-12-3x^2+12x-12}{(x+2)(x-2)^2(x-3)}$

$= \dfrac{4x^2-15x+6}{(x+2)(x-2)^2(x-3)}$

85. $\dfrac{9}{x^2+9x+14} - \dfrac{3x}{x^2+10x+21} + \dfrac{4}{x^2+5x+6} = \dfrac{9}{(x+2)(x+7)} - \dfrac{3x}{(x+3)(x+7)} + \dfrac{4}{(x+3)(x+2)}$

$\text{LCD} = (x+2)(x+7)(x+3)$

$\dfrac{9}{(x+2)(x+7)} \cdot \dfrac{(x+3)}{(x+3)} - \dfrac{3x}{(x+3)(x+7)} \cdot \dfrac{(x+2)}{(x+2)} + \dfrac{4}{(x+3)(x+2)} \cdot \dfrac{(x+7)}{(x+7)}$

$= \dfrac{9(x+3)}{(x+2)(x+7)(x+3)} - \dfrac{3x(x+2)}{(x+3)(x+7)(x+2)} + \dfrac{4(x+7)}{(x+3)(x+2)(x+7)}$

$= \dfrac{9x+27-3x^2-6x+4x+28}{(x+2)(x+7)(x+3)} = \dfrac{-3x^2+7x+55}{(x+2)(x+7)(x+3)}$

87. $\dfrac{5+x}{x^3-27}+\dfrac{x}{x^3+3x^2+9x}$

$=\dfrac{5+x}{(x-3)(x^2+3x+9)}+\dfrac{x}{x(x^2+3x+9)}$

$\text{LCD}=x(x-3)(x^2+3x+9)$

$\dfrac{(5+x)}{(x-3)(x^2+3x+9)}\cdot\dfrac{x}{x}+\dfrac{x}{x(x^2+3x+9)}\cdot\dfrac{(x-3)}{(x-3)}$

$=\dfrac{(5+x)(x)}{x(x-3)(x^2+3x+9)}+\dfrac{x(x-3)}{x(x-3)(x^2+3x+9)}$

$=\dfrac{5x+x^2+x^2-3x}{x(x-3)(x^2+3x+9)}=\dfrac{2x^2+2x}{x(x-3)(x^2+3x+9)}$

$=\dfrac{2x(x+1)}{x(x-3)(x^2+3x+9)}=\dfrac{2(x+1)}{(x-3)(x^2+3x+9)}$

Exercise Set 6.5

1. $\dfrac{\frac{1}{2}}{\frac{3}{4}}=\dfrac{1}{2}\cdot\dfrac{4}{3}=\dfrac{2}{3}$

3. $\dfrac{-\frac{4x}{9}}{-\frac{2x}{3}}=-\dfrac{4x}{9}\cdot-\dfrac{3}{2x}=\dfrac{12x}{18x}=\dfrac{2}{3}$

5. $\dfrac{\frac{1+x}{6}}{\frac{1+x}{3}}=\dfrac{1+x}{6}\cdot\dfrac{3}{1+x}=\dfrac{1}{2}$

7. $\dfrac{\frac{(y+1)(y-1)}{6}}{\frac{(y+1)(y+2)}{8}}=\dfrac{(y+1)(y-1)}{6}\cdot\dfrac{8}{(y+1)(y+2)}$

$=\dfrac{4(y-1)}{3(y+2)}$

9. $t=\dfrac{d}{r}$

$t=\dfrac{\frac{20x}{3}}{\frac{5x}{9}}=\dfrac{20x}{3}\cdot\dfrac{9}{5x}=12\text{ hours}$

11. $\dfrac{\frac{1}{2}+\frac{2}{3}}{\frac{5}{9}-\frac{5}{6}}=\dfrac{\frac{1}{2}\cdot\frac{3}{3}+\frac{2}{3}\cdot\frac{2}{2}}{\frac{5}{9}\cdot\frac{2}{2}-\frac{5}{6}\cdot\frac{3}{3}}=\dfrac{\frac{3}{6}+\frac{4}{6}}{\frac{10}{18}-\frac{15}{18}}=\dfrac{\frac{7}{6}}{-\frac{5}{18}}$

$=\dfrac{7}{6}\cdot-\dfrac{18}{5}=-\dfrac{21}{5}$

13. $\dfrac{2+\frac{7}{10}}{1+\frac{3}{5}}=\dfrac{10\left(2+\frac{7}{10}\right)}{10\left(1+\frac{3}{5}\right)}=\dfrac{10(2)+10\left(\frac{7}{10}\right)}{10(1)+10\left(\frac{3}{5}\right)}$

$=\dfrac{20+7}{10+6}=\dfrac{27}{16}$

15. $\dfrac{\frac{1}{3}}{\frac{1}{2}-\frac{1}{4}}=\dfrac{\frac{1}{3}}{\frac{1}{2}\cdot\frac{2}{2}-\frac{1}{4}}=\dfrac{\frac{1}{3}}{\frac{2}{4}-\frac{1}{4}}=\dfrac{\frac{1}{3}}{\frac{1}{4}}=\dfrac{1}{3}\cdot\dfrac{4}{1}=\dfrac{4}{3}$

17. $\dfrac{-\frac{2}{9}}{-\frac{14}{3}}=-\dfrac{2}{9}\cdot-\dfrac{3}{14}=\dfrac{6}{126}=\dfrac{1}{21}$

19. $\dfrac{-\frac{5}{12x^2}}{\frac{25}{16x^3}}=-\dfrac{5}{12x^2}\cdot\dfrac{16x^3}{25}=-\dfrac{80x^3}{300x^2}=-\dfrac{4x}{15}$

21. $\dfrac{\frac{m}{n}-1}{\frac{m}{n}+1}=\dfrac{n\left(\frac{m}{n}-1\right)}{n\left(\frac{m}{n}+1\right)}=\dfrac{n\left(\frac{m}{n}\right)+n(-1)}{n\left(\frac{m}{n}\right)+n(1)}=\dfrac{m-n}{m+n}$

23. $\dfrac{\frac{1}{5}-\frac{1}{x}}{\frac{7}{10}+\frac{1}{x^2}}=\dfrac{10x^2\left(\frac{1}{5}-\frac{1}{x}\right)}{10x^2\left(\frac{7}{10}+\frac{1}{x^2}\right)}$

$=\dfrac{10x^2\left(\frac{1}{5}\right)+10x^2\left(-\frac{1}{x}\right)}{10x^2\left(\frac{7}{10}\right)+10x^2\left(\frac{1}{x^2}\right)}=\dfrac{2x^2-10x}{7x^2+10}$

$=\dfrac{2x(x-5)}{7x^2+10}$

25. $\dfrac{1+\frac{1}{y-2}}{y+\frac{1}{y-2}}=\dfrac{(y-2)\left[1+\frac{1}{y-2}\right]}{(y-2)\left[y+\frac{1}{y-2}\right]}$

$=\dfrac{(y-2)(1)+(y-2)\cdot\frac{1}{y-2}}{(y-2)(y)+(y-2)\cdot\frac{1}{y-2}}=\dfrac{y-2+1}{y^2-2y+1}$

$=\dfrac{y-1}{(y-1)(y-1)}=\dfrac{1}{y-1}$

27. $\dfrac{\frac{4y-8}{16}}{\frac{6y-12}{4}}=\dfrac{4y-8}{16}\cdot\dfrac{4}{6y-12}$

$=\dfrac{4(y-2)}{16}\cdot\dfrac{4}{6(y-2)}=\dfrac{16}{96}=\dfrac{1}{6}$

29. $\dfrac{\frac{x}{y}+1}{\frac{x}{y}-1} = \dfrac{y\left(\frac{x}{y}+1\right)}{y\left(\frac{x}{y}-1\right)} = \dfrac{y\left(\frac{x}{y}\right)+y(1)}{y\left(\frac{x}{y}\right)+y(-1)}$

$= \dfrac{x+y}{x-y}$

31. $\dfrac{1}{2+\frac{1}{3}} = \dfrac{1}{2\cdot\frac{3}{3}+\frac{1}{3}} = \dfrac{1}{\frac{6}{3}+\frac{1}{3}} = \dfrac{1}{\frac{7}{3}} = 1\cdot\dfrac{3}{7} = \dfrac{3}{7}$

33. $\dfrac{\frac{ax+ab}{x^2-b^2}}{\frac{x+b}{x-b}} = \dfrac{ax+ab}{x^2-b^2}\cdot\dfrac{x-b}{x+b}$

$= \dfrac{a(x+b)}{(x+b)(x-b)}\cdot\dfrac{x-b}{x+b} = \dfrac{a}{x+b}$

35. $\dfrac{\frac{-3+y}{4}}{\frac{8+y}{28}} = \dfrac{-3+y}{4}\cdot\dfrac{28}{8+y} = \dfrac{7(-3+y)}{8+y}$

37. $\dfrac{3+\frac{12}{x}}{1-\frac{16}{x^2}} = \dfrac{x^2\left(3+\frac{12}{x}\right)}{x^2\left(1-\frac{16}{x^2}\right)} = \dfrac{x^2(3)+x^2\left(\frac{12}{x}\right)}{x^2(1)+x^2\left(-\frac{16}{x^2}\right)}$

$= \dfrac{3x^2+12x}{x^2-16} = \dfrac{3x(x+4)}{(x-4)(x+4)} = \dfrac{3x}{x-4}$

39. $\dfrac{2+\frac{6}{x}}{1-\frac{9}{x^2}} = \dfrac{x^2\left(2+\frac{6}{x}\right)}{x^2\left(1-\frac{9}{x^2}\right)} = \dfrac{x^2(2)+x^2\left(\frac{6}{x}\right)}{x^2(1)+x^2\left(-\frac{9}{x^2}\right)}$

$= \dfrac{2x^2+6x}{x^2-9} = \dfrac{2x(x+3)}{(x+3)(x-3)} = \dfrac{2x}{x-3}$

41. $\dfrac{\frac{8}{x+4}+2}{\frac{12}{x+4}-2} = \dfrac{(x+4)\left[\frac{8}{x+4}+2\right]}{(x+4)\left[\frac{12}{x+4}-2\right]}$

$= \dfrac{(x+4)\left(\frac{8}{x+4}\right)+(x+4)(2)}{(x+4)\left(\frac{12}{x+4}\right)+(x+4)(-2)} = \dfrac{8+2x+8}{12-2x-8}$

$= \dfrac{2x+16}{-2x+4} = \dfrac{2(x+8)}{-2(x-2)} = -\dfrac{x+8}{x-2}$

43. $\dfrac{\frac{s}{r}+\frac{r}{s}}{\frac{s}{r}-\frac{r}{s}} = \dfrac{rs\left(\frac{s}{r}+\frac{r}{s}\right)}{rs\left(\frac{s}{r}-\frac{r}{s}\right)} = \dfrac{rs\left(\frac{s}{r}\right)+rs\left(\frac{r}{s}\right)}{rs\left(\frac{s}{r}\right)+rs\left(-\frac{r}{s}\right)}$

$= \dfrac{s^2+r^2}{s^2-r^2}$

45. Answers may vary.

47. $\dfrac{\frac{1}{3}+\frac{3}{4}}{2} = \dfrac{\frac{1}{3}\cdot\frac{4}{4}+\frac{3}{4}\cdot\frac{3}{3}}{2} = \dfrac{\frac{4}{12}+\frac{9}{12}}{2} = \dfrac{\frac{13}{12}}{2}$

$= \dfrac{13}{12}\cdot\dfrac{1}{2} = \dfrac{13}{24}$

49. $\dfrac{1}{\frac{1}{R_1}+\frac{1}{R_2}} = \dfrac{1}{\frac{R_2}{R_1R_2}+\frac{R_1}{R_1R_2}} = \dfrac{1}{\frac{R_2+R_1}{R_1R_2}} = \dfrac{R_1R_2}{R_2+R_1}$

51. $3x+5=7$

$3x=2$

$x=\dfrac{2}{3}$

53. $2x^2-x-1=0$

$(2x+1)(x-1)=0$

$2x+1=0 \quad\text{or}\quad x-1=0$

$2x=-1 \quad\text{or}\quad x=1$

$x=-\dfrac{1}{2}$

$x=-\dfrac{1}{2},\ 1$

55. $\dfrac{2+x}{x+2} = \dfrac{2+x}{2+x} = 1$

57. $\dfrac{2-x}{x-2} = -\dfrac{x-2}{x-2} = -1$

59. $\dfrac{x^{-1}+2^{-1}}{x^{-2}-4^{-1}} = \dfrac{\frac{1}{x}+\frac{1}{2}}{\frac{1}{x^2}-\frac{1}{4}} = \dfrac{4x^2\left(\frac{1}{x}+\frac{1}{2}\right)}{4x^2\left(\frac{1}{x^2}-\frac{1}{4}\right)}$

$= \dfrac{4x+2x^2}{4-x^2} = \dfrac{2x(2+x)}{(2-x)(2+x)} = \dfrac{2x}{2-x}$

61. $\dfrac{x+y^{-1}}{\frac{x}{y}} = \dfrac{x+\frac{1}{y}}{\frac{x}{y}} = \dfrac{y\left(x+\frac{1}{y}\right)}{y\left(\frac{x}{y}\right)} = \dfrac{xy+1}{x}$

63. $\dfrac{y^{-2}}{1-y^{-2}} = \dfrac{\frac{1}{y^2}}{1-\frac{1}{y^2}} = \dfrac{y^2\left(\frac{1}{y^2}\right)}{y^2\left(1-\frac{1}{y^2}\right)} = \dfrac{1}{y^2-1}$

Section 6.6 Mental Math

1. $\dfrac{x}{5} = 2$

$5 \cdot \dfrac{x}{5} = 2 \cdot 5$

$x = 10$

3. $\dfrac{z}{6} = 6$

$6 \cdot \dfrac{z}{6} = 6 \cdot 6$

$z = 36$

Exercise Set 6.6

1. $\dfrac{x}{5} + 3 = 9$

$5\left(\dfrac{x}{5} + 3\right) = 5(9)$

$5\left(\dfrac{x}{5}\right) + 5(3) = 5(9)$

$x + 15 = 45$

$x = 45 - 15$

$x = 30$

3. $\dfrac{x}{2} + \dfrac{5x}{4} = \dfrac{x}{12}$

$12\left(\dfrac{x}{2} + \dfrac{5x}{4}\right) = 12\left(\dfrac{x}{12}\right)$

$12\left(\dfrac{x}{2}\right) + 12\left(\dfrac{5x}{4}\right) = 12\left(\dfrac{x}{12}\right)$

$6x + 15x = x$

$21x = x$

$21x - x = 0$

$20x = 0$

$x = \dfrac{0}{20} = 0$

5. $2 + \dfrac{10}{x} = x + 5$

$x\left(2 + \dfrac{10}{x}\right) = x(x + 5)$

$2x + 10 = x^2 + 5x$

$0 = x^2 + 3x - 10$

$0 = (x + 5)(x - 2)$

$x + 5 = 0 \quad$ or $\quad x - 2 = 0$

$x = -5 \quad$ or $\quad x = 2$

7. $\dfrac{a}{5} = \dfrac{a - 3}{2}$

$10\left(\dfrac{a}{5}\right) = 10\left(\dfrac{a - 3}{2}\right)$

$2a = 5(a - 3)$

$2a = 5a - 15$

$2a - 5a = -15$

$-3a = -15$

$a = \dfrac{-15}{-3} = 5$

9. $\dfrac{x - 3}{5} + \dfrac{x - 2}{2} = \dfrac{1}{2}$

$10\left(\dfrac{x - 3}{5} + \dfrac{x - 2}{2}\right) = 10\left(\dfrac{1}{2}\right)$

$10\left(\dfrac{x - 3}{5}\right) + 10\left(\dfrac{x - 2}{2}\right) = 10\left(\dfrac{1}{2}\right)$

$2(x - 3) + 5(x - 2) = 5$

$2x - 6 + 5x - 10 = 5$

$7x - 16 = 5$

$7x = 5 + 16$

$7x = 21$

$x = \dfrac{21}{7} = 3$

11. $\dfrac{20x}{3} + \dfrac{32x}{6} = 180$

$6\left(\dfrac{20x}{3} + \dfrac{32x}{6}\right) = 6 \cdot 180$

$40x + 32x = 1080$

$72x = 1080$

$x = 15$

$\dfrac{20x}{3} = \dfrac{20 \cdot 15}{3} = 100°$

$\dfrac{32x}{6} = \dfrac{32 \cdot 15}{6} = 80°$

13. $\dfrac{150}{x} + \dfrac{450}{x} = 90$

$x\left(\dfrac{150}{x} + \dfrac{450}{x}\right) = x \cdot 90$

$150 + 450 = 90x$

$600 = 90x$

$6.\overline{6} = x$

$\dfrac{150}{x} = \dfrac{150}{6.\overline{6}} = 22.5°$

$\dfrac{450}{x} = \dfrac{450}{6.\overline{6}} = 67.5°$

15. $\dfrac{9}{2a-5}=-2$

$(2a-5)\left(\dfrac{9}{2a-5}\right)=(2a-5)(-2)$

$9=-4a+10$

$9-10=-4a$

$-1=-4a$

$\dfrac{1}{4}=\dfrac{-1}{-4}=a$

17. $\dfrac{y}{y+4}+\dfrac{4}{y+4}=3$

$(y+4)\left[\dfrac{y}{y+4}+\dfrac{4}{y+4}\right]=(y+4)(3)$

$(y+4)\left(\dfrac{y}{y+4}\right)+(y+4)\left(\dfrac{4}{y+4}\right)=(y+4)(3)$

$y+4=3y+12$

$y-3y=12-4$

$-2y=8$

$y=\dfrac{8}{-2}=-4$

-4 is an extraneous solution. If $y=-4$, the denominator would equal zero.

19. $\dfrac{2x}{x+2}-2=\dfrac{x-8}{x-2}$

$(x+2)(x-2)\left[\dfrac{2x}{x+2}-2\right]=(x+2)(x-2)\left[\dfrac{x-8}{x-2}\right]$

$(x+2)(x-2)\left(\dfrac{2x}{x+2}\right)+(x+2)(x-2)(-2)$

$\qquad=(x+2)(x-8)$

$(x-2)(2x)+(x+2)(x-2)(-2)=(x+2)(x-8)$

$2x^2-4x+(x^2-4)(-2)=x^2-6x-16$

$2x^2-4x-2x^2+8=x^2-6x-16$

$-4x+8=x^2-6x-16$

$0=x^2-6x+4x-16-8$

$0=x^2-2x-24$

$0=(x-6)(x+4)$

$x-6=0 \qquad$ or $\qquad x+4=0$

$x=6 \qquad$ or $\qquad x=-4$

21. $\dfrac{4y}{y-4}+5=\dfrac{5y}{y-4}$

$(y-4)\left[\dfrac{4y}{y-4}+5\right]=(y-4)\left[\dfrac{5y}{y-4}\right]$

$(y-4)\left(\dfrac{4y}{y-4}\right)+(y-4)(5)=(y-4)\left[\dfrac{5y}{y-4}\right]$

$4y+5y-20=5y$

$4y+5y-5y=20$

$4y=20$

$y=\dfrac{20}{4}=5$

23. $\dfrac{7}{x-2}+1=\dfrac{x}{x+2}$

$(x-2)(x+2)\left[\dfrac{7}{x-2}+1\right]=(x-2)(x+2)\left[\dfrac{x}{x+2}\right]$

$(x-2)(x+2)\left(\dfrac{7}{x-2}\right)+(x-2)(x+2)(1)$

$\qquad=(x-2)(x)$

$(x+2)(7)=(x-2)(x+2)(1)=(x-2)(x)$

$7x+14+x^2-4=x^2-2x$

$x^2+7x+10=x^2-2x$

$x^2-x^2+7x+2x=-10$

$9x=-10$

$x=-\dfrac{10}{9}$

25. $\dfrac{x+1}{x+3}=\dfrac{2x^2-15x}{x^2+x-6}-\dfrac{x-3}{x-2}$

$\dfrac{x+1}{x+3}=\dfrac{2x^2-15x}{(x+3)(x-2)}-\dfrac{x-3}{x-2}$

$(x+3)(x-2)\left[\dfrac{x+1}{x+3}\right]$

$\qquad=(x+3)(x-2)\left[\dfrac{2x^2-15x}{(x+3)(x-2)}-\dfrac{x-3}{x-2}\right]$

$(x-2)(x+1)$

$\qquad=(x+3)(x-2)\left[\dfrac{2x^2-15x}{(x+3)(x-2)}\right]-(x+3)(x-2)\left(\dfrac{x-3}{x-2}\right)$

$(x-2)(x+1)=2x^2-15x-(x+3)(x-3)$

$x^2-x-2=2x^2-15x-(x^2-9)$

$x^2-x-2=2x^2-15x-x^2+9$

$x^2-x-2=x^2-15x+9$

$x^2-x^2-x+15x=9+2$

$14x=11$

$x=\dfrac{11}{14}$

27. $\dfrac{y}{2y+2}+\dfrac{2y-16}{4y+4}=\dfrac{2y-3}{y+1}$

$\dfrac{y}{2(y+1)}+\dfrac{2y-16}{4(y+1)}=\dfrac{2y-3}{y+1}$

$4(y+1)\left[\dfrac{y}{2(y+1)}+\dfrac{2y-16}{4(y+1)}\right]=4(y+1)\left(\dfrac{2y-3}{y+1}\right)$

$4(y+1)\left[\dfrac{y}{2(y+1)}\right]+4(y+1)\left[\dfrac{2y-16}{4(y+1)}\right]$
$=4(2y-3)$

$2y+2y-16=8y-12$
$4y-16=8y-12$
$4y-8y=-12+16$
$-4y=4$
$y=\dfrac{4}{-4}=-1$

-1 is an extraneous solution. If $y=-1$, the denominator would equal zero.

29. Expression;

$\dfrac{1}{x}+\dfrac{2}{3}=\dfrac{1}{x}\cdot\dfrac{3}{3}+\dfrac{2}{3}\cdot\dfrac{x}{x}$

$=\dfrac{3}{3x}+\dfrac{2x}{3x}=\dfrac{3+2x}{3x}$

31. Equation;

$\dfrac{1}{x}+\dfrac{2}{3}=\dfrac{3}{x}$

$3x\left(\dfrac{1}{x}+\dfrac{2}{3}\right)=3x\left(\dfrac{3}{x}\right)$

$3+2x=9$
$2x=6$
$x=3$

33. Expression;

$\dfrac{2}{x+1}-\dfrac{1}{x}=\dfrac{2}{(x+1)}\cdot\dfrac{x}{x}-\dfrac{1}{x}\cdot\dfrac{(x+1)}{(x+1)}$

$=\dfrac{2x}{x(x+1)}-\dfrac{x+1}{x(x+1)}=\dfrac{2x-(x+1)}{x(x+1)}$

$=\dfrac{2x-x-1}{x(x+1)}=\dfrac{x-1}{x(x+1)}$

35. Equation;

$\dfrac{2}{x+1}-\dfrac{1}{x}=1$

$x(x+1)\left(\dfrac{2}{x+1}-\dfrac{1}{x}\right)=x(x+1)\cdot1$

$2x-(x+1)=x(x+1)$
$2x-x-1=x^2+x$
$x-1=x^2+x$
$-1=x^2$
No solution

37. Answers may vary.

39. $\dfrac{2x}{7}-5x=9$

$7\left(\dfrac{2x}{7}-5x\right)=7(9)$

$7\left(\dfrac{2x}{7}\right)+7(-5x)=7(9)$

$2x-35x=63$
$-33x=63$
$x=-\dfrac{63}{33}=-\dfrac{21}{11}$

41. $\dfrac{2}{y}+\dfrac{1}{2}=\dfrac{5}{2y}$

$2y\left(\dfrac{2}{y}+\dfrac{1}{2}\right)=2y\left(\dfrac{5}{2y}\right)$

$2y\left(\dfrac{2}{y}\right)+2y\left(\dfrac{1}{2}\right)=2y\left(\dfrac{5}{2y}\right)$

$4+y=5$
$y=5-4$
$y=1$

43. $\dfrac{4x+10}{7}=\dfrac{8}{2}$

$14\left(\dfrac{4x+10}{7}\right)=14\left(\dfrac{8}{2}\right)$

$2(4x+10)=7(8)$
$8x+20=56$
$8x=56-20$
$8x=36$
$x=\dfrac{36}{8}=\dfrac{9}{2}$

45. $2 + \dfrac{3}{a-3} = \dfrac{a}{a-3}$

$(a-3)\left[2 + \dfrac{3}{a-3}\right] = (a-3)\left[\dfrac{3}{a-3}\right]$

$(a-3)(2) + (a-3)\left(\dfrac{3}{a-3}\right) = a$

$2a - 6 + 3 = a$
$2a - 3 = a$
$-3 = a - 2a$
$-3 = -a$
$3 = \dfrac{-3}{-1} = a$

3 is an extraneous solution. If $a = 3$, the denominator would equal zero.

47. $\dfrac{5}{x} + \dfrac{2}{3} = \dfrac{7}{2x}$

$6x\left(\dfrac{5}{x} + \dfrac{2}{3}\right) = 6x\left(\dfrac{7}{2x}\right)$

$6x\left(\dfrac{5}{x}\right) + 6x\left(\dfrac{2}{3}\right) = 21$

$30 + 4x = 21$
$4x = 21 - 30$
$4x = -9$
$x = -\dfrac{9}{4}$

49. $\dfrac{2a}{a+4} = \dfrac{3}{a-1}$

$(a+4)(a-1)\left[\dfrac{2a}{a+4}\right] = (a+4)(a-1)\left[\dfrac{3}{a-1}\right]$

$(a-1)(2a) = (a+4)(3)$
$2a^2 - 2a = 3a + 12$
$2a^2 - 2a - 3a - 12 = 0$
$2a^2 - 5a - 12 = 0$
$(2a+3)(a-4) = 0$
$2a + 3 = 0 \qquad \text{or} \qquad a - 4 = 0$
$2a = -3 \qquad \text{or} \qquad a = 4$
$a = -\dfrac{3}{2}$

51. $\dfrac{x+1}{3} - \dfrac{x-1}{6} = \dfrac{1}{6}$

$6\left[\dfrac{x+1}{3} - \dfrac{x-1}{6}\right] = 6\left[\dfrac{1}{6}\right]$

$6\left(\dfrac{x+1}{3}\right) - 6\left(\dfrac{x-1}{6}\right) = 1$

$2(x+1) - (x-1) = 1$

$2x + 2 - x + 1 = 1$
$x + 3 = 1$
$x = 1 - 3$
$x = -2$

53. $\dfrac{4r-1}{r^2+5r-14} + \dfrac{2}{r+7} = \dfrac{1}{r-2}$

$(r+7)(r-2)\left[\dfrac{4r-1}{(r+7)(r-2)} + \dfrac{2}{r+7}\right]$

$= (r+7)(r-2)\left(\dfrac{1}{r-2}\right)$

$(r+7)(r-2)\left(\dfrac{4r-1}{(r+7)(r-2)}\right) + (r+7)(r-2)\left(\dfrac{2}{r+7}\right)$

$= r + 7$

$4r - 1 + (r-2)(2) = r + 7$
$4r - 1 + 2r - 4 = r + 7$
$6r - 5 = r + 7$
$6r - r = 7 + 5$
$5r = 12$
$r = \dfrac{12}{5}$

55. $\dfrac{t}{t-4} = \dfrac{t+4}{6}$

$6(t-4)\left(\dfrac{t}{t-4}\right) = 6(t-4)\left(\dfrac{t+4}{6}\right)$

$6t = (t-4)(t+4)$
$6t = t^2 - 16$
$0 = t^2 - 6t - 16$
$0 = (t-8)(t+2)$
$t - 8 = 0 \qquad \text{or} \qquad t + 2 = 0$
$t = 8 \qquad \text{or} \qquad t = -2$

57. $\dfrac{x}{2x+6} + \dfrac{x+1}{3x+9} = \dfrac{2}{4x+12}$

$\dfrac{x}{2(x+3)} + \dfrac{x+1}{3(x+3)} = \dfrac{2}{4(x+3)}$

$12(x+3)\left[\dfrac{x}{2(x+3)} + \dfrac{x+1}{3(x+3)}\right] = 12(x+3)\left[\dfrac{2}{4(x+3)}\right]$

$12(x+3)\left[\dfrac{x}{2(x+3)}\right] + 12(x+3)\left[\dfrac{x+1}{3(x+3)}\right] = 6$

$6x + 4(x+1) = 6$
$6x + 4x + 4 = 6$
$10x + 4 = 6$
$10x = 2$
$x = \dfrac{2}{10} = \dfrac{1}{5}$

59. $\dfrac{D}{R} = T$

$R\left(\dfrac{D}{R}\right) = R(T)$

$D = RT$

$\dfrac{D}{T} = R$

61. $\dfrac{3}{x} = \dfrac{5y}{x+2}$

$x(x+2)\left(\dfrac{3}{x}\right) = x(x+2)\left(\dfrac{5y}{x+2}\right)$

$(x+2)(3) = x(5y)$

$3x + 6 = 5xy$

$\dfrac{3x+6}{5x} = y$

63. $\dfrac{3a+2}{3b-2} = -\dfrac{4}{2a}$

$2a(3b-2)\left(\dfrac{3a+2}{3b-2}\right) = 2a(3b-2)\left(-\dfrac{4}{2a}\right)$

$2a(3a+2) = (3b-2)(-4)$

$6a^2 + 4a = -12b + 8$

$6a^2 + 4a - 8 = -12b$

$-\dfrac{6a^2 + 4a - 8}{12} = b$

$-\dfrac{2(3a^2 + 2a - 4)}{12} = b$

$-\dfrac{3a^2 + 2a - 4}{6} = b$

65. $\dfrac{A}{BH} = \dfrac{1}{2}$

$2BH\left(\dfrac{A}{BH}\right) = 2BH\left(\dfrac{1}{2}\right)$

$2A = BH$

$\dfrac{2A}{H} = B$

67. $\dfrac{C}{\pi r} = 2$

$\pi r\left(\dfrac{C}{\pi r}\right) = \pi r(2)$

$C = 2\pi r$

$\dfrac{C}{2\pi} = r$

69. $\dfrac{1}{a} = \dfrac{1}{b} + \dfrac{1}{c}$

$abc\left(\dfrac{1}{a}\right) = abc\left(\dfrac{1}{b} + \dfrac{1}{c}\right)$

$bc = abc\left(\dfrac{1}{b}\right) + abc\left(\dfrac{1}{c}\right)$

$bc = ac + ab$

$bc = a(c + b)$

$\dfrac{bc}{c+b} = a$

71. $\dfrac{m^2}{6} - \dfrac{n}{3} = \dfrac{p}{2}$

$6\left[\dfrac{m^2}{6} - \dfrac{n}{3}\right] = 6\left[\dfrac{p}{2}\right]$

$6\left(\dfrac{m^2}{6}\right) - 6\left(\dfrac{n}{3}\right) = 3p$

$m^2 - 2n = 3p$

$m^2 - 3p = 2n$

$\dfrac{m^2 - 3p}{2} = n$

73.
$$\frac{5}{a^2+4a+3}+\frac{2}{a^2+a-6}-\frac{3}{a^2-a+2}=0$$
$$\frac{5}{(a+3)(a+1)}+\frac{2}{(a+3)(a-2)}-\frac{3}{(a-2)(a+1)}=0$$
$$(a+3)(a+1)(a-2)\left[\frac{5}{(a+3)(a+1)}+\frac{2}{(a+3)(a-2)}-\frac{3}{(a-2)(a+1)}\right]=(a+3)(a+1)(a-2)(0)$$
$$(a+3)(a+1)(a-2)\left[\frac{5}{(a+3)(a+1)}\right]+(a+3)(a+1)(a-2)\left[\frac{2}{(a+3)(a-2)}\right]-(a+3)(a+1)(a-2)\left[\frac{3}{(a-2)(a+1)}\right]=0$$
$$5(a-2)+2(a+1)-3(a+3)=0$$
$$5a-10+2a+2-3a-9=0$$
$$4a-17=0$$
$$4a=17$$
$$a=\frac{17}{4}$$

75.

77.

79. $x=2$; $y=-2$; $(2,0)$; $(0,-2)$

81. $x=-4$; $x=-2$; $x=3$; $y=4$
$(-4,0)$; $(-2,0)$; $(3,0)$; $(0,4)$

Exercise Set 6.7

1. $\dfrac{2}{15}$

3. $\dfrac{10}{12}=\dfrac{5}{6}$

5. 3 gallons = 12 quarts
$$\frac{5}{12}$$

7. 2 dollars = 40 nickels
$$\frac{4}{40}=\frac{1}{10}$$

9. 5 meters = 500 centimeters
$$\frac{175}{500}=\frac{7}{20}$$

11. 3 hours = 180 minutes
$$\frac{190}{180}=\frac{19}{18}$$

13. Answers may vary.

15. $\dfrac{2}{3}=\dfrac{x}{6}$
$$2\cdot6=3\cdot x$$
$$12=3x$$
$$4=\frac{12}{3}=x$$

17. $\dfrac{x}{10}=\dfrac{5}{9}$
$$x\cdot9=10\cdot5$$
$$9x=50$$
$$x=\frac{50}{9}$$

19. $\dfrac{4x}{6} = \dfrac{7}{2}$

$4x \cdot 2 = 6 \cdot 7$

$8x = 42$

$x = \dfrac{42}{8} = \dfrac{21}{4}$

21. $\dfrac{a}{25} = \dfrac{12}{10}$

$a \cdot 10 = 25 \cdot 12$

$10a = 300$

$a = \dfrac{300}{10} = 30$

23. $\dfrac{x-3}{x} = \dfrac{4}{7}$

$7 \cdot (x-3) = 4 \cdot x$

$7x - 21 = 4x$

$-21 = 4x - 7x$

$-21 = -3x$

$7 = \dfrac{-21}{-3} = x$

25. $\dfrac{5x+1}{x} = \dfrac{6}{3}$

$3 \cdot (5x+1) = 6 \cdot x$

$15x + 3 = 6x$

$3 = 6x - 15x$

$3 = -9x$

$-\dfrac{1}{3} = -\dfrac{3}{9} = x$

27. $\dfrac{x+1}{2x+3} = \dfrac{2}{3}$

$3 \cdot (x+1) = 2 \cdot (2x+3)$

$3x + 3 = 4x + 6$

$3 - 6 = 4x - 3x$

$-3 = x$

29. $\dfrac{9}{5} = \dfrac{12}{3x+2}$

$9 \cdot (3x+2) = 5 \cdot 12$

$27x + 18 = 60$

$27x = 60 - 18$

$27x = 42$

$x = \dfrac{42}{27} = \dfrac{14}{9}$

31. $\dfrac{3}{x+1} = \dfrac{5}{2x}$

$3 \cdot 2x = 5 \cdot (x+1)$

$6x = 5x + 5$

$6x - 5x = 5$

$x = 5$

33. $\dfrac{x+1}{x} = \dfrac{x+2}{x-2}$

$(x+1) \cdot (x-2) = x \cdot (x+2)$

$x^2 - x - 2 = x^2 + 2x$

$x^2 - x^2 - x - 2x = 2$

$-3x = 2$

$x = -\dfrac{2}{3}$

35. Yes; Answers may vary.

37. $\dfrac{5.79}{110} = 0.0526$

$\dfrac{13.99}{240} = 0.0583$

The best buy is 110 oz. for \$5.79.

39. $\dfrac{0.69}{6} = 0.115$

$\dfrac{0.90}{8} = 0.1125$

$\dfrac{1.89}{16} = 0.1181$

The best buy is 8 oz. for \$0.90.

41. $\dfrac{3}{100} = \dfrac{x}{4100}$

$3(4100) = 100x$

$12{,}300 = 100x$

$123 = x$

123 pounds

43. $\dfrac{28}{80} = \dfrac{x}{208}$

$28(208) = 80x$

$5824 = 80x$

$72.8 = x$

73 brown M&M's.

45. $\dfrac{110}{28.4} = \dfrac{x}{42.6}$

$110 \cdot 42.6 = 28.4 \cdot x$

$4686 = 28.4x$

$\dfrac{4686}{28.4} = x$

$165 \text{ cal} = x$

47. $\dfrac{35}{1} = \dfrac{x}{13.5}$

$x = 35(13.5)$

$x = 472.5 \text{ miles}$

49. $\dfrac{4045}{12} = 337 \text{ yds/game}$

51. $40 \cdot 9 = 360 \text{ sq. ft}$

53. $\dfrac{20 \cdot 12}{9} = 26 \text{ students}$

55. $\dfrac{343}{7} = \dfrac{x}{5}$

$7x = 1715$

$x = 245 \text{ miles}$

57. $\dfrac{8}{2} = \dfrac{36}{x}$

$8 \cdot x = 2 \cdot 36$

$8x = 72$

$x = 9 \text{ gallons}$

59. $\dfrac{\$153}{8 \text{ hours}} = \$19.125 / \text{hour} \approx \$19.13 / \text{hour}$

61. $\dfrac{\$35,063}{26 \text{ weeks}} = \$1348.58 \text{ per week}$

$\left(6 \text{ months} = \dfrac{1}{2} \text{ year} = \dfrac{1}{2} \times 52 \text{ weeks} = 26 \text{ weeks} \right)$

63. $2976 - 1568 = 1408 \text{ megawatts}$

65. $\dfrac{2976}{1000} \cdot (560,000) = 1,666,560 \text{ people}$

67. $\dfrac{a}{b} = \dfrac{c}{d}$

$ad = bc$

$x = ad, y = bc$

The ratio is 1.

69. $(0, 4), (2, 10)$

$m = \dfrac{10 - 4}{2 - 0} = \dfrac{6}{2} = 3; \text{ upward}$

71. $(-2, 7), (3, -2)$

$m = \dfrac{-2 - 7}{3 - (-2)} = -\dfrac{9}{5}; \text{ downward}$

73. $(0, -4), (2, -4)$

$m = \dfrac{-4 - (-4)}{2 - 0} = \dfrac{0}{2} = 0; \text{ horizontal line}$

Exercise Set 6.8

1. Let $x = $ number

Reciprocal $= \dfrac{1}{x}$

$3\left(\dfrac{1}{x}\right) = 9\left(\dfrac{1}{6}\right)$

$\dfrac{3}{x} = \dfrac{9}{6}$

$3 \cdot 6 = x \cdot 9$

$18 = 9x$

$2 = \dfrac{18}{9} = x$

3. Let $x = $ number

$\dfrac{3x + 3}{x + 1} = \dfrac{3}{2}$

$2 \cdot (2x + 3) = 3 \cdot (x + 1)$

$4x + 6 = 3x + 3$

$4x - 3x = 3 - 6$

$x = -3$

5.

	Time	In one hour
Experienced	4	$\dfrac{1}{4}$
Apprentice	5	$\dfrac{1}{5}$
Together	x	$\dfrac{1}{x}$

$\dfrac{1}{4} + \dfrac{1}{5} = \dfrac{1}{x}$

$20x\left(\dfrac{1}{4} + \dfrac{1}{5}\right) = 20x\left(\dfrac{1}{x}\right)$

$20x\left(\dfrac{1}{4}\right) + 20x\left(\dfrac{1}{5}\right) = 20$

$5x + 4x = 20$

$9x = 20$

$x = \dfrac{20}{9} = 2\dfrac{2}{9}$ hours

7.

	Time	In one minute
Belt	2	$\dfrac{1}{2}$
Smaller	6	$\dfrac{1}{6}$
Together	x	$\dfrac{1}{x}$

$\dfrac{1}{2} + \dfrac{1}{6} = \dfrac{1}{x}$

$6x\left(\dfrac{1}{2} + \dfrac{1}{6}\right) = 6x\left(\dfrac{1}{x}\right)$

$6x\left(\dfrac{1}{2}\right) + 6x\left(\dfrac{1}{6}\right) = 6$

$3x + x = 6$

$4x = 6$

$x = \dfrac{6}{4} = \dfrac{3}{2} = 1\dfrac{1}{2}$ minutes

9.

	distance	= rate · time	
Trip	12	r	x
Return Trip	18	r	$x+1$

$\dfrac{12}{r} = x$

$\dfrac{18}{r} = x+1$

$\dfrac{18}{r} = \dfrac{12}{r} + 1$

$r\left(\dfrac{18}{r}\right) = r\left(\dfrac{12}{r} + 1\right)$

$18 = 12 + r$

$6 = r$

6 mph

11.

	distance	= rate · time	
First part	20	r	$\dfrac{20}{r}$
Cooldown	16	$r-2$	$\dfrac{16}{r-2}$

$\dfrac{20}{r} = \dfrac{16}{r-2}$

$20(r-2) = 16r$

$20r - 40 = 16r$

$20r - 16r = 40$

$4r = 40$

$r = \dfrac{40}{4} = 10$

The first part is at 10 mph, and the cooldown at 8 mph.

13. $\dfrac{4}{12} = \dfrac{x}{18}$

$12 \cdot x = 4 \cdot 18$

$12x = 72$

$x = \dfrac{72}{12} = 6$

15. $\dfrac{x}{3.75} = \dfrac{12}{9}$

$x \cdot 9 = 3.75 \cdot 12$

$9x = 45$

$x = \dfrac{45}{9} = 5$

17. Let x = number

$\dfrac{1}{4} = \dfrac{x}{8}$

$1 \cdot 8 = 4 \cdot x$

$8 = 4x$

$2 = \dfrac{8}{4} = x$

19.

	Time	In one hour
Marcus	6	$\dfrac{1}{6}$
Tony	4	$\dfrac{1}{4}$
Together	x	$\dfrac{1}{x}$

$$\frac{1}{6} + \frac{1}{4} = \frac{1}{x}$$

$$12x\left(\frac{1}{6} + \frac{1}{4}\right) = 12x\left(\frac{1}{x}\right)$$

$$12x\left(\frac{1}{6}\right) + 12x\left(\frac{1}{4}\right) = 12$$

$$2x + 3x = 12$$

$$5x = 12$$

$$x = \frac{12}{5} = 2.4 \text{ hours}$$

$$2.4 \text{ hours } \cdot \$45 / \text{hour} = \$108$$

21. Let x = speed of car in still air.

	distance	= rate	· time
With the wind	11	$x + 3$	$\dfrac{11}{x+3}$
Into wind	10	$x - 3$	$\dfrac{10}{x-3}$

$$\frac{10}{x-3} = \frac{11}{x+3}$$
$$10 \cdot (x+3) = 11 \cdot (x-3)$$
$$10x + 30 = 11x - 33$$
$$30 + 33 = 11x - 10x$$
$$63 \text{ mph} = x$$

23.
$$\frac{10}{16} = \frac{y}{34}$$
$$16 \cdot y = 10 \cdot 34$$
$$16y = 340$$
$$y = \frac{340}{16} = 21.25$$

25. Let x = number

$$\frac{2}{x-3} - \frac{4}{x+3} = 8\left(\frac{1}{x^2 - 9}\right)$$

$$\frac{2}{x-3} - \frac{4}{x+3} = \frac{8}{x^2 - 9}$$

$$(x-3)(x+3)\left[\frac{2}{x-3} - \frac{4}{x+3}\right]$$

$$= (x+3)(x-3)\left[\frac{8}{(x+3)(x-3)}\right]$$

$$(x-3)(x+3)\left(\frac{2}{x-3}\right) - (x-3)(x+3)\left(\frac{4}{x+3}\right) = 8$$

$$2(x+3) - 4(x-3) = 8$$
$$2x + 6 - 4x + 12 = 8$$
$$-2x + 18 = 8$$
$$-2x = 8 - 18$$
$$-2x = -10$$
$$x = \frac{-10}{-2} = 5$$

27. Let x = speed of plane in still air

	distance	= rate	· time
With the wind	630	$x + 35$	$\dfrac{630}{x+35}$
Against the wind	455	$x - 35$	$\dfrac{455}{x-35}$

$$\frac{630}{x+35} = \frac{455}{x-35}$$
$$630(x-35) = 455(x+35)$$
$$630x - 22{,}050 = 455x + 15{,}925$$
$$630x - 455x = 15{,}925 + 22{,}050$$
$$175x = 37{,}975$$
$$x = \frac{37{,}975}{175} = 217 \text{ mph}$$

29. Let x = speed of the wind

	distance	= rate	· time
With the wind	48	$16 + x$	$\dfrac{48}{16+x}$
Against the wind	16	$16 - x$	$\dfrac{16}{16-x}$

$$\frac{48}{16+x} = \frac{16}{16-x}$$
$$48 \cdot (16-x) = 16 \cdot (16+x)$$
$$768 - 48x = 256 + 16x$$
$$768 - 256 = 16x + 48x$$
$$512 = 64x$$
$$\frac{512}{64} = x$$
$$8 \text{ mph} = x$$

$$\frac{1}{20} + \frac{1}{15} + \frac{1}{x} = \frac{1}{6}$$
$$60x\left(\frac{1}{20}\right) + 60x\left(\frac{1}{15}\right) + 60x\left(\frac{1}{x}\right) = 60x\left(\frac{1}{6}\right)$$
$$3x + 4x + 60 = 10x$$
$$7x + 60 = 10x$$
$$60 = 10x - 7x$$
$$60 = 3x$$
$$20 \text{ hours} = x$$

31.

	Time	In one hour
1st custodian	3	$\frac{1}{3}$
2nd custodian	x	$\frac{1}{x}$
Together	$1\frac{1}{2} = \frac{3}{2}$	$\frac{1}{\frac{3}{2}} = \frac{2}{3}$

$$\frac{1}{3} + \frac{1}{x} = \frac{2}{3}$$
$$3x\left(\frac{1}{3} + \frac{1}{x}\right) = 3x\left(\frac{2}{3}\right)$$
$$3x\left(\frac{1}{3}\right) + 3x\left(\frac{1}{x}\right) = 2x$$
$$x + 3 = 2x$$
$$3 = 2x - x$$
$$3 \text{ hours} = x$$

33.

	Time	In one hour
First pipe	20	$\frac{1}{20}$
Second pipe	15	$\frac{1}{15}$
Third pipe	x	$\frac{1}{x}$
Together	6	$\frac{1}{6}$

35. $\frac{x}{11} = \frac{2}{5}$; $\frac{y}{14} = \frac{2}{5}$
$$x \cdot 5 = 11 \cdot 2 \qquad\quad y \cdot 5 = 14 \cdot 2$$
$$5x = 22 \qquad\qquad 5y = 28$$
$$x = \frac{22}{5} = 4.4' \qquad y = \frac{28}{5} = 5.6'$$

37.

	Time	In one hour
First cook	6	$\frac{1}{6}$
Second Cook	7	$\frac{1}{7}$
Third cook	x	$\frac{1}{x}$
Together	2	$\frac{1}{2}$

$$\frac{1}{6} + \frac{1}{7} + \frac{1}{x} = \frac{1}{2}$$
$$42x\left(\frac{1}{6} + \frac{1}{7} + \frac{1}{x}\right) = 42x\left(\frac{1}{2}\right)$$
$$42x\left(\frac{1}{6}\right) + 42x\left(\frac{1}{7}\right) + 42x\left(\frac{1}{x}\right) = 21x$$
$$7x + 6x + 42 = 21x$$
$$13x + 42 = 21x$$
$$42 = 21x - 13x$$
$$42 = 8x$$
$$5\frac{1}{4} \text{ hours} = \frac{42}{8} = x$$

39. Let x = time for faster pump
then $3x$ = time for slower pump

	Time	In one minute
Faster pump	x	$\dfrac{1}{x}$
Slower pump	$3x$	$\dfrac{1}{3x}$
Together	21	$\dfrac{1}{21}$

$$\frac{1}{x} + \frac{1}{3x} = \frac{1}{21}$$

$$21x\left(\frac{1}{x} + \frac{1}{3x}\right) = 21x\left(\frac{1}{21}\right)$$

$$21x\left(\frac{1}{x}\right) + 21x\left(\frac{1}{3x}\right) = x$$

$$21 + 7 = x$$
$$28 = x$$

The faster pump takes 28 minutes and the slower pump takes 84 minutes.

41. Age when son was born.
$$\frac{1}{6}x + \frac{1}{12}x + \frac{1}{7}x + 5$$

$$= \frac{1}{6}(84) + \frac{1}{12}(84) + \frac{1}{7}(84) + 5$$

$$= 14 + 7 + 12 + 5$$
$$= 38 \text{ years}$$

Age of son when he died.
$$\frac{1}{2}x = \frac{1}{2}(84) = 42 \text{ years}$$

43. Answers may vary.

45. $-x + 3y = 6$

If $x = 0$, then
$-0 + 3y = 6$
$3y = 6$
$y = 2$

If $y = 0$, then
$-x + 3(0) = 6$
$-x = 6$
$x = -6$

If $x = -3$, then
$-(-3) + 3y = 6$
$3 + 3y = 6$
$3y = 3$
$y = 1$

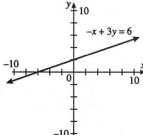

47. $y = 2x$

If $x = 0$, then
$y = 2(0)$
$y = 0$

If $y = 0$, then
$0 = 2x$
$0 = x$

If $x = 1$, then
$y = 2(1)$
$y = 2$

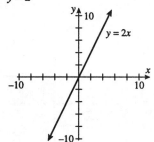

49. $y - x = -5$

If $x = 0$, then
$y - 0 = -5$
$y = -5$

If $y = 0$, then
$0 - x = -5$
$-x = -5$
$x = 5$

If $x = 1$, then
$y - 1 = -5$
$y = -4$

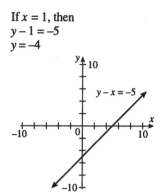

Chapter 6 - Review

1. $\dfrac{x+5}{x^2-4} = \dfrac{x+5}{(x+2)(x-2)}$

Undefined if $x = 2$ or $x = -2$

2. $\dfrac{5x+9}{4x^2-4x-15}$

$4x^2 - 4x - 15 = 0$
$(2x+3)(2x-5) = 0$
$2x + 3 = 0$ or $2x - 5 = 0$
$\dfrac{2x}{2} = -\dfrac{3}{2}$ or $\dfrac{2x}{2} = \dfrac{5}{2}$
$x = -\dfrac{3}{2}$ or $x = \dfrac{5}{2}$

3. $\dfrac{z^2-z}{z+xy}$; $x = 5, y = 7, z = -2$

$\dfrac{(-2)^2-(-2)}{-2+(5)(7)} = \dfrac{4+2}{-2+35} = \dfrac{6}{33} = \dfrac{2}{11}$

4. If $x = 5, y = 7$ and $z = -2$

$\dfrac{x^2+xy-z^2}{x+y+z} = \dfrac{5^2+(5)(7)-(-2)^2}{5+7+(-2)}$

$= \dfrac{25+35-4}{10} = \dfrac{56}{10} = \dfrac{28}{5}$

5. $\dfrac{x+2}{x^2-3x-10} = \dfrac{x+2}{(x-5)(x+2)} = \dfrac{1}{x-5}$

6. $\dfrac{x+4}{x^2+5x+4} = \dfrac{x+4}{(x+4)(x+1)} = \dfrac{1}{x+1}$

7. $\dfrac{x^3-4x}{x^2+3x+2} = \dfrac{x(x^2-4)}{(x+2)(x+1)}$

$= \dfrac{x(x+2)(x-2)}{(x+2)(x+1)} = \dfrac{x(x-2)}{x+1}$

8. $\dfrac{5x^2-125}{x^2+2x-15} = \dfrac{5(x^2-25)}{(x+5)(x-3)}$

$= \dfrac{5(x-5)(x+5)}{(x+5)(x-3)} = \dfrac{5(x-5)}{(x-3)}$

9. $\dfrac{x^2-x-6}{x^2-3x-10} = \dfrac{(x-3)(x+2)}{(x-5)(x+2)} = \dfrac{x-3}{x-5}$

10. $\dfrac{x^2-2x}{x^2+2x-8} = \dfrac{x(x-2)}{(x+4)(x-2)} = \dfrac{x}{x+4}$

11. $\dfrac{x^2+6x+5}{2x^2+11x+5} = \dfrac{(x+5)(x+1)}{(2x+1)(x+5)} = \dfrac{x+1}{2x+1}$

12. $\dfrac{x^2+xa+xb+ab}{x^2-xc+bx-bc} = \dfrac{x(x+a)+b(x+a)}{x(x-c)+b(x-c)}$

$= \dfrac{(x+a)(x+b)}{(x-c)(x+b)} = \dfrac{x+a}{x-c}$

13. $\dfrac{x^2+5x-2x-10}{x^2-3x-2x+6} = \dfrac{x(x+5)-2(x+5)}{x(x-3)-2(x-3)}$

$= \dfrac{(x+5)(x-2)}{(x-3)(x-2)} = \dfrac{x+5}{x-3}$

14. $\dfrac{x^2-9}{9-x^2} = \dfrac{(x-3)(x+3)}{(3-x)(3+x)} = \dfrac{(x-3)(x+3)}{-(x-3)(3+x)}$

$= \dfrac{1}{-1} = -1$

15. $\dfrac{4-x}{x^3-64} = -\dfrac{x-4}{x^3-64} = -\dfrac{x-4}{(x-4)(x^2+4x+16)}$

$= -\dfrac{1}{x^2+4x+16}$

16. $\dfrac{15x^3 \cdot y^2}{z} \cdot \dfrac{z}{5xy^3} = \dfrac{3 \cdot 5x^2y^2 \cdot x}{z} \cdot \dfrac{z}{5xy^2 \cdot y} = \dfrac{3x^2}{y}$

17. $\dfrac{-y^3}{8} \cdot \dfrac{9x^2}{y^3} = -\dfrac{9x^2y^3}{8y^3} = -\dfrac{9x^2}{8}$

18. $\dfrac{x^2-9}{x^2-4}\cdot\dfrac{x-2}{x+3}=\dfrac{(x-3)(x+3)}{(x-2)(x+2)}\cdot\dfrac{(x-2)}{(x+3)}=\dfrac{x-3}{x+2}$

19. $\dfrac{2x+5}{x-6}\cdot\dfrac{2x}{-x+6}=\dfrac{2x+5}{x-6}\cdot-\dfrac{2x}{x-6}=-\dfrac{2x(2x+5)}{(x-6)^2}$

20. $\dfrac{x^2-5x-24}{x^2-x-12}+\dfrac{x^2-10x+16}{x^2+x-6}$

$=\dfrac{(x-8)(x+3)}{(x-4)(x+3)}\cdot\dfrac{(x+3)(x-2)}{(x-8)(x-2)}=\dfrac{x+3}{x-4}$

21. $\dfrac{4x+4y}{xy^2}+\dfrac{3x+3y}{x^2y}=\dfrac{4x+4y}{xy^2}\cdot\dfrac{x^2y}{3x+3y}$

$=\dfrac{4(x+y)}{xy^2}\cdot\dfrac{x^2y}{3(x+y)}=\dfrac{4x^2y}{3xy^2}=\dfrac{4x}{3y}$

22. $\dfrac{x^2+x-42}{x-3}\cdot\dfrac{(x-3)^2}{x+7}$

$=\dfrac{(x+7)(x-6)}{x-3}\cdot\dfrac{(x-3)(x-3)}{x+7}=(x-6)(x-3)$

23. $\dfrac{2a+2b}{3}\cdot\dfrac{a-b}{a^2-b^2}=\dfrac{2(a+b)}{3}\cdot\dfrac{(a-b)}{(a+b)(a-b)}=\dfrac{2}{3}$

24. $\dfrac{x^2-9x+14}{x^2-5x+6}\cdot\dfrac{x+2}{x^2-5x-14}$

$=\dfrac{(x-7)(x-2)}{(x-3)(x-2)}\cdot\dfrac{(x+2)}{(x-7)(x+2)}=\dfrac{1}{x-3}$

25. $(x-3)\cdot\dfrac{x}{x^2+3x-18}=(x-3)\cdot\dfrac{x}{(x+6)(x-3)}$

$=\dfrac{x}{x+6}$

26. $\dfrac{2x^2-9x+9}{8x-12}+\dfrac{x^2-3x}{2x}$

$=\dfrac{(2x-3)(x-3)}{2\cdot2(2x-3)}\cdot\dfrac{2x}{x(x-3)}=\dfrac{1}{2}$

27. $\dfrac{x^2-y^2}{x^2+xy}+\dfrac{3x^2-2xy-y^2}{3x^2+6x}$

$=\dfrac{x^2-y^2}{x^2+xy}\cdot\dfrac{3x^2+6x}{3x^2-2xy-y^2}$

$=\dfrac{(x+y)(x-y)}{x(x+y)}\cdot\dfrac{3x(x+2)}{(3x+y)(x-y)}=\dfrac{3(x+2)}{3x+y}$

28. $\dfrac{x^2-y^2}{8x^2-16xy+8y^2}+\dfrac{x+y}{4x-y}$

$=\dfrac{(x-y)(x+y)}{2\cdot2\cdot2(x-y)(x-y)}\cdot\dfrac{4x-y}{x+y}=\dfrac{4x-y}{8(x-y)}$

29. $\dfrac{x-y}{4}+\dfrac{y^2-2y-xy+2x}{16x+24}$

$=\dfrac{x-y}{4}\cdot\dfrac{16x+24}{y^2-2y-xy+2x}$

$=\dfrac{x-y}{4}\cdot\dfrac{8(2x+3)}{y(y-2)-x(y-2)}$

$=\dfrac{x-y}{4}\cdot\dfrac{8(2x+3)}{(y-2)(y-x)}$

$=-\dfrac{(y-x)}{4}\cdot\dfrac{8(2x+3)}{(y-2)(y-x)}=-\dfrac{2(2x+3)}{y-2}$

30. $\dfrac{y-3}{4x+3}+\dfrac{9-y^2}{4x^2-x-3}=\dfrac{y-3}{4x+3}\cdot\dfrac{(4x+3)(x-1)}{-1(y-3)(y+3)}$

$=\dfrac{x-1}{-(y+3)}=\dfrac{x-1}{-y-3}$

31. $\dfrac{5x-4}{3x-1}+\dfrac{6}{3x-1}=\dfrac{5x-4+6}{3x-1}=\dfrac{5x+2}{3x-1}$

32. $\dfrac{4x-5}{3x^2}-\dfrac{2x+5}{3x^2}=\dfrac{4x-5-2x-5}{3x^2}=\dfrac{2x-10}{3x^2}$

33. $\dfrac{9x+7}{6x^2}-\dfrac{3x+4}{6x^2}=\dfrac{9x+7-(3x+4)}{6x^2}$

$=\dfrac{9x+7-3x-4}{6x^2}=\dfrac{6x+3}{6x^2}=\dfrac{3(2x+1)}{6x^2}$

$=\dfrac{2x+1}{2x^2}$

34. $\dfrac{x+4}{2x},\ \dfrac{3}{7x}$

LCD $=14x$

35. $x^2-5x-24=(x-8)(x+3)$

$x^2+11x+24=(x+8)(x+3)$

LCD $=(x-8)(x+8)(x+3)$

36. $\dfrac{x+2}{x^2+11x+18}=\dfrac{(x+2)(x-5)}{(x+2)(x+9)(x-5)}$

$=\dfrac{x^2-3x-10}{(x+2)(x-5)(x+9)}$

37. $\dfrac{3x-5}{x^2+4x+4} = \dfrac{3x-5}{(x+2)^2} \cdot \dfrac{(x+3)}{(x+3)}$

$= \dfrac{(3x-5)(x+3)}{(x+2)^2(x+3)} = \dfrac{3x^2+4x-15}{(x+2)^2(x+3)}$

38. $\dfrac{4}{5x^2} - \dfrac{6}{y} = \dfrac{4y-30x^2}{5x^2 y}$

39. $\dfrac{2}{x-3} - \dfrac{4}{x-1}$; LCD $= (x-3)(x-1)$

$\dfrac{2}{(x-3)} \cdot \dfrac{(x-1)}{(x-1)} - \dfrac{4}{(x-1)} \cdot \dfrac{(x-3)}{(x-3)}$

$= \dfrac{2(x-1)}{(x-3)(x-1)} - \dfrac{4(x-3)}{(x-1)(x-3)}$

$= \dfrac{2(x-1)-4(x-3)}{(x-3)(x-1)} = \dfrac{2x-2-4x+12}{(x-3)(x-1)}$

$= \dfrac{-2x+10}{(x-3)(x-1)}$

40. $\dfrac{x+7}{x+3} - \dfrac{x-3}{x+7} = \dfrac{(x+7)^2 - (x-3)(x+3)}{(x+3)(x+7)}$

$= \dfrac{x^2+14x+49-x^2+9}{(x+3)(x+7)} = \dfrac{14x+58}{(x+3)(x+7)}$

41. $\dfrac{4}{x+3} - 2$; LCD $= x+3$

$\dfrac{4}{x+3} - 2 \cdot \dfrac{(x+3)}{(x+3)} = \dfrac{4}{x+3} - \dfrac{2(x+3)}{x+3}$

$= \dfrac{4-2(x+3)}{x+3} = \dfrac{4-2x-6}{x+3} = \dfrac{-2x-2}{x+3}$

42. $\dfrac{3}{x^2+2x-8} + \dfrac{2}{x^2-3x+2}$

$= \dfrac{3}{(x+4)(x-2)} + \dfrac{2}{(x-2)(x-1)}$

$= \dfrac{3x-3+2x+8}{(x+4)(x-2)(x-1)} = \dfrac{5x+5}{(x+4)(x-2)(x-1)}$

43. $\dfrac{2x-5}{6x+9} - \dfrac{4}{2x^2+3x} = \dfrac{2x-5}{3(2x+3)} - \dfrac{4}{x(2x+3)}$

LCD $= 3x(2x+3)$

$\dfrac{2x-5}{3(2x+3)} \cdot \dfrac{x}{x} - \dfrac{4}{x(2x+3)} \cdot \dfrac{3}{3}$

$= \dfrac{x(2x-5)}{3x(2x+3)} - \dfrac{4(3)}{3x(2x+3)} = \dfrac{x(2x-5)-4(3)}{3x(2x+3)}$

$= \dfrac{2x^2-5x-12}{3x(2x+3)} = \dfrac{(2x+3)(x-4)}{3x(2x+3)} = \dfrac{x-4}{3x}$

44. $\dfrac{x-1}{x^2-2x+1} - \dfrac{x+1}{x-1} = \dfrac{x-1}{(x-1)(x-1)} - \dfrac{x+1}{x-1}$

$= \dfrac{x-1-(x-1)(x+1)}{(x-1)(x-1)} = \dfrac{x-1-x^2+1}{(x-1)(x-1)}$

$= \dfrac{-x^2+x}{(x-1)(x-1)} = \dfrac{-1(x)(x-1)}{(x-1)(x-1)} = \dfrac{-x}{x-1}$

45. $\dfrac{x-1}{x^2+4x+4} + \dfrac{x-1}{x+2} = \dfrac{x-1}{(x+2)^2} + \dfrac{x-1}{x+2}$

LCD $= (x+2)^2$

$\dfrac{x-1}{(x+2)^2} + \dfrac{(x-1)}{(x+2)} \cdot \dfrac{(x+2)}{(x+2)}$

$= \dfrac{x-1}{(x+2)^2} + \dfrac{(x-1)(x+2)}{(x+2)^2} = \dfrac{x-1+(x-1)(x+2)}{(x+2)^2}$

$= \dfrac{x-1+x^2+x-2}{(x+2)^2} = \dfrac{x^2+2x-3}{(x+2)^2}$

46. $P = 2l + 2w$

$P = 2\left(\dfrac{x}{8}\right) + 2\left(\dfrac{x+2}{4x}\right)$

$P = \dfrac{x}{4} + \dfrac{x+2}{2x}$

$P = \dfrac{x}{4} \cdot \dfrac{x}{x} + \dfrac{(x+2)}{(2x)} \cdot \dfrac{2}{2}$

$P = \dfrac{x^2}{4x} + \dfrac{2(x+2)}{4x}$

$P = \dfrac{x^2+2x+4}{4x}$

$A = lw$

$A = \left(\dfrac{x}{8}\right)\left(\dfrac{x+2}{4x}\right)$

$A = \dfrac{x(x+2)}{8 \cdot 4x}$

$A = \dfrac{x+2}{32}$

47. $P = \dfrac{3x}{4x-4} + \dfrac{2x}{3x-3} + \dfrac{x}{x-1}$

$P = \dfrac{3x}{4(x-1)} \cdot \dfrac{3}{3} + \dfrac{2x}{3(x-1)} \cdot \dfrac{4}{4} + \dfrac{x}{x-1} \cdot \dfrac{12}{12}$

$P = \dfrac{9x}{12(x-1)} + \dfrac{8x}{12(x-1)} + \dfrac{12x}{12(x-1)}$

$P = \dfrac{9x+8x+12x}{12(x-1)}$

$P = \dfrac{29x}{12(x-1)}$

$A = \dfrac{1}{2}bh$

$A = \dfrac{1}{2}\left(\dfrac{x}{x-1}\right)\left(\dfrac{6y}{5}\right)$

$A = \dfrac{6xy}{2 \cdot 5(x-1)}$

$A = \dfrac{3xy}{5(x-1)}$

48. $\dfrac{\frac{5x}{27}}{-\frac{10xy}{21}} = \dfrac{5x}{27} \cdot \dfrac{-21}{10xy} = \dfrac{5x(3)(-7)}{3 \cdot 9 \cdot 2 \cdot 5xy} = -\dfrac{7}{18y}$

49. $\dfrac{\frac{8x}{x^2-9}}{\frac{4}{x+3}} = \dfrac{8x}{x^2-9} \cdot \dfrac{x+3}{4} = \dfrac{8x(x+3)}{(x+3)(x-3)4} = \dfrac{2x}{x-3}$

50. $\dfrac{\frac{3}{5}+\frac{2}{7}}{\frac{1}{5}+\frac{5}{6}} = \dfrac{210\left(\frac{3}{5}+\frac{2}{7}\right)}{210\left(\frac{1}{5}+\frac{5}{6}\right)} = \dfrac{210\left(\frac{3}{5}\right)+210\left(\frac{2}{7}\right)}{210\left(\frac{1}{5}\right)+210\left(\frac{5}{6}\right)}$

$= \dfrac{126+60}{42+175} = \dfrac{186}{217} = \dfrac{6 \cdot 31}{7 \cdot 31} = \dfrac{6}{7}$

51. $\dfrac{\frac{2}{a}+\frac{1}{2a}}{a+\frac{a}{2}} = \dfrac{2a\left(\frac{2}{a}+\frac{1}{2a}\right)}{2a\left(a+\frac{a}{2}\right)} = \dfrac{4+1}{2a^2+a^2} = \dfrac{5}{3a^2}$

52. $\dfrac{3-\frac{1}{y}}{2-\frac{1}{y}} = \dfrac{y\left(3-\frac{1}{y}\right)}{y\left(2-\frac{1}{y}\right)} = \dfrac{y(3)-y\left(\frac{1}{y}\right)}{y(2)-y\left(\frac{1}{y}\right)} = \dfrac{3y-1}{2y-1}$

53. $\dfrac{2+\frac{1}{x^2}}{\frac{1}{x}+\frac{2}{x^2}} = \dfrac{x^2\left(2+\frac{1}{x^2}\right)}{x^2\left(\frac{1}{x}+\frac{2}{x^2}\right)} = \dfrac{x^2(2)+x^2\left(\frac{1}{x^2}\right)}{x^2\left(\frac{1}{x}\right)+x^2\left(\frac{2}{x^2}\right)}$

$= \dfrac{2x^2+1}{x+2}$

54. $\dfrac{\frac{1}{a}+\frac{1}{b}}{\frac{1}{ab}} = \dfrac{ab\left(\frac{1}{a}+\frac{1}{b}\right)}{ab\left(\frac{1}{ab}\right)} = \dfrac{b+a}{1} = b+a$

55. $\dfrac{\frac{6}{x+2}+4}{\frac{8}{x+2}-4} = \dfrac{\frac{6}{x+2}+\frac{4(x+2)}{x+2}}{\frac{8}{x+2}-\frac{4(x+2)}{x+2}} = \dfrac{\frac{6+4(x+2)}{x+2}}{\frac{8-4(x+2)}{x+2}}$

$= \dfrac{6+4(x+2)}{x+2} \cdot \dfrac{x+2}{8-4(x+2)} = \dfrac{6+4x+8}{8-4x-8}$

$= \dfrac{14+4x}{-4x} = \dfrac{2(7+2x)}{2(-2x)} = -\dfrac{7+2x}{2x}$

56. $\dfrac{x+4}{9} = \dfrac{5}{9}$

$9x+36 = 45$

$\dfrac{9x}{9} = \dfrac{9}{9}$

$x = 1$

57. $\dfrac{n}{10} = 9 - \dfrac{n}{5}$

$10\left(\dfrac{n}{10}\right) = 10\left(9-\dfrac{n}{5}\right)$

$n = 10(9) - 10\left(\dfrac{n}{5}\right)$

$n = 90 - 2n$

$n + 2n = 90$

$3n = 90$

$n = \dfrac{90}{3} = 30$

58. $\dfrac{5y-3}{7} = \dfrac{15y-2}{28}$

$28(5y-3) = 7(15y-2)$

$140y - 84 = 105y - 14$

$\dfrac{35y}{35} = \dfrac{70}{35}$

$y = 2$

59. $\dfrac{2}{x+1} - \dfrac{1}{x-2} = -\dfrac{1}{2}$

$2(x+1)(x-2)\left[\dfrac{2}{x+1} - \dfrac{1}{x-2}\right] = 2(x+1)(x-2)\left(-\dfrac{1}{2}\right)$

$2(x+1)(x-2)\left(\dfrac{2}{x+1}\right) - 2(x+1)(x-2)\left(\dfrac{1}{x-2}\right)$

$\qquad\qquad = -(x+1)(x-2)$

$4(x-2) - 2(x+1) = -(x+1)(x-2)$

$4x - 8 - 2x - 2 = -(x^2 - x - 2)$

$2x - 10 = -x^2 + x + 2$

$x^2 + 2x - x - 10 - 2 = 0$

$x^2 + x - 12 = 0$

$(x+4)(x-3) = 0$

$x+4 = 0 \qquad$ or $\qquad x-3 = 0$

$x = -4 \qquad\quad$ or $\qquad x = 3$

60. $\dfrac{1}{a+3} + \dfrac{1}{a-3} = -\dfrac{5}{a^2-9}$

$\dfrac{1}{a+3} + \dfrac{1}{a-3} = -\dfrac{5}{(a-3)(a+3)}$

$[(a-3)(a+3)]\left[\dfrac{1}{a+3} + \dfrac{1}{a-3}\right]$

$\qquad\qquad = (a-3)(a+3)\left[-\dfrac{5}{(a-3)(a+3)}\right]$

$a - 3 + a + 3 = -5$

$\dfrac{2a}{2} = -\dfrac{5}{2} \Rightarrow a = -\dfrac{5}{2}$

61. $\dfrac{y}{2y+2} + \dfrac{2y-16}{4y+4} = \dfrac{y-3}{y+1}$

$\dfrac{y}{2(y+1)} + \dfrac{2(y-8)}{4(y+1)} = \dfrac{y-3}{y+1}$

$\dfrac{y}{2(y+1)} + \dfrac{y-8}{2(y+1)} = \dfrac{y-3}{y+1}$

$2(y+1)\left[\dfrac{y}{2(y+1)} + \dfrac{y-8}{2(y+1)}\right] = 2(y+1)\left[\dfrac{y-3}{y+1}\right]$

$2(y+1)\left[\dfrac{y}{2(y+1)}\right] + 2(y+1)\left[\dfrac{y-8}{2(y+1)}\right] = 2(y-3)$

$y + y - 8 = 2y - 6$

$2y - 8 = 2y - 6$

$2y - 2y = -6 + 8$

$0 = 2 \qquad\qquad$ False

No solution.

62. $\dfrac{4}{x+3} + \dfrac{8}{x^2-9} = 0$

$\Rightarrow \dfrac{4}{x+3} + \dfrac{8}{(x-3)(x+3)} = 0$

$(x-3)(x+3)\left[\dfrac{4}{x+3} + \dfrac{8}{(x-3)(x+3)}\right] = 0(x-3)(x+3)$

$4x - 12 + 8 = 0$

$4x - 4 = 0$

$\dfrac{4x}{4} = \dfrac{4}{4} \Rightarrow x = 1$

63. $\dfrac{2}{x-3} - \dfrac{4}{x+3} = \dfrac{8}{x^2-9}$

$\dfrac{2}{x-3} - \dfrac{4}{x+3} = \dfrac{8}{(x+3)(x-3)}$

$(x+3)(x-3)\left[\dfrac{2}{x-3} - \dfrac{4}{x+3}\right]$

$\qquad = (x+3)(x-3)\left[\dfrac{8}{(x+3)(x-3)}\right]$

$(x+3)(x-3)\left(\dfrac{2}{x-3}\right) - (x+3)(x-3)\left(\dfrac{4}{x+3}\right) = 8$

$2(x+3) - 4(x-3) = 8$

$2x + 6 - 4x + 12 = 8$

$-2x + 18 = 8$

$-2x = 8 - 18$

$-2x = -10$

$x = \dfrac{-10}{-2} = 5$

64. $\dfrac{x-3}{x+1} - \dfrac{x-6}{x+5} = 0$

$(x+1)(x+5)\left[\dfrac{x-3}{x+1} - \dfrac{x-6}{x+5}\right] = 0(x+1)(x+5)$

$(x-3)(x+5) - (x-6)(x+1) = 0$

$x^2 + 2x - 15 - x^2 + 5x + 6 = 0$

$7x - 9 = 0$

$7x = 9$

$\dfrac{7x}{7} = \dfrac{9}{7}$

$x = \dfrac{9}{7}$

65. $x + 5 = \dfrac{6}{x}$

$$x(x+5) = x\left(\dfrac{6}{x}\right)$$

$$x^2 + 5x = 6$$

$$x^2 + 5x - 6 = 0$$

$$(x+6)(x-1) = 0$$

$x + 6 = 0$ or $x - 1 = 0$

$x = -6$ or $x = 1$

66. $\dfrac{4A}{b \cdot 5} = x^2$

$$\dfrac{5bx^2}{5x^2} = \dfrac{4A}{5x^2}$$

$$b = \dfrac{4A}{5x^2}$$

67. $\dfrac{x}{7} + \dfrac{y}{8} = 10$

$$56\left(\dfrac{x}{7} + \dfrac{y}{8}\right) = 56(10)$$

$$56\left(\dfrac{x}{7}\right) + 56\left(\dfrac{y}{8}\right) = 560$$

$$8x + 7y = 560$$

$$7y = 560 - 8x$$

$$y = \dfrac{560 - 8x}{7}$$

68. $\dfrac{1}{5}$

69. $\dfrac{4}{6} = \dfrac{2}{3}$

70. $\dfrac{x}{2} = \dfrac{12}{4}$

$$\dfrac{4x}{4} = \dfrac{24}{4}$$

$$x = 6$$

71. $\dfrac{20}{1} = \dfrac{x}{25}$

$$1 \cdot x = 20 \cdot 25$$

$$x = 500$$

72. $\dfrac{32}{100} = \dfrac{100}{x}$

$$\dfrac{32x}{32} = \dfrac{10,000}{32}$$

$$x = 312.5$$

73. $\dfrac{20}{2} = \dfrac{c}{5}$

$$2 \cdot c = 20 \cdot 5$$

$$2c = 100$$

$$c = \dfrac{100}{2} = 50$$

74. $\dfrac{2}{x-1} = \dfrac{3}{x+3}$

$$2x + 6 = 3x - 3$$

$$9 = x$$

75. $\dfrac{4}{y-3} = \dfrac{2}{y-3}$

$$4 \cdot (y-3) = 2 \cdot (y-3)$$

$$4y - 12 = 2y - 6$$

$$4y - 2y = -6 + 12$$

$$2y = 6$$

$$y = \dfrac{6}{2} = 3$$

3 is an extraneous solution.

If $y = 3$, the denominator would be zero.

76. $\dfrac{y+2}{y} = \dfrac{5}{3}$

$$3y + 6 = 5y$$

$$\dfrac{6}{2} = \dfrac{2y}{2}$$

$$3 = y$$

77. $\dfrac{x-3}{3x+2} = \dfrac{2}{6}$

$$6 \cdot (x-3) = 2 \cdot (3x+2)$$

$$6x - 18 = 6x + 4$$

$$6x - 6x = 4 + 18$$

$$0 = 22 \qquad \text{False}$$

No solution.

78. $\dfrac{1.29}{10} = 0.129$

$$\dfrac{2.15}{16} = 0.134$$

The best buy is 10 oz. for $1.29.

79. $\dfrac{0.89}{8} = 0.111$

$\dfrac{1.63}{15} = 0.109$

$\dfrac{2.36}{20} = 0.118$

The best buy is 15 oz. for $1.63.

80. $\dfrac{300}{20} = \dfrac{x}{45}$

$\dfrac{20x}{20} = \dfrac{13,500}{20}$

$x = 675$ parts

81. $\dfrac{90.00}{8} = \dfrac{x}{3}$

$3 \cdot (90.00) = 8x$

$270.00 = 8x$

$\dfrac{270.00}{8} = x$

$\$33.75 = x$

82. $\dfrac{100}{35} = \dfrac{x}{55}$

$\dfrac{35x}{35} = \dfrac{5500}{35}$

$x = \dfrac{5500}{35}$

$x = 157\dfrac{1}{7}$ or 157 letters

83. Let $x =$ number

$5\left(\dfrac{1}{x}\right) = \dfrac{3}{2}\left(\dfrac{1}{x}\right) + \dfrac{7}{6}$

$\dfrac{5}{x} = \dfrac{3}{2x} + \dfrac{7}{6}$

$6x\left(\dfrac{5}{x}\right) = 6x\left(\dfrac{3}{2x} + \dfrac{7}{6}\right)$

$30 = 6x\left(\dfrac{3}{2x}\right) + 6x\left(\dfrac{7}{6}\right)$

$30 = 9 + 7x$

$30 - 9 = 7x$

$21 = 7x$

$3 = \dfrac{21}{7} = x$

84. $\dfrac{1}{x} = \dfrac{1}{4-x}$

$x = 4 - x$

$\dfrac{2x}{2} = \dfrac{4}{2} \Rightarrow x = 2$

85.

	distance = rate · time		
Slower car	60	$x - 10$	$\dfrac{60}{x-10}$
Faster car	90	x	$\dfrac{90}{x}$

$\dfrac{60}{x-10} = \dfrac{90}{x}$

$60 \cdot x = 90 \cdot (x - 10)$

$60x = 90x - 900$

$60x - 90x = -900$

$-30x = -900$

$x = \dfrac{-900}{-30} = 30$

The speed of the faster car is 30 mph and the speed of the slower car is 20 mph.

86.

	distance = rate · time		
Upstream	48	$x - 4$	$\dfrac{48}{x-4}$
Downstream	72	$x + 4$	$\dfrac{72}{x+4}$

$\dfrac{48}{x-4} = \dfrac{72}{x+4}$

$48(x + 4) = 72(x - 4)$

$48x + 192 = 72x - 288$

$480 = 24x$

$\dfrac{480}{24} = \dfrac{24x}{24}$

$20 \text{ mph} = x$

87.

	Time	In one hour
Mark	7	$\dfrac{1}{7}$
Maria	x	$\dfrac{1}{x}$
Together	5	$\dfrac{1}{5}$

$$\frac{1}{7}+\frac{1}{x}=\frac{1}{5}$$

$$35x\left(\frac{1}{7}+\frac{1}{x}\right)=35x\left(\frac{1}{5}\right)$$

$$35x\left(\frac{1}{7}\right)+35x\left(\frac{1}{x}\right)=7x$$

$$5x+35=7x$$
$$35=7x-5x$$
$$35=2x$$

$$17\frac{1}{2}\text{ hours}=\frac{35}{2}=x$$

88. $\dfrac{1}{20}+\dfrac{1}{15}=\dfrac{1}{x}$

$$15x+20x=(20)(15)$$

$$\frac{35x}{35}=\frac{300}{35}$$

$$x=8\frac{4}{7}\text{ days}$$

Chapter 6 - Test

1. $\dfrac{x+5}{x^2+4x+3}=\dfrac{x+5}{(x+3)(x+1)}$

Undefined if $x=-3$ or $x=-1$

2. a. $C=\dfrac{100x+3000}{x}$

$$C=\frac{100(200)+3000}{200}=\$115$$

b. $C=\dfrac{100x+3000}{x}$

$$C=\frac{100(1000)+3000}{1000}=\$103$$

3. $\dfrac{3x-6}{5x-10}=\dfrac{3(x-2)}{5(x-2)}=\dfrac{3}{5}$

4. $\dfrac{x+10}{x^2-100}=\dfrac{x+10}{(x+10)(x-10)}=\dfrac{1}{x-10}$

5. $\dfrac{x+6}{x^2+12x+36}=\dfrac{x+6}{(x+6)(x+6)}=\dfrac{1}{x+6}$

6. $\dfrac{x+3}{x^3+27}=\dfrac{x+3}{(x+3)(x^2-3x+9)}=\dfrac{1}{x^2-3x+9}$

7. $\dfrac{2m^3-2m^2-12m}{m^2-5m+6}$

$$=\frac{2m(m^2-m-6)}{(m-3)(m-2)}$$

$$=\frac{2m(m-3)(m+2)}{(m-3)(m-2)}$$

$$=\frac{2m(m+2)}{m-2}$$

8. $\dfrac{ay+3a+2y+6}{ay+3a+5y+15}$

$$=\frac{a(y+3)+2(y+3)}{a(y+3)+5(y+3)}$$

$$=\frac{(y+3)(a+2)}{(y+3)(a+5)}$$

$$=\frac{a+2}{a+5}$$

9. $\dfrac{y-x}{x^2-y^2}$

$$=-\frac{x-y}{x^2-y^2}$$

$$=-\frac{x-y}{(x+y)(x-y)}$$

$$=-\frac{1}{x+y}$$

10. $\dfrac{x^2-13x+42}{x^2+10x+21}+\dfrac{x^2-4}{x^2+x-6}$

$$=\frac{x^2-13x+42}{x^2+10x+21}\cdot\frac{x^2+x-6}{x^2-4}$$

$$=\frac{(x-6)(x-7)}{(x+7)(x+3)}\cdot\frac{(x+3)(x-2)}{(x-2)(x+2)}$$

$$=\frac{(x-6)(x-7)}{(x+7)(x+2)}$$

11. $\dfrac{3}{x-1} \cdot (5x - 5) = \dfrac{3}{x-1} \cdot 5(x-1) = 15$

12. $\dfrac{y^2 - 5y + 6}{2y + 4} \cdot \dfrac{y+2}{2y-6}$

$= \dfrac{(y-3)(y-2)}{2(y+2)} \cdot \dfrac{y+2}{2(y-3)}$

$= \dfrac{y-2}{4}$

13. $\dfrac{5}{2x+5} - \dfrac{6}{2x+5} = \dfrac{5-6}{2x+5} = \dfrac{-1}{2x+5}$

14. $\dfrac{5a}{a^2 - a - 6} - \dfrac{2}{a-3}$

$= \dfrac{5a}{(a-3)(a+2)} - \dfrac{2}{a-3}$

LCD $= (a-3)(a+2)$

$\dfrac{5a}{(a-3)(a+2)} - \dfrac{2}{(a-3)} \cdot \dfrac{a+2}{a+2}$

$= \dfrac{5a}{(a-3)(a+2)} - \dfrac{2(a+2)}{(a-3)(a+2)}$

$= \dfrac{5a - 2(a+2)}{(a-3)(a+2)}$

$= \dfrac{5a - 2a - 4}{(a-3)(a+2)}$

$= \dfrac{3a - 4}{(a-3)(a+2)}$

15. $\dfrac{6}{x^2 - 1} + \dfrac{3}{x+1}$

$= \dfrac{6}{(x+1)(x-1)} + \dfrac{3}{x+1}$

LCD $= (x+1)(x-1)$

$\dfrac{6}{(x+1)(x-1)} + \dfrac{3}{(x+1)} \cdot \dfrac{x-1}{x-1}$

$= \dfrac{6}{(x+1)(x-1)} + \dfrac{3(x-1)}{(x+1)(x-1)}$

$= \dfrac{6 + 3(x-1)}{(x+1)(x-1)}$

$= \dfrac{6 + 3x - 3}{(x+1)(x-1)}$

$= \dfrac{3x + 3}{(x+1)(x-1)} = \dfrac{3(x+1)}{(x+1)(x-1)} = \dfrac{3}{x-1}$

16. $\dfrac{x^2 - 9}{x^2 - 3x} + \dfrac{xy + 5x + 3y + 15}{2x + 10}$

$= \dfrac{x^2 - 9}{x^2 - 3x} \cdot \dfrac{2x + 10}{xy + 5x + 3y + 15}$

$= \dfrac{(x+3)(x-3)}{x(x-3)} \cdot \dfrac{2(x+5)}{x(y+5) + 3(y+5)}$

$= \dfrac{(x+3)(x-3)}{x(x-3)} \cdot \dfrac{2(x+5)}{(y+5)(x+3)} = \dfrac{2(x+5)}{x(y+5)}$

17. $\dfrac{x+2}{x^2 + 11x + 18} + \dfrac{5}{x^2 - 3x - 10}$

$= \dfrac{x+2}{(x+9)(x+2)} + \dfrac{5}{(x-5)(x+2)}$

LCD $= (x+9)(x+2)(x-5)$

$\dfrac{(x+2)}{(x+9)(x+2)} \cdot \dfrac{x-5}{x-5} + \dfrac{5}{(x-5)(x+2)} \cdot \dfrac{x+9}{x+9}$

$= \dfrac{(x+2)(x-5)}{(x+9)(x+2)(x-5)} + \dfrac{5(x+9)}{(x-5)(x+2)(x+9)}$

$= \dfrac{x^2 - 3x - 10}{(x+9)(x+2)(x-5)} + \dfrac{5x + 45}{(x+9)(x+2)(x-5)}$

$= \dfrac{x^2 - 3x - 10 + 5x + 45}{(x+9)(x+2)(x-5)}$

$= \dfrac{x^2 + 2x + 35}{(x+9)(x+2)(x-5)}$

18. $\dfrac{4y}{y^2 + 6y + 5} - \dfrac{3}{y^2 + 5y + 4}$

$= \dfrac{4y}{(y+5)(y+1)} - \dfrac{3}{(y+4)(y+1)}$

LCD $= (y+5)(y+1)(y+4)$

$\dfrac{4y}{(y+5)(y+1)} \cdot \dfrac{(y+4)}{(y+4)} - \dfrac{3}{(y+4)(y+1)} \cdot \dfrac{(y+5)}{(y+5)}$

$= \dfrac{4y(y+4)}{(y+5)(y+1)(y+4)} - \dfrac{3(y+5)}{(y+4)(y+1)(y+5)}$

$= \dfrac{4y(y+4) - 3(y+5)}{(y+5)(y+1)(y+4)} = \dfrac{4y^2 + 16y - 3y - 15}{(y+5)(y+1)(y+4)}$

$= \dfrac{4y^2 + 13y - 15}{(y+5)(y+1)(y+4)}$

19. $\dfrac{4}{y} - \dfrac{5}{3} = \dfrac{-1}{5}$

$15y\left(\dfrac{4}{y} - \dfrac{5}{3}\right) = 15y\left(-\dfrac{1}{5}\right)$

$15y\left(\dfrac{4}{y}\right) + 15y\left(-\dfrac{5}{3}\right) = -3y$

$60 - 25y = -3y$

$60 = -3y + 25y$

$60 = 22y$

$\dfrac{30}{11} = \dfrac{60}{22} = y$

20. $\dfrac{5}{y+1} = \dfrac{4}{y+2}$

$5 \cdot (y+2) = 4 \cdot (y+1)$

$5y + 10 = 4y + 4$

$5y - 4y = 4 - 10$

$y = -6$

21. $\dfrac{a}{a-3} = \dfrac{3}{a-3} - \dfrac{3}{2}$

$2(a-3)\left(\dfrac{a}{a-3}\right) = 2(a-3)\left(\dfrac{3}{a-3} - \dfrac{3}{2}\right)$

$2(a-3)\left(\dfrac{a}{a-3}\right) = 2(a-3)\left(\dfrac{3}{a-3}\right) + 2(a-3)\left(-\dfrac{3}{2}\right)$

$2a = 6 - 3(a-3)$

$2a = 6 - 3a + 9$

$2a = 15 - 3a$

$2a + 3a = 15$

$5a = 15$

$\dfrac{5a}{5} = \dfrac{15}{5}$

$a = 3$

3 is an extraneous solution. If $x = 3$, the denominator would be zero.

22. $\dfrac{10}{x^2 - 25} = \dfrac{3}{x+5} + \dfrac{1}{x-5}$

$\dfrac{10}{(x+5)(x-5)} = \dfrac{3}{x+5} + \dfrac{1}{x-5}$

$(x+5)(x-5)\left[\dfrac{10}{(x+5)(x-5)}\right]$

$\quad = (x+5)(x-5)\left[\dfrac{3}{x+5} + \dfrac{1}{x-5}\right]$

$10 = (x+5)(x-5)\left(\dfrac{3}{x+5}\right) + (x+5)(x-5)\left(\dfrac{1}{x-5}\right)$

$10 = 3(x-5) + 1(x+5)$

$10 = 3x - 15 + x + 5$

$10 = 4x - 10$

$10 + 10 = 4x$

$20 = 4x$

$5 = \dfrac{20}{4} = x$

5 is an extraneous solution. If $x = 5$, the denominator would be zero.

23. $\dfrac{\frac{5x^2}{yz^2}}{\frac{10x}{z^3}} = \dfrac{5x^2}{yz^2} \cdot \dfrac{z^3}{10x} = \dfrac{xz}{2y}$

24. $\dfrac{\frac{b}{a} - \frac{a}{b}}{\frac{b}{a} + \frac{b}{a}} = \dfrac{ab\left(\frac{b}{a} - \frac{a}{b}\right)}{ab\left(\frac{b}{a} + \frac{b}{a}\right)} = \dfrac{ab\left(\frac{b}{a}\right) - ab\left(\frac{a}{b}\right)}{ab\left(\frac{b}{a}\right) + ab\left(\frac{b}{a}\right)}$

$= \dfrac{b^2 - a^2}{b^2 + b^2} = \dfrac{b^2 - a^2}{2b^2}$

25. $\dfrac{5 - \frac{1}{y^2}}{\frac{1}{y} + \frac{2}{y^2}} = \dfrac{\frac{5y^2}{y^2} - \frac{1}{y^2}}{\frac{y}{y^2} + \frac{2}{y^2}} = \dfrac{\frac{5y^2-1}{y^2}}{\frac{y+2}{y^2}} = \dfrac{5y^2-1}{y^2} \cdot \dfrac{y^2}{y+2}$

$= \dfrac{5y^2 - 1}{y+2}$

26. $\dfrac{3}{85} = \dfrac{x}{510}$

$\dfrac{3(510)}{85} = x$

$18 = x$

27. Let $x = $ number

$x + 5\left(\dfrac{1}{x}\right) = 6$

$x + \dfrac{5}{x} = 6$

$x\left(x + \dfrac{5}{x}\right) = x(6)$

$x(x) + x\left(\dfrac{5}{x}\right) = 6x$

$x^2 + 5 = 6x$

$x^2 - 6x + 5 = 0$

$(x-1)(x-5) = 0$

$x - 1 = 0 \quad$ or $\quad x - 5 = 0$

$x = 1 \quad\quad$ or $\quad x = 5$

28. Let x = speed of boat in still water

	distance = rate · time		
downstream	16	$x+2$	$\dfrac{16}{x+2}$
upstream	14	$x-2$	$\dfrac{14}{x-2}$

$$\frac{16}{x+2} = \frac{14}{x-2}$$
$$16 \cdot (x-2) = 14 \cdot (x+2)$$
$$16x - 32 = 14x + 28$$
$$16x - 14x = 28 + 32$$
$$2x = 60$$
$$x = \frac{60}{2} = 30 \text{ mph}$$

29.

	Time	In one hour
First pipe	12	$\dfrac{1}{12}$
Second pipe	15	$\dfrac{1}{15}$
Together	x	$\dfrac{1}{x}$

$$\frac{1}{12} + \frac{1}{15} = \frac{1}{x}$$
$$60x\left(\frac{1}{12} + \frac{1}{15}\right) = 60x\left(\frac{1}{x}\right)$$
$$60x\left(\frac{1}{12}\right) + 60x\left(\frac{1}{15}\right) = 60x\left(\frac{1}{x}\right)$$
$$5x + 4x = 60$$
$$9x = 60$$
$$x = \frac{60}{9} = \frac{20}{3} = 6\frac{2}{3} \text{ hours}$$

30. $\dfrac{1.19}{6} = 0.198$

$\dfrac{2.15}{10} = 0.215$

$\dfrac{3.25}{16} = 0.203$

The best buy is 6 oz. for $1.19.

Chapter 6 - Cumulative Review

1. a. $\dfrac{15}{x} = 4$

   b. $12 - 3 = x$

   c. $4x + 17 = 21$

2. a. $-3 + (-7) = -10$

   b. $5 + (+12) = 17$

   c. $(-1) + (-20) = -21$

   d. $-2 + (-10) = -12$

3. a. Commutative property for multiplication

   b. Distributive property

   c. Associative property for addition

4. $3 - x = 7$
$$3 - 3 - x = 7 - 3$$
$$-x = 4$$
$$\frac{-x}{-1} = \frac{4}{-1}$$
$$x = -4$$

5. $2x + 7 = x - 3$
$$2x - x + 7 = x - x - 3$$
$$x + 7 = -3$$
$$x + 7 - 7 = -3 - 7$$
$$x = -10$$

6. $P = 2L + 2W$
$$P - 2L = 2L - 2L + 2W$$
$$P - 2L = 2W$$
$$\frac{P - 2L}{2} = \frac{2W}{2}$$
$$\frac{P - 2L}{2} = W$$

7. $x + 4 \le -6$
$$x + 4 - 4 \le -6 - 4$$
$$x \le -10$$

8. $y = 3x$
If $x = 0$, then
$y = 3(0)$
$y = 0$

If $x = 1$, then
$y = 3(1)$
$y = 3$

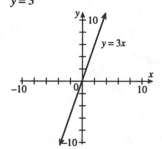

9. a. $\dfrac{x^5}{x^2} = x^{5-2} = x^3$

b. $\dfrac{4^7}{4^3} = 4^{7-3} = 4^4 = 256$

c. $\dfrac{(-3)^5}{(-3)^2} = (-3)^{5-2} = (-3)^3 = -27$

d. $\dfrac{2x^5 y^2}{xy} = 2x^{5-1} y^{2-1} = 2x^4 y$

10. a. $5x(2x^3 + 6) = 5x(2x^3) + 5x(6) = 10x^4 + 30x$

b. $-3x^2(5x^2 + 6x - 1)$
$= -3x^2(5x^2) - 3x^2(6x) - 3x^2(-1)$
$= -15x^4 - 18x^3 + 3x^2$

c. $(3n^2 - 5n + 4)(2n)$
$= 3n^2(2n) - 5n(2n) + 4(2n)$
$= 6n^3 - 10n^2 + 8n$

11. a. $\dfrac{y}{y^{-2}} = y^{1+2} = y^3$

b. $\dfrac{p^{-4}}{q^{-9}} = \dfrac{q^9}{p^4}$

c. $\dfrac{x^{-5}}{x^7} = \dfrac{1}{x^{7+5}} = \dfrac{1}{x^{12}}$

12.
$$x^2 + 1 \overline{\smash{\big)}\, 2x^4 - x^3 + 3x^2 + x - 1}$$

quotient: $2x^2 - x + 1$

$\underline{2x^4 \qquad + 2x^2}$
$\qquad -x^3 + x^2 + x$
$\qquad \underline{-x^3 \qquad - x}$
$\qquad\qquad x^2 + 2x - 1$
$\qquad\qquad \underline{x^2 \qquad + 1}$
$\qquad\qquad\qquad 2x - 2$

Answer: $2x^2 - x + 1 + \dfrac{2x - 2}{x^2 + 1}$

13. $x^2 + 7x + 12 = (x + 3)(x + 4)$

14. $x^3 + 8 = (x + 2)(x^2 - 2x + 4)$

15. $2x^3 - 4x^2 - 30x = 0$
$2x(x^2 - 2x - 15) = 0$
$2x(x - 5)(x + 3) = 0$ or
$2x = 0$ or $x - 5 = 0$ or $x + 3 = 0$
$x = 0$ or $x = 5$ or $x = -3$

16. $\dfrac{x^2 + x}{3x} \cdot \dfrac{6}{5x + 5} = \dfrac{x(x+1)}{3x} \cdot \dfrac{6}{5(x+1)} = \dfrac{2}{5}$

17. $\dfrac{3x^2 + 2x}{x - 1} - \dfrac{10x - 5}{x - 1} = \dfrac{3x^2 + 2x - (10x - 5)}{x - 1}$
$= \dfrac{3x^2 + 2x - 10x + 5}{x - 1} = \dfrac{3x^2 - 8x + 5}{x - 1}$
$= \dfrac{(3x - 5)(x - 1)}{x - 1} = 3x - 5$

18. $\dfrac{6x}{x^2 - 4} - \dfrac{3}{x + 2}$

$= \dfrac{6x}{(x + 2)(x - 2)} - \dfrac{3}{x + 2} \cdot \dfrac{x - 2}{x - 2}$

$= \dfrac{6x - 3(x - 2)}{(x + 2)(x - 2)}$

$= \dfrac{6x - 3x + 6}{(x + 2)(x - 2)}$

$= \dfrac{3x + 6}{(x + 2)(x - 2)}$

$= \dfrac{3(x + 2)}{(x + 2)(x - 2)}$

$= \dfrac{3}{x - 2}$

19. $\dfrac{\frac{1}{z} - \frac{1}{2}}{\frac{1}{3} - \frac{z}{6}}$

$= \dfrac{\frac{1}{z} \cdot \frac{2}{2} - \frac{1}{2} \cdot \frac{z}{z}}{\frac{1}{3} \cdot \frac{2}{2} - \frac{z}{6}}$

$= \dfrac{\frac{2}{2z} - \frac{z}{2z}}{\frac{2}{6} - \frac{z}{6}}$

$= \dfrac{\frac{2-z}{2z}}{\frac{2-z}{6}} = \dfrac{2-z}{2z} \cdot \dfrac{6}{2-z} = \dfrac{3}{z}$

20. $\dfrac{t-4}{2} - \dfrac{t-3}{9} = \dfrac{5}{18}$

$\dfrac{t-4}{2} \cdot \dfrac{9}{9} - \dfrac{t-3}{9} \cdot \dfrac{2}{2} = \dfrac{5}{18}$

$\dfrac{9(t-4)}{18} - \dfrac{2(t-3)}{18} = \dfrac{5}{18}$

$\dfrac{9t - 36 - 2t + 6}{18} = \dfrac{5}{18}$

$\dfrac{7t - 30}{18} = \dfrac{5}{18}$

$7t - 30 = 5$

$7t = 35$

$t = 5$

21.

	Hours to Complete Total Job	Part of Job Completed in 1 hr
Sam	3	$\dfrac{1}{3}$
Frank	7	$\dfrac{1}{7}$
Together	x	$\dfrac{1}{x}$

$\dfrac{1}{3} + \dfrac{1}{7} = \dfrac{1}{x}$

$21x\left(\dfrac{1}{3}\right) + 21x\left(\dfrac{1}{7}\right) = 21x\left(\dfrac{1}{x}\right)$

$7x + 3x = 21$

$10x = 21$

$x = \dfrac{21}{10}$ or $2\dfrac{1}{10}$ hours

Chapter 7

Section 7.1 Mental Math

1. $y = 2x - 1$; $m = 2$; $(0, -1)$

3. $y = x + \dfrac{1}{3}$; $m = 1$; $\left(0, \dfrac{1}{3}\right)$

5. $y = \dfrac{5}{7}x - 4$; $m = \dfrac{5}{7}$; $(0, -4)$

Exercise Set 7.1

1. $2x + y = 4$
$y = -2x + 4$
$m = -2$; $(0, 4)$

3. $x + 9y = 1$
$9y = -x + 1$
$y = -\dfrac{1}{9}x + \dfrac{1}{9}$
$m = -\dfrac{1}{9}$; $\left(0, \dfrac{1}{9}\right)$

5. $4x - 3y = 12$
$-3y = -4x + 12$
$y = \dfrac{-4x}{-3} + \dfrac{12}{-3}$
$y = \dfrac{4}{3}x - 4$
$m = \dfrac{4}{3}$; $(0, -4)$

7. $x + y = 0$
$y = -x$
$m = -1$; $(0, 0)$

9. $y = -3$
$m = 0$; $(0, -3)$

11. $-x + 5y = 20$
$5y = x + 20$
$y = \dfrac{x}{5} + \dfrac{20}{5}$
$y = \dfrac{1}{5}x + 4$
$m = \dfrac{1}{5}$; $(0, 4)$

13. $m = 2$; y-intercept $= (0, 1)$; B

15. $m = -3$; y-intercept $= (0, -2)$; D

17. $x - 3y = -6$
$-3y = -x - 6$
$y = \dfrac{1}{3}x + 2$
$m = \dfrac{1}{3}$
$3x - y = 0$
$-y = -3x + 0$
$y = 3x$
$m = 3$
Neither; the slopes are not the same, and their product is not -1.

19. $2x - 7y = 1$
$-7y = -2x + 1$
$y = \dfrac{2}{7}x - \dfrac{1}{7}$
$m = \dfrac{2}{7}$
$2y = 7x - 2$
$y = \dfrac{7}{2}x - 1$
$m = \dfrac{7}{2}$
Neither; the slopes are not the same, and their product is not -1.

21. $10 + 3x = 5y$
$y = \dfrac{3}{5}x + 2$
$m = \dfrac{3}{5}$
$5x + 3y = 1$
$3y = -5x + 1$
$y = -\dfrac{5}{3}x + \dfrac{1}{3}$
$m = -\dfrac{5}{3}$
Perpendicular; the product of the two slopes is -1.

23. $6x = 5y + 1$
$-5y = -6x + 1$
$y = \dfrac{6}{5}x - \dfrac{1}{5}$
$m = \dfrac{6}{5}$; y-intercept $= \left(0, -\dfrac{1}{5}\right)$
$-12x + 10y = 1$
$10y = 12x + 1$
$y = \dfrac{6}{5}x + \dfrac{1}{10}$
$m = \dfrac{6}{5}$; y-intercept $= \left(0, \dfrac{1}{10}\right)$
Parallel; the slopes are the same and their y-intercepts are different.

25. Answers may vary.

27. $y = -x + 1$

29. $y = 2x + \dfrac{3}{4}$

31. $y = \dfrac{2}{7}x$

33. $y = \dfrac{2}{3}x + 5$
y-intercept $= (0, 5)$
slope $= \dfrac{2}{3}$
another point $= (3, 7)$

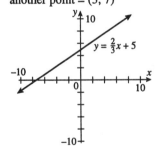

35. $y = -\dfrac{3}{5}x - 2$
y-intercept $= (0, -2)$
slope $= -\dfrac{3}{5}$
another point $= (5, -5)$

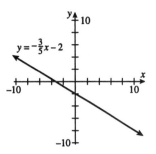

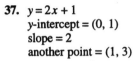

37. $y = 2x + 1$
y-intercept $= (0, 1)$
slope $= 2$
another point $= (1, 3)$

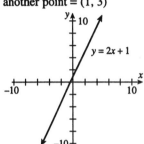

39. $y = -5x$
y-intercept $= (0, 0)$
slope $= -5$
another point $= (1, -5)$

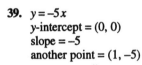

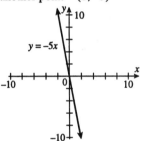

41. $4x + y = 6$
$y = -4x + 6$
y-intercept $= (0, 6)$
slope $= -4$
another point $= (1, 2)$

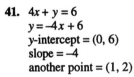

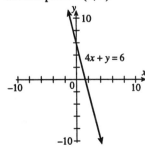

43. $x - y = -2$
$-y = -x - 2$
$y = x + 2$
y-intercept = (0, 2)
slope = 1
another point = (1, 3)

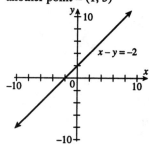

45. $3x + 5y = 10$
$5y = -3x + 10$
$y = -\dfrac{3}{5}x + 2$
y-intercept = (0, 2)
slope = $-\dfrac{3}{5}$
another point = (5, -1)

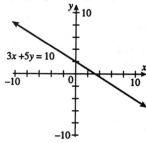

47. $4x - 7y = -14$
$-7y = -4x - 14$
$y = \dfrac{4}{7}x + 2$
y-intercept = (0, 2)
slope = $\dfrac{4}{7}$
another point = (7, 6)

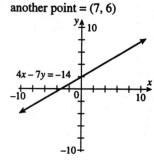

49. Answers may vary.

51. (0, 273) and (100, 373)
$m = \dfrac{373 - 273}{100 - 0} = \dfrac{100}{100} = 1$
intercept = (0, 273)
$K = C + 273$

53. $y = -1$

55. $x = 2$

57. $y - 2 = 3(x - 6)$
$y - 2 = 3x - 18$
$-3x + y - 2 = -18$
$-3x + y = -16$
$3x - y = 16$

59. $y + 4 = -6(x + 1)$
$y + 4 = -6x - 6$
$6x + y + 4 = -6$
$6x + y = -10$

Section 7.2 Mental Math

1. $y - 8 = 3(x - 4)$
$m = 3$; answers may vary. One example is (4, 8).

3. $y + 3 = -2(x - 10)$
$m = -2$; answers may vary. One example is (10, -3).

5. $y = \dfrac{2}{5}(x + 1)$
$m = \dfrac{2}{5}$; answers may vary. One example is (-1, 0).

Exercise Set 7.2

1. Slope 6; through (2, 2)

$$y - y_1 = m(x - x_1)$$
$$y - 2 = 6(x - 2)$$
$$y - 2 = 6x - 12$$
$$-2 = 6x - y - 12$$
$$-2 + 12 = 6x - y$$
$$10 = 6x - y$$
$$6x - y = 10$$

3. Slope –8; through (–1, –5)

$$y - y_1 = m(x - x_1)$$
$$y - (-5) = -8[x - (-1)]$$
$$y + 5 = -8(x + 1)$$
$$y + 5 = -8x - 8$$
$$8x + y + 5 = -8$$
$$8x + y = -8 - 5$$
$$8x + y = -13$$

5. Slope $\dfrac{1}{2}$; through (5, –6)

$$y - y_1 = m(x - x_1)$$
$$y - (-6) = \frac{1}{2}(x - 5)$$
$$y + 6 = \frac{1}{2}(x - 5)$$
$$2(y + 6) = 2\left[\frac{1}{2}(x - 5)\right]$$
$$2y + 12 = x - 5$$
$$12 = x - 2y - 5$$
$$12 + 5 = x - 2y$$
$$17 = x - 2y$$
$$x - 2y = 17$$

7. Through (3, 2) and (5, 6)

$$m = \frac{6 - 2}{5 - 3} = \frac{4}{2} = 2$$
$$y - y_1 = m(x - x_1)$$
$$y - 2 = 2(x - 3)$$
$$y - 2 = 2x - 6$$
$$-2 = 2x - y - 6$$
$$-2 + 6 = 2x - y$$
$$4 = 2x - y$$
$$2x - y = 4$$

9. Through (–1, 3) and (–2, –5)

$$m = \frac{-5 - 3}{-2 - (-1)} = \frac{-8}{-2 + 1} = \frac{-8}{-1} = 8$$
$$y - y_1 = m(x - x_1)$$
$$y - 3 = 8[x - (-1)]$$
$$y - 3 = 8(x + 1)$$
$$y - 3 = 8x + 8$$
$$-3 = 8x - y + 8$$
$$-3 - 8 = 8x - y$$
$$-11 = 8x - y$$
$$8x - y = -11$$

11. Through (2, 3) and (–1, –1)

$$m = \frac{-1 - 3}{-1 - 2} = \frac{-4}{-3} = \frac{4}{3}$$
$$y - y_1 = m(x - x_1)$$
$$y - 3 = \frac{4}{3}(x - 2)$$
$$3(y - 3) = 3\left[\frac{4}{3}(x - 2)\right]$$
$$3(y - 3) = 4(x - 2)$$
$$3y - 9 = 4x - 8$$
$$-9 = 4x - 3y - 8$$
$$-9 + 8 = 4x - 3y$$
$$-1 = 4x - 3y$$
$$4x - 3y = -1$$

13. Vertical line through (0, 2)
    On a vertical line, the *x*-coordinate stays the same.
    $$x = 0$$

15. Horizontal line through (–1, 3)
    On a horizontal line, the *y*-coordinate stays the same.
    $$y = 3$$

17. Vertical line through $\left(-\dfrac{7}{3}, -\dfrac{2}{5}\right)$

    On a vertical line, the *x*-coordinate stays the same.
    $$x = -\frac{7}{3}$$

19. Parallel to *y* = 5, through (1, 2)
    $$y = 5$$
    $$y = 0x + 5$$
    $$m = 0$$
    Slope of the given line is 0, so slope of the parallel line is the same, 0.
    $$y - y_1 = m(x - x_1)$$
    $$y - 2 = 0(x - 1)$$
    $$y - 2 = 0$$
    $$y = 2$$

21. Perpendicular to $x = -3$, through $(-2, 5)$
$x = -3$ is a vertical line, therefore the perpendicular would be a horizontal line with slope = 0.
On a horizontal line, the y-coordinate stays the same.
$y = 5$

23. Parallel to $x = 0$, through $(6, -8)$
$x = 0$ is a vertical line, therefore the parallel line would be a vertical line.
On a vertical line, the x-coordinate stays the same.
$x = 6$

25. **a.** $(1, 32)$ and $(3, 96)$
$$m = \frac{96 - 32}{3 - 1} = \frac{64}{2} = 32$$
$$s - 32 = 32(t - 1)$$
$$s = 32t$$

   b. $s = 32(4) = 128$
   128 ft/sec

27. Slope $-\dfrac{1}{2}$; through $\left(0, \dfrac{5}{3}\right)$
$$y - y_1 = m(x - x_1)$$
$$y - \frac{5}{3} = -\frac{1}{2}(x - 0)$$
$$y - \frac{5}{3} = -\frac{1}{2}x$$
$$6\left(y - \frac{5}{3}\right) = 6\left(-\frac{1}{2}x\right)$$
$$6y - 10 = -3x$$
$$3x + 6y - 10 = 0$$
$$3x + 6y = 10$$

29. Slope 1; through $(-7, 9)$
$$y - y_1 = m(x - x_1)$$
$$y - 9 = 1[x - (-7)]$$
$$y - 9 = 1(x + 7)$$
$$y - 9 = x + 7$$
$$-9 = x - y + 7$$
$$-9 - 7 = x - y$$
$$-16 = x - y$$
$$x - y = -16$$

31. Through $(10, 7)$ and $(7, 10)$
$$m = \frac{10 - 7}{7 - 10} = \frac{3}{-3} = -1$$
$$y - y_1 = m(x - x_1)$$
$$y - 7 = -1(x - 10)$$
$$y - 7 = -x + 10$$
$$x + y - 7 = 10$$
$$x + y = 10 + 7$$
$$x + y = 17$$

33. Through $(6, 7)$ parallel to x-axis.
The x-axis is a horizontal line, so the parallel line is horizontal.
On a horizontal line, the y-coordinate stays the same.
$y = 7$

35. Slope $-\dfrac{4}{7}$, through $(-1, -2)$
$$y - y_1 = m(x - x_1)$$
$$y - (-2) = -\frac{4}{7}[x - (-1)]$$
$$y + 2 = -\frac{4}{7}(x + 1)$$
$$7(y + 2) = 7\left[-\frac{4}{7}(x + 1)\right]$$
$$7(y + 2) = -4(x + 1)$$
$$7y + 14 = -4x - 4$$
$$4x + 7y + 14 = -4$$
$$4x + 7y = -4 - 14$$
$$4x + 7y = -18$$

37. Through $(-8, 1)$ and $(0, 0)$
$$m = \frac{0 - 1}{0 - (-8)} = \frac{-1}{8}$$
$$y - y_1 = m(x - x_1)$$
$$y - 0 = -\frac{1}{8}(x - 0)$$
$$y = -\frac{1}{8}x$$
$$8(y) = 8\left(-\frac{1}{8}x\right)$$
$$8y = -x$$
$$x + 8y = 0$$

39. Through $(0, 0)$ with slope 3.
$$y - y_1 = m(x - x_1)$$
$$y - 0 = 3(x - 0)$$
$$y = 3x$$
$$0 = 3x - y$$
$$3x - y = 0$$

41. Through $(-6, -6)$ and $(0, 0)$
$$m = \frac{0 - (-6)}{0 - (-6)} = \frac{6}{6} = 1$$
$$y - y_1 = m(x - x_1)$$
$$y - 0 = 1(x - 0)$$
$$y = x$$
$$0 = x - y$$
$$x - y = 0$$

43. Slope -5, y-intercept 7
$$y = mx + b$$
$$y = -5x + 7$$
$$5x + y = 7$$

45. Through $(-1, 5)$ and $(0, -6)$
$$m = \frac{-6 - 5}{0 - (-1)} = \frac{-11}{1} = -11$$
$$y - y_1 = m(x - x_1)$$
$$y - (-6) = -11(x - 0)$$
$$y + 6 = -11x$$
$$11x + y + 6 = 0$$
$$11x + y = -6$$

47. With undefined slope, through $\left(-\dfrac{3}{4}, 1\right)$

When the slope is undefined, the line is a vertical line. On a vertical line, the x-coordinate stays the same.
$$x = -\frac{3}{4}$$

49. Through $(-2, -3)$, perpendicular to the y-axis. The y-axis is a vertical line, so the perpendicular is a horizontal line. On a horizontal line, the y-coordinate stays the same.
$$y = -3$$

51. Slope 7, through $(1, 3)$
$$y - y_1 = m(x - x_1)$$
$$y - 3 = 7(x - 1)$$
$$y - 3 = 7x - 7$$
$$-3 = 7x - y - 7$$
$$-3 + 7 = 7x - y$$
$$4 = 7x - y$$
$$7x - y = 4$$

53. $(10, 63)$ and $(15, 94)$
Let $x =$ radius and $y =$ circumference
$$m = \frac{94 - 63}{15 - 10} = \frac{31}{5}$$
$$y - y_1 = m(x - x_1)$$
$$y - 63 = \frac{31}{5}(x - 10)$$
$$5(y - 63) = 5\left[\frac{31}{5}(x - 10)\right]$$
$$5(y - 63) = 31(x - 10)$$
$$5y - 315 = 31x - 310$$
$$-315 = 31x - 5y - 310$$
$$-315 + 310 = 31x - 5y$$
$$-5 = 31x - 5y$$
$$31x - 5y = -5$$

55. a. $(2, 2600)$ and $(5, 2000)$
$$m = \frac{2000 - 2600}{5 - 2} = \frac{-600}{3} = -200$$
$$V - 2600 = -200(t - 2)$$
$$V = -200t + 3000$$

b. $V = -200(10) + 3000$
$$V = 1000$$
$$\$1000$$

57. Answers may vary.

59. a. Slope $= 3$, point $(-1, 2)$
$$y - 2 = 3(x + 1)$$
$$y - 2 = 3x + 3$$
$$-2 = 3x - y + 3$$
$$-5 = 3x - y \text{ or } 3x - y = -5$$

b. Slope $= -\dfrac{1}{3}$, point $(-1, 2)$
$$y - 2 = -\frac{1}{3}(x + 1)$$
$$-3(y - 2) = -3\left(-\frac{1}{3}\right)(x + 1)$$
$$-3y + 6 = x + 1$$
$$6 = x + 3y + 1$$
$$5 = x + 3y \text{ or } x + 3y = 5$$

61. $3x + 2y = 7$
$$2y = -3x + 7$$
$$y = -\frac{3}{2}x + \frac{7}{2}$$

a. Slope $= -\dfrac{3}{2}$, point $(3, -5)$
$$y + 5 = -\frac{3}{2}(x - 3)$$
$$2(y + 5) = 2\left(-\frac{3}{2}\right)(x - 3)$$
$$2y + 10 = -3x + 9$$
$$3x + 2y + 10 = 9$$
$$3x + 2y = -1$$

b. Slope $= \dfrac{2}{3}$, point $(3, -5)$
$$y + 5 = \frac{2}{3}(x - 3)$$
$$3(y + 5) = 3\left(\frac{2}{3}\right)(x - 3)$$
$$3y + 15 = 2x - 6$$
$$15 = 2x - 3y - 6$$
$$21 = 2x - 3y \text{ or } 2x - 3y = 21$$

63. $y = 2x - 6$

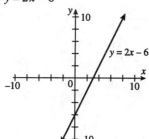

65. $x + 3y = 5$

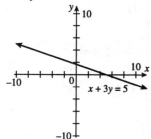

67. $y = -2$

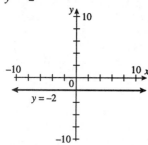

Exercise Set 7.3

For Exercises 1 through 8, graph ordered pair solutions and connect with a smooth curve.

1. $y = x^2 + 2$

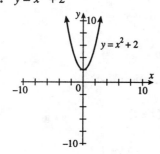

3. $y = |x| - 1$

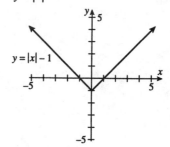

5. $y = -x^2$

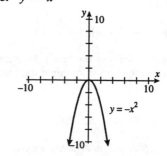

7. $y = |x + 5|$

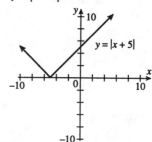

9. C

11. D

13. A

15. $(0, 1); (0, -1); (-2, 0); (2, 0)$

17. $(2, 0); \left(\dfrac{2}{3}, -2\right)$

19. (2, any real number)

21. There is no such point.

23. $(2, -1)$

25. Linear

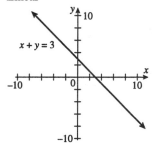

$x + y = 3$

27. Linear

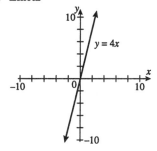

$y = 4x$

29. Linear

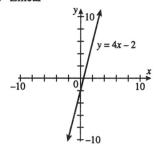

$y = 4x - 2$

31. Not linear

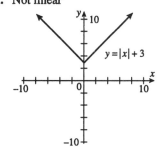

$y = |x| + 3$

33. Linear

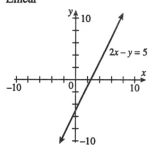

$2x - y = 5$

35. Not linear

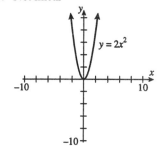

$y = 2x^2$

37. Not linear

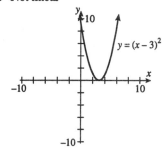

$y = (x - 3)^2$

39. Linear

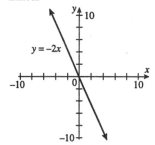

$y = -2x$

41. Linear

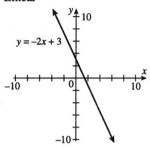

$y = -2x + 3$

43. Not Linear

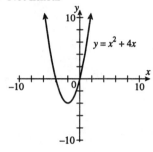

$y = x^2 + 4x$

45. $y = x^3$

x	-3	-2	-1	0	1	2	3
y	-27	-8	-1	0	1	8	27

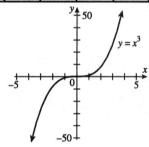

$y = x^3$

47. $x^2 - 3x + 1 = (2)^2 - 3(2) + 1 = 4 - 6 + 1 = -1$

49. $x^2 - 3x + 1 = (-1)^2 - 3(-1) + 1 = 1 + 3 + 1 = 5$

51. No

53. Yes

Exercise Set 7.4

1. $\{(2, 4), (0, 0), (-7, 10), (10, -7)\}$
The domain is $\{-7, 0, 2, 10\}$.
The range is $\{-7, 0, 4, 10\}$.

3. $\{(0, -2), (1, -2), (5, -2)\}$
The domain is $\{0, 1, 5\}$.
The range is $\{-2\}$.

5. $\{(1, 1), (2, 2), (-3, -3), (0, 0)\}$
Yes, it is a function.

7. $\{(-1, 0), (-1, 6), (-1, 8)\}$
No, the x-value -1 has more than one y value.

9. $y = x + 1$
Yes, it is a function. Each x value will produce exactly one y value.

11. $x = 2y^2$
No, it is not a function.
The ordered pairs $(2, 1)$ and $(2, -1)$ make the equation true. Therefore, an x value has more than one y value, so it is not a function.

13. $y - x = 7$
Yes, it is a function. Each x value will produce exactly one y value.

15. $y = \dfrac{1}{x}$
Yes, it is a function.
Each x value ($x \ne 0$), will produce exactly one y value.

17. $x = 5$
No, it is not a function.
$x = 5$ is a vertical line that would have ordered pairs such as $(5, 1)$, and $(5, 3)$ where the x value would have more than one y value.

19. $y = x^3$
Yes, it is a function.

21. $y < 2x + 1$
No, it is not a function. A vertical line would pass through more than one point in the shaded portion of the graph.

23. $y = x + 3$
Yes, it is a function.

25. Yes

27. No

29. No

31. Yes

33. $f(x) = 2x - 5$

   a. $f(-2) = 2(-2) - 5 = -4 - 5 = -9$

   b. $f(0) = 2(0) - 5 = 0 - 5 = -5$

   c. $f(3) = 2(3) - 5 = 6 - 5 = 1$

35. $f(x) = x^2 + 2$

   a. $f(-2) = (-2)^2 + 2 = 4 + 2 = 6$

   b. $f(0) = (0)^2 + 2 = 0 + 2 = 2$

   c. $f(3) = (3)^2 + 2 = 9 + 2 = 11$

37. $f(x) = x^3$

   a. $f(-2) = (-2)^3 = -8$

   b. $f(0) = (0)^3 = 0$

   c. $f(3) = (3)^3 = 27$

39. $f(x) = |x|$

   a. $f(-2) = |-2| = 2$

   b. $f(0) = |0| = 0$

   c. $f(3) = |3| = 3$

41. $h(x) = 5x$

   a. $h(-1) = 5(-1) = -5$

   b. $h(0) = 5(0) = 0$

   c. $h(4) = 5(4) = 20$

43. $h(x) = 2x^2 + 3$

   a. $h(-1) = 2(-1)^2 + 3 = 2(1) + 3 = 2 + 3 = 5$

   b. $h(0) = 2(0)^2 + 3 = 2(0) + 3 = 0 + 3 = 3$

   c. $h(4) = 2(4)^2 + 3 = 2(16) + 3 = 32 + 3 = 35$

45. $h(x) = -x^2 - 2x + 3$

   a. $h(-1) = -(-1)^2 - 2(-1) + 3 = -1 + 2 + 3 = 4$

   b. $h(0) = -0^2 - 2(0) + 3 = 0 - 0 + 3 = 3$

   c. $h(4) = -4^2 - 2(4) + 3 = -16 - 8 + 3 = -21$

47. $h(x) = 6$

   a. $h(-1) = 6$

   b. $h(0) = 6$

   c. $h(4) = 6$

49.

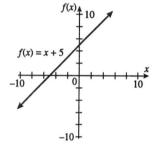

51.

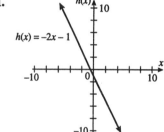

53.

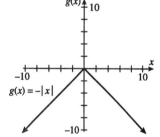

55.

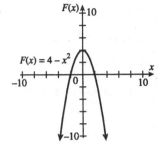

$F(x) = 4 - x^2$

57.

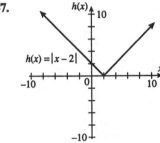

$h(x) = |x - 2|$

59. 5:20 A.M.

61. Answers may vary.

63. $f(x) = 3x - 7$
The domain is all real numbers.

65. $h(x) = \dfrac{1}{x+5}$
The domain is all real numbers except –5.

67. $g(x) = |x + 1|$
The domain is all real numbers.

69. Domain: all real numbers
Range: $y \geq -4$

71. Domain: all real numbers
Range: all real numbers

73. Domain: all real numbers
Range: $y = 2$

75. 9 P.M.

77. January 1 and December 1

79. Yes; it passes the vertical line test.

81. Answers may vary.

83. $H(x) = 2.59x + 47.24$

    a. $H(46) = 2.59(46) + 47.24 = 166.38$ cm

    b. $H(39) = 2.59(39) + 47.24 = 148.25$ cm

85. $\dfrac{3}{x} + \dfrac{5}{x} + \dfrac{3}{2x} = \dfrac{6+10+3}{2x} = \dfrac{19}{2x}$ meters

87. $(-2, 1)$

89. $(-3, -1)$

91. $f(x) = 2x + 7$

    a. $f(2) = 2(2) + 7$
        $f(2) = 4 + 7$
        $f(2) = 11$

    b. $f(a) = 2(a) + 7$
        $f(a) = 2a + 7$

    c. $f(a + 2) = 2(a + 2) + 7$
        $f(a + 2) = 2a + 4 + 7$
        $f(a + 2) = 2a + 11$

93. $h(x) = x^2 + 7$

    a. $h(3) = (3)^2 + 7$
        $h(3) = 9 + 7$
        $h(3) = 16$

    b. $h(a) = (a)^2 + 7$
        $h(a) = a^2 + 7$

    c. $h(a - 3) = (a - 3)^2 + 7$
        $h(a - 3) = a^2 + 2(a)(-3) + (3)^2 + 7$
        $h(a - 3) = a^2 - 6a + 9 + 7$
        $h(a - 3) = a^2 - 6a + 16$

Chapter 7 - Review

1. $3x + y = 7$
$y = -3x + 7$
$m = -3; (0, 7)$

2. $x - 6y = -1$
$-6y = -x - 1$
$y = \dfrac{1}{6}x + \dfrac{1}{6}$
$m = \dfrac{1}{6}; \left(0, \dfrac{1}{6}\right)$

3. $y = 2$
$m = 0; (0, 2)$

4. $x = -5$
undefined slope; no y-intercept

5. $x - y = -6$
$-y = -x - 6$

5. $x - y = -6$
$-y = -x - 6$
$y = x + 6$
$m = 1$
$x + y = 3$
$y = -x + 3$
$m = -1$
perpendicular

6. $3x + y = 7$
$y = -3x + 7$
$m = -3$; y-intercept $= (0, 7)$
$-3x - y = 10$
$y = -3x - 10$
$m = -3$; y-intercept $= (0, -10)$
parallel

7. $y = 4x + \dfrac{1}{2}$
$m = 4$
$4x + 2y = 1$
$2y = -4x + 1$
$y = -2x + \dfrac{1}{2}$
$m = -2$
neither

8. $y = -5x + \dfrac{1}{2}$

9. $y = \dfrac{2}{3}x + 6$

10. $y = -3x$
Slope $= -3$
y-intercept $= (0, 0)$

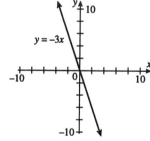

11. $y = 3x - 1$
Slope $= 3$
y-intercept $= (0, -1)$

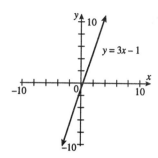

12. $-x + 2y = 8$
$2y = x + 8$
$y = \dfrac{1}{2}x + 4$
Slope $= \dfrac{1}{2}$
y-intercept $= (0, 4)$

13. $5x - 3y = 15$
$-3y = -5x + 15$
$y = \dfrac{5}{3}x - 5$
Slope $= \dfrac{5}{3}$
y-intercept $= (0, -5)$

14. D

15. C

16. A

17. B

18. Slope 4, through (2, 0)
$$y - y_1 = m(x - x_1)$$
$$y - 0 = 4(x - 2)$$
$$y = 4x - 8$$
$$0 = 4x - y - 8$$
$$8 = 4x - y$$
$$4x - y = 8$$

19. $m = -3, (0, -5)$
$$y + 5 = -3(x - 0)$$
$$y + 5 = -3x$$
$$3x + y = -5$$

20. Slope $\dfrac{1}{2}$, through $\left(0, -\dfrac{7}{2}\right)$
$$y - y_1 = m(x - x_1)$$
$$y - \left(-\dfrac{7}{2}\right) = \dfrac{1}{2}(x - 0)$$
$$y + \dfrac{7}{2} = \dfrac{1}{2}x$$
$$2\left(y + \dfrac{7}{2}\right) = 2\left(\dfrac{1}{2}x\right)$$
$$2y + 7 = x$$
$$7 = x - 2y$$
$$x - 2y = 7$$

21. $m = 0, (-2, -3)$
$$y + 3 = 0(x + 2)$$
$$y = -3$$

22. With 0 slope, through the origin. So $m = 0$ through (0, 0). Since the slope is 0, the line is a horizontal line. On a horizontal line, the y-coordinate stays the same.
$$y = 0$$

23. $m = -6, (2, -1)$
$$y + 1 = -6(x - 2)$$
$$y + 1 = -6x + 12$$
$$6x + y = 11$$

24. Slope 12, $\left(\dfrac{1}{2}, 5\right)$
$$y - y_1 = m(x - x_1)$$
$$y - 5 = 12\left(x - \dfrac{1}{2}\right)$$
$$y - 5 = 12x - 6$$
$$-5 = 12x - y - 6$$
$$-5 + 6 = 12x - y$$
$$1 = 12x - y$$
$$12x - y = 1$$

25. (0, 6) and (6, 0)
$$m = \dfrac{0 - 6}{6 - 0} = \dfrac{-6}{6} = -1$$
$$y - 6 = -1(x - 0)$$
$$y - 6 = -x$$
$$x + y = 6$$

26. Through (0, –4) and (–8, 0)
$$m = \dfrac{0 - (-4)}{-8 - 0} = \dfrac{4}{-8} = -\dfrac{1}{2}$$
$$y - y_1 = m(x - x_1)$$
$$y - (-4) = -\dfrac{1}{2}(x - 0)$$
$$y + 4 = -\dfrac{1}{2}x$$
$$2(y + 4) = 2\left(-\dfrac{1}{2}x\right)$$
$$2y + 8 = -x$$
$$x + 2y + 8 = 0$$
$$x + 2y = -8$$

27. Vertical line through (5, 7)
$$x = 5$$

28. Horizontal line, through (–6, 8)
On a horizontal line, the y-coordinate stays the same.
$$y = 8$$

29. (6, 0)
perpendicular to $y = 8$
The line has no slope, thus it is a vertical line.
$$x = 6$$

30. Through (10, 12), perpendicular to $x = -2$
$x = -2$ is a vertical line, so the perpendicular is a horizontal line. On a horizontal line, the y-coordinate stays the same.
$$y = 12$$

31. a. $m = -3$; point (5, 0)
$$y - 0 = -3(x - 5)$$
$$y = -3x + 15$$
$$3x + y = 15$$

b. $m = \dfrac{1}{3}$; point (5, 0)
$$y - 0 = \dfrac{1}{3}(x - 5)$$
$$y = \dfrac{1}{3}x - \dfrac{5}{3}$$
$$3y = x - 5$$
$$x - 3y = 5$$

32.

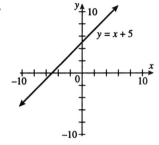

33.

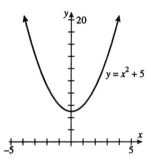

34.

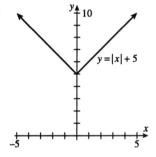

35.

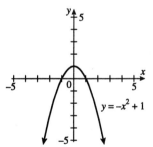

36.

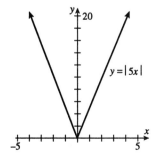

37.

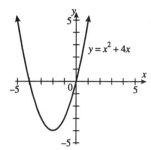

38. $(-5, 4)$

39. $(-1, -3)$

40. $(-5, 4)$

41. $(6, -1)$

42. No

43. Yes

44. Yes

45. Yes

46. No

47. Yes

48. No

49. No

50. Yes

51. $f(x) = -2x + 6$

   a. $f(0) = -2(0) + 6 = 0 + 6 = 6$

   b. $f(-2) = -2(-2) + 6 = 4 + 6 = 10$

   c. $f\left(\dfrac{1}{2}\right) = -2\left(\dfrac{1}{2}\right) + 6 = -1 + 6 = 5$

52. $h(x) = -5 - 3x$

   a. $h(2) = -5 - 3(2)$
        $h(2) = -5 - 6$
        $h(2) = -11$

   b. $h(-3) = -5 - 3(-3)$
        $h(-3) = -5 + 9$
        $h(-3) = 4$

   c. $h(0) = -5 - 3(0)$
        $h(0) = -5 - 0$
        $h(0) = -5$

53. $g(x) = x^2 + 12x$

   a. $g(3) = 3^2 + 12(3) = 9 + 36 = 45$

   b. $g(-5) = (-5)^2 + 12(-5) = 25 - 60 = -35$

   c. $g(0) = 0^2 + 12(0) = 0$

54. $h(x) = 6 - |x|$

   a. $h(-1) = 6 - |-1|$
        $h(-1) = 6 - 1$
        $h(-1) = 5$

   b. $h(1) = 6 - |1|$
        $h(1) = 6 - 1$
        $h(1) = 5$

   c. $h(-4) = 6 - |-4|$
        $h(-4) = 6 - 4$
        $h(-4) = 2$

55.   All real numbers

56.   All real numbers except 2

57.   Domain: $-3 \le x \le 5$
     Range: $-4 \le y \le 2$

58.   Domain: all real numbers
     Range: $y \ge 0$

59.   Domain: $x = 3$
     Range: all real numbers

60.   Domain: all real numbers
     Range: $y \le 2$

Chapter 7 - Test

1. $7x - 3y = 2$
     $-3y = -7x + 2$
     $y = \dfrac{7}{3}x - \dfrac{2}{3}$
     $m = \dfrac{7}{3}; \left(0, -\dfrac{2}{3}\right)$

2. $y = 2x - 6$
     $m = 2$
     $-4x = 2y$
     $y = -2x$
     $m = -2$
     neither

3. $y - 2 = -\dfrac{1}{4}(x - 2)$

     $-4(y - 2) = -4\left(-\dfrac{1}{4}\right)(x - 2)$

     $-4y + 8 = x - 2$
     $10 = x + 4y$ or $x + 4y = 10$

4.   Through the origin and $(6, -7)$;
     $(0, 0)$ and $(6, -7)$
     $m = \dfrac{-7 - 0}{6 - 0} = -\dfrac{7}{6}$
     $y - y_1 = m(x - x_1)$
     $y - 0 = -\dfrac{7}{6}(x - 0)$
     $y = -\dfrac{7}{6}x$
     $6(y) = 6\left(-\dfrac{7}{6}x\right)$
     $6y = -7x$
     $7x + 6y = 0$

5.   Through $(2, -5)$ and $(1, 3)$
     $m = \dfrac{3 - (-5)}{1 - 2} = \dfrac{3 + 5}{-1} = \dfrac{8}{-1} = -8$
     $y - y_1 = m(x - x_1)$
     $y - (-5) = -8(x - 2)$
     $y + 5 = -8x + 16$
     $8x + y + 5 = 16$
     $8x + y = 16 - 5$
     $8x + y = 11$

6.   Through $(-5, -1)$, parallel to $x = 7$;
     $x = 7$ is a vertical line, so the parallel would be the
     same, a vertical line.
     On a vertical line, the x-coordinate stays the same.
     $x = -5$

7. Slope $\frac{1}{8}$, through (0, 12)

$y - y_1 = m(x - x_1)$

$y - 12 = \frac{1}{8}(x - 0)$

$y - 12 = \frac{1}{8}x$

$8(y - 12) = 8\left(\frac{1}{8}x\right)$

$8y - 96 = x$

$-96 = x - 8y$

$x - 8y = -96$

8.

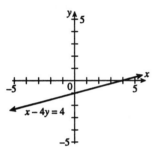

9.

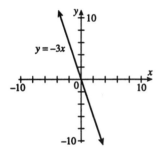

10.

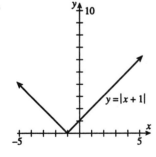

11.

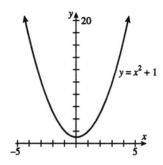

12. Yes

13. Yes

14. Yes

15. No

16. No

17. $f(x) = 2x - 4$

  a. $f(-2) = 2(-2) - 4$
      $f(-2) = -4 - 4$
      $f(-2) = -8$

  b. $f(0.2) = 2(0.2) - 4$
      $f(0.2) = 0.4 - 4$
      $f(0.2) = -3.6$

  c. $f(0) = 2(0) - 4$
      $f(0) = -4$

18. $h(x) = x^3 - x$

  a. $h(-1) = (-1)^3 - (-1)$
      $h(-1) = -1 + 1$
      $h(-1) = 0$

  b. $h(0) = 0^3 - 0$
      $h(0) = 0 - 0$
      $h(0) = 0$

  c. $h(4) = 4^3 - 4$
      $h(4) = 64 - 4$
      $h(4) = 60$

19. $g(x) = 6$

  a. $g(0) = 6$

  b. $g(a) = 6$

  c. $g(242) = 6$

20. All real numbers except -1

21. Domain: all real numbers
Range: $y \le 4$

22. Domain: all real numbers
Range: all real numbers

Chapter 7 - Cumulative Review

1. a. $3 + (-7) + (-8) = 3 - 7 - 8 = -12$

  b. $[7 + (-10)] + [-2 + (-4)] = (7 - 10) + (-2 - 4)$
  $= (-3) + (-6) = -9$

2. a. The Lion King; $313 million

  b. $217 - 145 = 72 million

3. a. $2x + 6$

  b. $\dfrac{x-4}{7}$

  c. $5 + 3(x + 1) = 5 + 3x + 3 = 3x + 8$

4. $\dfrac{5}{2}x = 15$
$\left(\dfrac{2}{5}\right)\dfrac{5}{2}x = \left(\dfrac{2}{5}\right)15$
$x = 6$

5. $2x < -4$
$\dfrac{2x}{2} < \dfrac{-4}{2}$
$x < -2$

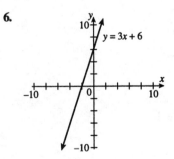

6.

7. undefined slope

8. a. $-2t^2 + 3t + 6$; 2; trinomial

  b. $15x - 10$; 1; binomial

  c. $7x + 3x^3 + 2x^2 - 1$; 3; none of these

  d. $7x^2y - 6xy$; 3; binomial

9. $x^2 + 4x - 12 = (x + 6)(x - 2)$

10. $x^2 - 9x = -20$
$x^2 - 9x + 20 = 0$
$(x - 5)(x - 4) = 0$
$x - 5 = 0$ or $x - 4 = 0$
$x = 5$ or $x = 4$

11. $\dfrac{2x^2 - 11x + 5}{5x - 25} + \dfrac{4x - 2}{10}$
$= \dfrac{(2x - 1)(x - 5)}{5(x - 5)} \cdot \dfrac{10}{2(2x - 1)}$
$= \dfrac{2x - 1}{5} \cdot \dfrac{5}{2x - 1} = 1$

12. $\dfrac{4b}{9a} \cdot \dfrac{3ab}{3ab} = \dfrac{12ab^2}{27a^2b}$

13. $1 + \dfrac{m}{m+1} = \dfrac{m+1}{m+1} + \dfrac{m}{m+1} = \dfrac{m+1+m}{m+1} = \dfrac{2m+1}{m+1}$

14. $\dfrac{\frac{x+1}{y}}{\frac{x}{y}+2} = \dfrac{\frac{x+1}{y}}{\frac{x+2y}{y}} = \dfrac{x+1}{y} \cdot \dfrac{y}{x+2y} = \dfrac{x+1}{x+2y}$

15. $3 - \dfrac{6}{x} = x + 8$
$x\left(3 - \dfrac{6}{x}\right) = x(x + 8)$
$3x - 6 = x^2 + 8x$
$0 = x^2 + 5x + 6$
$0 = (x + 3)(x + 2)$
$x + 3 = 0$ or $x + 2 = 0$
$x = -3$ or $x = -2$

16. $5x + y = 2$
$y = -5x + 2$
$m = -5$; $(0, 2)$

17. $x = -1$

18.

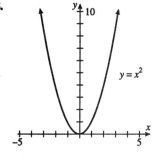

19. a. $(-1, 2)$

    b. $(-2, 0)$

    c. $\left(-2\dfrac{2}{3}, -1\right), (1, -1), (3, -1)$

20. $g(x) = x^2 - 3$

    a. $g(2) = (2)^2 - 3 = 4 - 3 = 1$

    b. $g(-2) = (-2)^2 - 3 = 4 - 3 = 1$

    c. $g(0) = (0)^2 - 3 = 0 - 3 = -3$

Chapter 8

Section 8.1 Mental Math

1. consistent, independent

3. consistent, dependent

5. inconsistent, independent

7. consistent, independent

Exercise Set 8.1

1. $\begin{cases} x + y = 8 \\ 3x + 2y = 21 \end{cases}$

   a. $(2, 4)$
   $x + y = 8;$
   $2 + 4 \: ? \: 8$
   $6 \neq 8$

   $3x + 2y = 21$
   $3(2) + 2(4) \: ? \: 21$
   $6 + 8 \: ? \: 21$
   $14 \neq 21$

   no

   b. $(5, 3)$

   $x + y = 8$
   $5 + 3 \: ? \: 8$
   $8 = 8$

   $3x + 2y = 21$
   $3(5) + 2(3) \: ? \: 21$
   $15 + 6 \: ? \: 21$
   $21 = 21$

   yes

   c. $(1, 9)$

   $x + y = 8$
   $1 + 9 \: ? \: 8$
   $10 \neq 8$

   $3x + 2y = 21$
   $3(1) + 2(9) \: ? \: 21$
   $3 + 18 \: ? \: 21$
   $21 = 21$

   no

3. $\begin{cases} 3x - y = 5 \\ x + 2y = 11 \end{cases}$

   a. $(2, -1)$

   $3x - y = 5$
   $3(2) - (-1) \: ? \: 5$
   $6 + 1 \: ? \: 5$
   $7 \neq 5$

   $x + 2y = 11$
   $2 + 2(-1) \: ? \: 11$
   $2 - 2 \: ? \: 11$
   $0 \neq 11$

   no

   b. $(3, 4)$

   $3x - y = 5$
   $3(3) - 4 \: ? \: 5$
   $9 - 4 \: ? \: 5$
   $5 = 5$

   $x + 2y = 11$
   $3 + 2(4) \: ? \: 11$
   $3 + 8 \: ? \: 11$
   $11 = 11$

   yes

   c. $(0, -5)$

   $3x - y = 5$
   $3(0) - (-5) \: ? \: 5$
   $0 + 5 \: ? \: 5$
   $5 = 5$

   $x + 2y = 11$
   $0 + 2(-5) \: ? \: 11$
   $-10 \: ? \: 11$
   $-10 \neq 11$

   no

5. $\begin{cases} 2y = 4x \\ 2x - y = 0 \end{cases}$

   a. $(-3, -6)$

      $2y = 4x$
      $2(-6) \; ? \; 4(-3)$
      $-12 = -12$

      $2x - y = 0$
      $2(-3) - (-6) \; ? \; 0$
      $-6 + 6 \; ? \; 0$
      $0 = 0$

      yes

   b. $(0, 0)$

      $2y = 4x$
      $2(0) \; ? \; 4(0)$
      $0 = 0$

      $2x - y = 0$
      $2(0) - 0 \; ? \; 0$
      $0 = 0$

      yes

   c. $(1, 2)$

      $2y = 4x$
      $2(2) \; ? \; 4(1)$
      $4 = 4$

      $2x - y = 0$
      $2(1) - 2 \; ? \; 0$
      $2 - 2 \; ? \; 0$
      $0 = 0$

      yes

7. Answers may vary.

9. consistent, independent

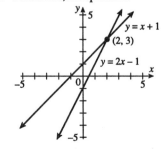

11. consistent, independent

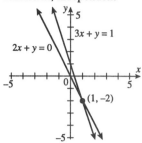

13. consistent, independent

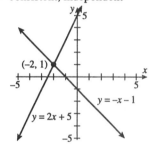

15. consistent, independent

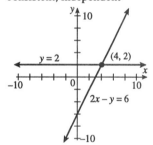

17. inconsistent, independent

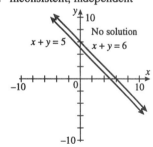

19. consistent, dependent

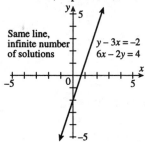

Same line, infinite number of solutions

$y - 3x = -2$
$6x - 2y = 4$

21. consistent, independent

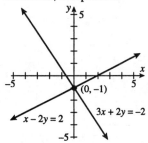

$x - 2y = 2$
$(0, -1)$
$3x + 2y = -2$

23. consistent, independent

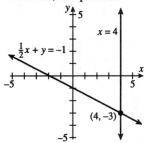

$x = 4$
$\frac{1}{2}x + y = -1$
$(4, -3)$

25. consistent, independent

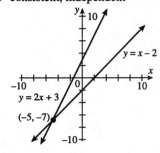

$y = x - 2$
$y = 2x + 3$
$(-5, -7)$

27. consistent, independent

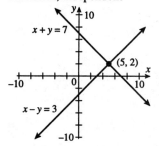

$x + y = 7$
$(5, 2)$
$x - y = 3$

29. Answers may vary.

31. $\begin{cases} 4x + y = 24 \\ x + 2y = 2 \end{cases}$

$4x + y = 24$ $x + 2y = 2$
$y = -4x + 24$ $2y = -x + 2$

$m = -4$ $\dfrac{2y}{2} = \dfrac{-x}{2} + \dfrac{2}{2}$

$\qquad\qquad\qquad y = -\dfrac{1}{2}x + 1$

$\qquad\qquad\qquad m = -\dfrac{1}{2}$

a. Lines intersecting at a single point

b. One solution

33. $\begin{cases} 2x + y = 0 \\ 2y = 6 - 4x \end{cases}$

$2x + y = 0$ $2y = 6 - 4x$

$y = -2x$ $\dfrac{2y}{2} = \dfrac{6}{2} - \dfrac{4x}{2}$

$m = -2, b = 0$ $y = 3 - 2x$
$\qquad\qquad\qquad y = -2x + 3$
$\qquad\qquad\qquad m = -2, b = 3$

a. The lines are parallel.

b. No solutions

35. $\begin{cases} 6x - y = 4 \\ \frac{1}{2}y = -2 + 3x \end{cases}$

$6x - y = 4$ | $\frac{1}{2}y = -2 + 3x$

$-y = -6x + 4$ | $2\left(\frac{1}{2}y\right) = 2(-2 + 3x)$

$\frac{-y}{-1} = \frac{-6x}{-1} + \frac{4}{-1}$ | $y = -4 + 6x$

$y = 6x - 4$ | $y = 6x - 4$

$m = 6,\ b = -4$ | $m = 6,\ b = -4$

a. Identical lines

b. Infinite number of solutions

37. $\begin{cases} x = 5 \\ y = -2 \end{cases}$

$x = 5$ | $y = -2$

m is undefined. | $m = 0$

$b =$ none | $b = -2$

a. Lines intersecting at a single point

b. One solution

39. $\begin{cases} 3y - 2x = 3 \\ x + 2y = 9 \end{cases}$

$3y - 2x = 3$ | $x + 2y = 9$

$3y = 2x + 3$ | $2y = -x + 9$

$\frac{3y}{3} = \frac{2x}{3} + \frac{3}{3}$ | $\frac{2y}{2} = \frac{-y}{2} + \frac{9}{2}$

$y = \frac{2}{3}x + 1$ | $y = -\frac{1}{2}x + \frac{9}{2}$

$m = \frac{2}{3},\ b = 1$ | $m = -\frac{1}{2},\ b = \frac{9}{2}$

a. The lines intersect at one point.

b. One solution

41. $\begin{cases} 6y + 4x = 6 \\ 3y - 3 = -2x \end{cases}$

$6y + 4x = 6$ | $3y - 3 = -2x$

$6y = -4x + 6$ | $3y = -2x + 3$

$\frac{6y}{6} = -\frac{4}{6}x + \frac{6}{6}$ | $\frac{3y}{3} = -\frac{2}{3}x + \frac{3}{3}$

$y = -\frac{2}{3}x + 1$ | $y = -\frac{2}{3}x + 1$

$m = -\frac{2}{3},\ b = 1$ | $m = -\frac{2}{3},\ b = 1$

a. Identical lines

b. Infinite number of solutions

43. $\begin{cases} x + y = 4 \\ x + y = 3 \end{cases}$

$x + y = 4$ | $x + y = 3$

$y = -x + 4$ | $y = -x + 3$

$m = -1,\ b = 4$ | $m = -1,\ b = 3$

a. The lines are parallel.

b. No solution

45. Answers may vary.

47. 1984, 1988

49. a. (4, 9)

b. Yes

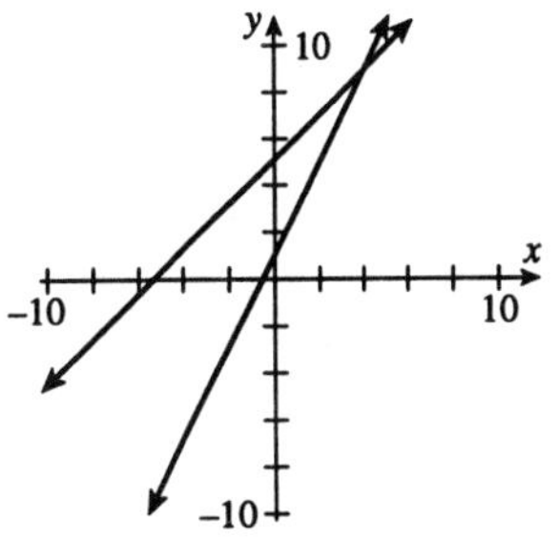

51. $-2x + 3(x + 6) = 17$
$-2x + 3x + 18 = 17$
$x + 18 = 17$
$x = -1$

53. $-y + 12\left(\frac{y-1}{4}\right) = 3$
$-y + 3(y - 1) = 3$
$-y + 3y - 3 = 3$
$2y - 3 = 3$
$2y = 6$
$y = 3$

55. $3z - (4z - 2) = 9$
$3z - 4z + 2 = 9$
$-z + 2 = 9$
$-z = 7$
$z = -7$

Exercise Set 8.2

1. $\begin{cases} x + y = 3 \\ \quad x = 2y \end{cases}$

Replace x with $2y$.
$2y + y = 3$
$3y = 3$
$y = 1$
To find x, use $y = 1$ in the equation
$x = 2y$
$x = 2(1)$
$x = 2$
The solution is $(2, 1)$.

3. $\begin{cases} x + y = 6 \\ \quad y = -3x \end{cases}$

Replace y with $-3x$.
$x + (-3x) = 6$
$-2x = 6$
$x = \dfrac{6}{-2} = -3$
To find y, use $x = -3$ in the equation
$y = -3x$
$y = -3(-3)$
$y = 9$
The solution is $(-3, 9)$.

5. $\begin{cases} 3x + 2y = 16 \\ \quad\quad x = 3y - 2 \end{cases}$

Replace x with $3y - 2$.
$3(3y - 2) + 2y = 16$
$9y - 6 + 2y = 16$
$11y - 6 = 16$
$11y = 16 + 6$
$11y = 22$
$y = \dfrac{22}{11} = 2$
To find x, use $y = 2$ in the equation
$x = 3y - 2$
$x = 3(2) - 2$
$x = 6 - 2$
$x = 4$
The solution is $(4, 2)$.

7. $\begin{cases} x + 2y = 6 \\ 2x + 3y = 8 \end{cases}$

Solve the first equation for x.
$x + 2y = 6$
$x = -2y + 6$
Replace x with $-2y + 6$ in the second equation.
$2(-2y + 6) + 3y = 8$
$-4y + 12 + 3y = 8$
$-y + 12 = 8$
$-y = +8 - 12$
$-y = -4$
$y = 4$
To find x, use $y = 4$ in the equation
$x = -2y + 6$
$x = -2(4) + 6$
$x = -8 + 6$
$x = -2$
The solution is $(-2, 4)$.

9. $\begin{cases} 2x - 5y = 1 \\ 3x + y = -7 \end{cases}$

Solve the second equation for y.
$3x + y = -7$
$y = -3x - 7$
Replace y with $-3x - 7$ in the first equation.
$2x - 5(-3x - 7) = 1$
$2x + 15x + 35 = 1$
$17x + 35 = 1$
$17x = 1 - 35$
$17x = -34$
$x = \dfrac{-34}{17} = -2$
To find y, use $x = -2$ in the equation
$y = -3x - 7$
$y = -3(-2) - 7$
$y = 6 - 7$
$y = -1$
The solution is $(-2, -1)$.

11. $\begin{cases} \quad\quad 2y = x + 2 \\ 6x - 12y = 0 \end{cases}$

Solve the first equation for x.
$2y = x + 2$
$2y - 2 = x$
Replace x with $2y - 2$ in the second equation.
$6(2y - 2) - 12y = 0$
$12y - 12 - 12y = 0$
$-12 = 0$
This is a contradiction, $-12 \neq 0$. Therefore, there is no solution.

13. $\begin{cases} \dfrac{1}{3}x - y = 2 \\ x - 3y = 6 \end{cases}$

Solve the first equation for y.

$\dfrac{1}{3}x - y = 2$

$\dfrac{1}{3}x = 2 + y$

$\dfrac{1}{3}x - 2 = y$

Replace y with $\dfrac{1}{3}x - 2$ in the second equation.

$x - 3\left(\dfrac{1}{3}x - 2\right) = 6$

$x - x + 6 = 6$

$6 = 6$

This is an identity.
The equations are dependent and there are infinite solutions.

15. Answers may vary.

17. $\begin{cases} 3x - 4y = 10 \\ x = 2y \end{cases}$

Replace x with $2y$ in the first equation.

$3(2y) - 4y = 10$

$6y - 4y = 10$

$2y = 10$

$y = \dfrac{10}{2} = 5$

To find x, use $y = 5$ in the equation.

$x = 2y$

$x = 2(5)$

$x = 10$

The solution is (10, 5).

19. $\begin{cases} y = 3x + 1 \\ 4y - 8x = 12 \end{cases}$

Replace y with $3x + 1$ in the second equation.

$4(3x + 1) - 8x = 12$

$12x + 4 - 8x = 12$

$4x + 4 = 12$

$4x = 12 - 4$

$4x = 8$

$x = \dfrac{8}{4} = 2$

To find y, use $x = 2$ in the equation

$y = 3x + 1$

$y = 3(2) + 1$

$y = 6 + 1$

$y = 7$

The solution is (2, 7).

21. $\begin{cases} 4x + y = 11 \\ 2x + 5y = 1 \end{cases}$

Solve the first equation for y.

$4x + y = 11$

$y = -4x + 11$

Replace y with $-4x + 11$ in the second equation.

$2x + 5(-4x + 11) = 1$

$2x - 20x + 55 = 1$

$-18x + 55 = 1$

$-18x = 1 - 55$

$-18x = -54$

$x = \dfrac{-54}{-18} = 3$

To find y, use $x = 3$ in the equation

$4x + y = 11$

$4(3) + y = 11$

$12 + y = 11$

$y = 11 - 12$

$y = -1$

The solution is (3, −1).

23. $\begin{cases} 2x - 3y = -9 \\ 3x = y + 4 \end{cases}$

Solve the second equation for y.

$3x = y + 4$

$3x - 4 = y$

Replace y with $3x - 4$ in the first equation.

$2x - 3(3x - 4) = -9$

$2x - 9x + 12 = -9$

$-7x + 12 = -9$

$-7x = -9 - 12$

$-7x = -21$

$x = \dfrac{-21}{-7} = 3$

To find y, use $x = 3$ in the equation,

$3x - 4 = y$

$3(3) - 4 = y$

$9 - 4 = y$

$5 = y$

The solution is (3, 5).

25. $\begin{cases} 6x - 3y = 5 \\ x + 2y = 0 \end{cases}$

Solve the second equation for x.

$x + 2y = 0$

$x = -2y$

Replace x with $-2y$ in the first equation.

$6(-2y) - 3y = 5$

$-12y - 3y = 5$

$-15y = 5$

$y = \dfrac{5}{-15} = -\dfrac{1}{3}$

To find x, use $y = -\dfrac{1}{3}$ in the equation

$x = -2y$

$x = -2\left(-\dfrac{1}{3}\right)$

$x = \dfrac{2}{3}$

The solution is $\left(\dfrac{2}{3}, -\dfrac{1}{3}\right)$.

27. $\begin{cases} 3x - y = 1 \\ 2x - 3y = 10 \end{cases}$

Solve the first equation for y.

$3x - y = 1$

$-y = -3x + 1$

$y = \dfrac{-3}{-1}x + \dfrac{1}{-1}$

$y = 3x - 1$

Replace y with $3x - 1$ in the second equation.

$2x - 3(3x - 1) = 10$

$2x - 9x + 3 = 10$

$-7x + 3 = 10$

$-7x = 10 - 3$

$-7x = 7$

$x = \dfrac{7}{-7} = -1$

To find y, use $x = -1$ in the equation

$y = 3x - 1$

$y = 3(-1) - 1$

$y = -3 - 1$

$y = -4$

The solution is $(-1, -4)$.

29. $\begin{cases} -x + 2y = 10 \\ -2x + 3y = 18 \end{cases}$

Solve the first equation for x.

$-x + 2y = 10$

$2y = 10 + x$

$2y - 10 = x$

Replace x with $2y - 10$ in the second equation.

$-2(2y - 10) + 3y = 18$

$-4y + 20 + 3y = 18$

$-y + 20 = 18$

$-y = 18 - 20$

$-y = -2$

$y = 2$

To find x, use $y = 2$ in the equation

$2y - 10 = x$

$2(2) - 10 = x$

$4 - 10 = x$

$-6 = x$

The solution is $(-6, 2)$.

31. $\begin{cases} 5x + 10y = 20 \\ 2x + 6y = 10 \end{cases}$

Solve the second equation for x.

$2x + 6y = 10$

$2x = 10 - 6y$

$x = \dfrac{10}{2} - \dfrac{6}{2}y$

$x = 5 - 3y$

Replace x with $5 - 3y$ in the first equation.

$5(5 - 3y) + 10y = 20$

$25 - 15y + 10y = 20$

$25 - 5y = 20$

$-5y = 20 - 25$

$-5y = -5$

$y = \dfrac{-5}{-5} = 1$

To find x, use $y = 1$ in the equation

$x = 5 - 3y$

$x = 5 - 3(1)$

$x = 5 - 3$

$x = 2$

The solution is $(2, 1)$.

33. $\begin{cases} 3x + 6y = 9 \\ 4x + 8y = 16 \end{cases}$

Solve the first equation for x.

$3x + 6y = 9$

$3x = -6y + 9$

$x = \dfrac{-6y}{3} + \dfrac{9}{3}$

$x = -2y + 3$

Replace x with $-2y + 3$ in the second equation.

$4(-2y + 3) + 8y = 16$

$-8y + 12 + 8y = 16$

$12 = 16$

There is a contradiction, $12 \neq 16$. There is no solution.

35. $\begin{cases} y = 2x + 9 \\ y = 7x + 10 \end{cases}$

Replace y with $2x + 9$ in the second equation.

$2x + 9 = 7x + 10$

$9 = 7x - 2x + 10$

$9 = 5x + 10$

$9 - 10 = 5x$

$-1 = 5x$

$-\dfrac{1}{5} = x$

To find y, use $x = -\dfrac{1}{5}$ in the equation

$y = 2x + 9$

$y = 2\left(-\dfrac{1}{5}\right) + 9$

$y = -\dfrac{2}{5} + 9$

$y = -\dfrac{2}{5} + \dfrac{45}{5}$

$y = \dfrac{43}{5}$

The solution is $\left(-\dfrac{1}{5}, \dfrac{43}{5}\right)$

37. $-5y + 6y = 3x + 2(x - 5) - 3x + 5$

$-5y + 6y = 3x + 2x - 10 - 3x + 5$

$y = 2x - 5$

$4(x + y) - x + y = -12$

$4x + 4y - x + y = -12$

$3x + 5y = -12$

$3x + 5(2x - 5) = -12$

$3x + 10x - 25 = -12$

$13x - 25 = -12$

$13x = 13$

$x = 1$

$y = 2x - 5$

$y = 2(1) - 5$

$y = 2 - 5$

$y = -3$

$(1, -3)$

39. $y = 5.1x + 14.56$

$\quad y = -2x - 3.9$

$5.1x + 14.56 = -2x - 3.9$

$7.1x = -18.46$

$x = -2.6$

$y = -2x - 3.9$

$y = -2(-2.6) - 3.9$

$y = 5.2 - 3.9$

$y = 1.3$

$(-2.6, 1.3)$

41. $3x + 2y = 14.05$

$\quad 5x + y = 18.5$

$5(3x + 2y) = 5(14.05)$

$-3(5x + y) = -3(18.5)$

$15x + 10y = 70.25$

$\underline{-15x - 3y = -55.5}$

$\qquad\quad 7y = 14.75$

$\qquad\quad\; y = 2.11$

$3x + 2y = 14.05$

$3x + 2(2.11) = 14.05$

$3x + 4.22 = 14.05$

$3x = 9.83$

$x = 3.28$

$(3.28, 2.11)$

43. $3x + 2y = 6$

$\quad -2(3x + 2y) = -2(6)$

$\quad -6x - 4y = -12$

45. $-4x + y = 3$

$\quad 3(-4x + y) = 3(3)$

$\quad -12x + 3y = 9$

47. $\quad 3n + 6m$

$\quad \underline{+ 2n - 6m}$

$\qquad 5n$

49. $\quad -5a - 7b$

$\quad \underline{+ 5a - 8b}$

$\qquad -15b$

Exercise Set 8.3

1. $\begin{cases} 3x+y=5 \\ 6x-y=4 \end{cases}$

$3x+y=5$
$6x-y=4$
$\overline{9x=9}$
$x=1$
To find y, use $x = 1$ in the equation,
$3x+y=5$
$3(1)+y=5$
$3+y=5$
$y=5-3$
$y=2$
The solution is $(1, 2)$.

3. $\begin{cases} x-2y=8 \\ -x+5y=-17 \end{cases}$

$x-2y=8$
$-x+5y=-17$
$\overline{3y=-9}$
$y=\dfrac{-9}{3}=-3$
To find x, use $y = -3$ in the equation
$x-2y=8$
$x-2(-3)=8$
$x+6=8$
$x=8-6$
$x=2$
The solution is $(2, -3)$.

5. $\begin{cases} 3x+2y=11 \\ 5x-2y=29 \end{cases}$

$3x+2y=11$
$5x-2y=29$
$\overline{8x=40}$
$x=\dfrac{40}{8}=5$
To find y, use $x = 5$ in the equation
$3x+2y=11$
$3(5)+2y=11$
$15+2y=11$
$2y=11-15$
$2y=-4$
$y=\dfrac{-4}{2}=-2$
The solution is $(5, -2)$.

7. $\begin{cases} 3x+y=-11 \\ 6x-2y=-2 \end{cases}$
Multiply 1st equation by 2.
$2(3x+y)=2(-11)$
$6x+2y=-22$
Add the two equations.
$6x+2y=-22$
$6x-2y=-2$
$\overline{12x=-24}$
$x=\dfrac{-24}{12}=-2$
To find y, use $x = -2$ in the equation
$3x+y=-11$
$3(-2)+y=-11$
$-6+y=-11$
$y=-11+6$
$y=-5$
The solution is $(-2, -5)$.

9. $\begin{cases} x+5y=18 \\ 3x+2y=-11 \end{cases}$
Multiply 1st equation by -3.
$-3(x+5y)=-3(18)$
$-3x-15y=-54$
Add the two equations.
$-3x-15y=-54$
$3x+2y=-11$
$\overline{-13y=-65}$
$y=\dfrac{-65}{-13}=5$
To find x, use $y = 5$ in the equation
$x+5y=18$
$x+5(5)=18$
$x+25=18$
$x=18-25$
$x=-7$
The solution is $(-7, 5)$.

11. $\begin{cases} 2x-5y=4 \\ 3x-2y=4 \end{cases}$
Multiply the 1st equation by -3.
Multiply 2nd equation by 2.
$-3(2x-5y)=-3(4)$
$2(3x-2y)=2(4)$

$-6x+15y=-12$
$6x-4y=8$
$\overline{11y=-4}$
$y=-\dfrac{4}{11}$

To find x, use $y = -\dfrac{4}{11}$ in the equation

$2x - 5y = 4$

$2x - 5\left(-\dfrac{4}{11}\right) = 4$

$2x + \dfrac{20}{11} = 4$

$2x = 4 - \dfrac{20}{11}$

$2x = \dfrac{44}{11} - \dfrac{20}{11}$

$2x = \dfrac{24}{11}$

$\dfrac{1}{2}(2x) = \dfrac{1}{2}\left(\dfrac{24}{11}\right)$

$x = \dfrac{12}{11}$

The solution is $\left(\dfrac{12}{11},\ -\dfrac{4}{11}\right)$.

13. $\begin{cases} 2x + 3y = 0 \\ 4x + 6y = 3 \end{cases}$

Multiply 1st equation by –2.
$-2(2x + 3y) = -2(0)$
$-4x - 6y = 0$
Add the equations.
$\quad -4x - 6y = 0$
$\quad\ \ 4x + 6y = 3$
$\quad\overline{\qquad\quad 0 = 3}$

This is a contradiction, $0 \neq 3$. There is no solution.

15. $\begin{cases} \dfrac{x}{3} + \dfrac{y}{6} = 1 \\ \dfrac{x}{2} - \dfrac{y}{4} = 0 \end{cases}$

Multiply 1st equation by 6.
Multiply 2nd equation by 4.

$6\left(\dfrac{x}{3} + \dfrac{y}{6}\right) = 6(1)$

$4\left(\dfrac{x}{2} - \dfrac{y}{4}\right) = 4(0)$

Add the equations.
$\quad 2x + y = 6$
$\quad 2x - y = 0$
$\quad\overline{4x\qquad = 6}$
$\quad\ x \quad = \dfrac{6}{4} = \dfrac{3}{2}$

To find y, use $x = \dfrac{3}{2}$ in the equation

$2x + y = 6$

$2\left(\dfrac{3}{2}\right) + y = 6$

$3 + y = 6$
$y = 6 - 3$
$y = 3$

The solution is $\left(\dfrac{3}{2},\ 3\right)$.

17. $\begin{cases} x - \dfrac{y}{3} = -1 \\ -\dfrac{x}{2} + \dfrac{y}{8} = \dfrac{1}{4} \end{cases}$

Multiply 1st equation by 3.
Multiply 2nd equation by 8.

$3\left(x - \dfrac{y}{3}\right) = 3(-1)$

$8\left(-\dfrac{x}{2} + \dfrac{y}{8}\right) = 8\left(\dfrac{1}{4}\right)$

Add the resulting equations.
$\quad 3x - y = -3$
$\quad -4x + y = \ \ 2$
$\quad\overline{\ -x \qquad = -1}$
$\quad\ x \qquad = \dfrac{-1}{-1} = 1$

To find y, use $x = 1$ in the equation
$-4x + y = 2$
$-4(1) + y = 2$
$-4 + y = 2$
$y = 2 + 4$
$y = 6$
The solution is $(1, 6)$.

19. Answers may vary.

21. $\begin{cases} x + y = 6 \\ x - y = 6 \end{cases}$

$\quad x + y = \ \ 6$
$\quad x - y = \ \ 6$
$\quad\overline{2x \qquad = 12}$
$\quad\ x \quad\ = \ \ 6$

To find y, use $x = 6$ in the equation
$x + y = 6$
$6 + y = 6$
$y = 6 - 6$
$y = 0$
The solution is $(6, 0)$.

23. $\begin{cases} 3x+y=4 \\ 9x+3y=6 \end{cases}$

Multiply 1st equation by –3.

$-3(3x+y)=-3(4)$

$-9x-3y=-12$

Add the equations.

$\begin{array}{r} -9x-3y=-12 \\ 9x+3y=6 \\ \hline 0=-6 \end{array}$

This is a contradiction, $0 \neq -6$. There is no solution.

25. $\begin{cases} 3x-2y=7 \\ 5x+4y=8 \end{cases}$

Multiply 1st equation by 2.

$2(3x-2y)=2(7)$

$6x-4y=14$

Add the equations.

$\begin{array}{r} 6x-4y=14 \\ 5x+4y=8 \\ \hline 11x=22 \end{array}$

$x=\dfrac{22}{11}=2$

To find y, use $x=2$ in the equation

$3x-2y=7$

$3(2)-2y=7$

$6-2y=7$

$-2y=7-6$

$-2y=1$

$y=-\dfrac{1}{2}$

The solution is $\left(2,\ -\dfrac{1}{2}\right)$.

27. $\begin{cases} \dfrac{2}{3}x+4y=-4 \\ 5x+6y=18 \end{cases}$

Multiply 1st equation by 3.

$3\left(\dfrac{2}{3}x+4y\right)=3(-4)$

$2x+12y=-12$

Since addition now would not eliminate a variable, multiply the second equation by –2.

$-2(5x+6y)=-2(18)$

$-10x-12y=-36$

Add the equations.

$\begin{array}{r} 2x+12y=-12 \\ -10x-12y=-36 \\ \hline -8x=-48 \end{array}$

$x=\dfrac{-48}{-8}=6$

To find y, use $x=6$ in the equation

$5x+6y=18$

$5(6)+6y=18$

$30+6y=18$

$6y=18-30$

$6y=-12$

$y=-\dfrac{12}{6}=-2$

The solution is $(6,-2)$.

29. $\begin{cases} 4x-6y=8 \\ 6x-9y=12 \end{cases}$

Multiply 1st equation by 3.

Multiply 2nd equation by –2.

$3(4x-6y)=3(8)$

$-2(6x-9x)=-2(12)$

Add the resulting equations.

$\begin{array}{r} 12x-18y=24 \\ -12x+18y=-24 \\ \hline 0=0 \end{array}$

This is an identity.

The equations are dependent.

There are an infinite number of solutions.

31. $\begin{cases} \dfrac{x}{3}-y=2 \\ -\dfrac{x}{2}+\dfrac{3y}{2}=-3 \end{cases}$

Multiply 1st equation by 3.

Multiply 2nd equation by 2.

$3\left(\dfrac{x}{3}-y\right)=3(2)$

$2\left(-\dfrac{x}{2}+\dfrac{3y}{2}\right)=2(-3)$

Add the equations.

$\begin{array}{r} x-3y=6 \\ -x+3y=-6 \\ \hline 0=0 \end{array}$

This is an identity.

The equations are dependent.

There are an infinite number of solutions.

33. $\begin{cases} 8x = -11y - 16 \\ 2x + 3y = -4 \end{cases}$

Multiply the 2nd equation by –4.
$-4(2x + 3y) = -4(-4)$
$-8x - 12y = 16$
$8x = -11y - 16 \Rightarrow 8x + 11y = -16$
Add the equations.
$\begin{array}{r} -8x - 12y = 16 \\ 8x + 11y = -16 \\ \hline -y = 0 \\ y = 0 \end{array}$

To find x, use $y = 0$ in the equation
$2x + 3y = -4$
$2x + 3(0) = -4$
$2x = -4$
$x = \dfrac{-4}{2} = -2$
The solution is (–2, 0).

35. a. $x + y = 5$
$3(x + y) = 3(5)$
$3x + 3y = 15$
$b = 15$

b. Any real number except 15

37. $\begin{cases} 2x - 3y = -11 \\ y = 4x - 3 \end{cases}$

Replace y with $4x - 3$ in the 1st equation.
$2 - 3(4x - 3) = -11$
$2x - 12x + 9 = -11$
$-10x + 9 = -11$
$-10x = -11 - 9$
$-10x = -20$
$x = \dfrac{-20}{-10} = 2$
To find y, use $x = 2$ in the equation
$y = 4x - 3$
$y = 4(2) - 3$
$y = 8 - 3$
$y = 5$
The solution is (2, 5).

39. $\begin{cases} x + 2y = 1 \\ 3x + 4y = -1 \end{cases}$

Multiply the 1st equation by –3.
$-3(x + 2y) = -3(1)$
$-3x - 6y = -3$

Add the equations.
$\begin{array}{r} -3x - 6y = -3 \\ 3x + 4y = -1 \\ \hline -2y = -4 \end{array}$
$y = \dfrac{-4}{-2} = 2$

To find x, use $y = 2$ in the equation
$x + 2y = 1$
$x + 2(2) = 1$
$x + 4 = 1$
$x = 1 - 4$
$x = -3$
The solution is (–3, 2).

41. $\begin{cases} 2y = x + 6 \\ 3x - 2y = -6 \end{cases}$

Solve the 1st equation for x.
$2y = x + 6$
$2y - 6 = x$
Replace x with $2y - 6$ in the second equation.
$3(2y - 6) - 2y = -6$
$6y - 18 - 2y = -6$
$4y - 18 = -6$
$4y = -6 + 18$
$4y = 12$
$y = \dfrac{12}{4} = 3$
To find x, use $y = 3$ in the equation.
$2y - 6 = x$
$2(3) - 6 = x$
$6 - 6 = x$
$0 = x$
The solution is (0, 3).

43. $\begin{cases} y = 2x - 3 \\ y = 5x - 18 \end{cases}$

Replace y with $2x - 3$ in the second equation.
$2x - 3 = 5x - 18$
$2x - 5x - 3 = -18$
$-3x - 3 = -18$
$-3x = -18 + 3$
$-3x = -15$
$x = \dfrac{-15}{-3} = 5$
To find y, use $x = 5$ in the equation.
$y = 2x - 3$
$y = 2(5) - 3$
$y = 10 - 3$
$y = 7$
The solution is (5, 7).

45. $\begin{cases} x + \dfrac{1}{6}y = \dfrac{1}{2} \\ 3x + 2y = 3 \end{cases}$

Multiply 1st equation by 6.

$6\left(x + \dfrac{1}{6}y \right) = 6\left(\dfrac{1}{2} \right)$

$6x + y = 3$

Since addition now would not eliminate a variable, multiply this equation by –2.

$-2(6x + y) = -2(3)$

$-12x - 2y = -6$

Add the equations.

$\begin{array}{r} -12x - 2y = -6 \\ 3x + 2y = 3 \\ \hline -9x = -3 \end{array}$

$x = \dfrac{-3}{-9} = \dfrac{1}{3}$

To find y, use $x = \dfrac{1}{3}$ in the equation.

$3x + 2y = 3$

$3\left(\dfrac{1}{3} \right) + 2y = 3$

$1 + 2y = 3$

$2y = 3 - 1$

$2y = 2$

$y = \dfrac{2}{2} = 1$

The solution is $\left(\dfrac{1}{3}, \ 1 \right)$.

47. $\begin{cases} \dfrac{x+2}{2} = \dfrac{y+11}{3} \\ \dfrac{x}{2} = \dfrac{2y+16}{6} \end{cases}$

Multiply 1st equation by 6.
Multiply 2nd equation by 6.

$6\left(\dfrac{x+2}{2} \right) = 6\left(\dfrac{y+11}{3} \right)$

$3(x + 2) = 2(y + 11)$ Simplifying

$3x + 6 = 2y + 22$

$3x - 2y = 22 - 6$

$3x - 2y = 16$

$6\left(\dfrac{x}{2} \right) = 6\left(\dfrac{2y+16}{6} \right)$

$3x = 2y + 16$ Simplifying

$3x - 2y = 16$

$-1(3x - 2y) = -1(16)$ Multiply by –1

$-3x + 2y = -16$

Add the equations.

$\begin{array}{r} 3x - 2y = 16 \\ -3x + 2y = -16 \\ \hline 0 = 0 \end{array}$

This is an identity.
The equations are dependent.
There are an infinite number of solutions.

49. $3(2x + 3y) = 3(14)$

$-2(3x - 4y) = -2(-69.1)$

$\begin{array}{r} 6x + 9y = 42 \\ -6x + 8y = 138.2 \\ \hline 17y = 180.2 \\ y = 10.6 \end{array}$

$2x + 3y = 14$

$2x + 3(10.6) = 14$

$2x + 31.8 = 14$

$2x = -17.8$

$x = -8.9$

$(-8.9, 10.6)$

51. $2x + 6 = x - 3$

53. $20 - 3x = 2$

55. Let x = number

$4(x + 6) = 2x$

Exercise Set 8.4

1. Let x = one number
y = other number
$\begin{cases} x + y = 15 \\ x - y = 7 \end{cases}$

3. Let x = larger account
y = smaller account
$\begin{cases} x + y = 6500 \\ x = y + 800 \end{cases}$

5. Let x = length
y = width
$\begin{cases} 30 = 2x + 2y \\ x = y + 3 \end{cases}$

Replace x with $y + 3$ in the 1st equation.

$30 = 2(y + 3) + 2y$

$30 = 2y + 6 + 2y$

$30 = 4y + 6$

$30 - 6 = 4y$

$24 = 4y$

$6 = \dfrac{24}{4} = y$

To find x, use $y = 6$ in the equation

$x = y + 3$

$x = 6 + 3$

$x = 9$

The correct answer is **c**,
length = 9 feet;
width = 6 feet.

7. Let x = speed of the plane
    y = speed of the wind

$\begin{cases} 4(x - y) = 400 \\ 2(x + y) = 400 \end{cases}$

$\dfrac{4(x - y)}{4} = \dfrac{400}{4}$ **Simplify the equation.**

$\dfrac{2(x + y)}{2} = \dfrac{400}{2}$

Add the equations.

$\begin{array}{r} x - y = 100 \\ x + y = 200 \\ \hline 2x = 300 \\ x = 150 \end{array}$

To find the y, use $x = 150$ in the equation

$x + y = 200$

$150 + y = 200$

$y = 200 - 150$

$y = 50$

The correct answer is **a**; Plane 150 mph and
Wind 50 mph.

9. Let x = number of dimes
    y = number of quarters

$\begin{cases} x + y = 100 \\ 10x + 25y = 1300 \end{cases}$

Multiply 1st equation by -10.

$-10(x + y) = -10(100)$

$-10x - 10y = -1000$

Add the equations.

$\begin{array}{r} -10x - 10y = -1000 \\ 10x + 25y = 1300 \\ \hline 15y = 300 \\ y = \dfrac{300}{15} = 20 \end{array}$

To find x, use $y = 20$ in the equation

$x + y = 100$

$x + 20 = 100$

$x = 100 - 20$

$x = 80$

The correct answer is **a**; 80 dimes, 20 quarters.

11. Let x = one number
    y = other number

$\begin{cases} x + y = 83 \\ x - y = 17 \end{cases}$

$\begin{array}{r} x + y = 83 \\ x - y = 17 \\ \hline 2x = 100 \\ x = \dfrac{100}{2} = 50 \end{array}$

To find y, use $x = 50$ in the equation

$x + y = 83$

$50 + y = 83$

$y = 83 - 50$

$y = 33$

The numbers are 50 and 33.

13. Let x = adult price
    y = children's price

$\begin{cases} 3x + 4y = 159 \\ 2x + 3y = 112 \end{cases}$

Multiply 1st equation by 2.
Multiply 2nd equation by -3.

$2(3x + 4y) = 2(159)$

$-3(2x + 3y) = -3(112)$

$\begin{array}{r} 6x + 8y = 318 \\ -6x - 9y = -336 \\ \hline -y = -18 \\ y = \dfrac{-18}{-1} = 18 \end{array}$

To find x, use $y = 18$ in the equation

$3x + 4y = 159$

$3x + 4(18) = 159$

$3x + 72 = 159$

$3x = 159 - 72$

$3x = 87$

$x = \dfrac{87}{3} = 29$

Adult's price is \$29, Children's price is \$18.

15. Let x = number of quarters
    y = number of nickels

$\begin{cases} x + y = 80 \\ 25x + 5y = 1460 \end{cases}$

Solve the 1st equation for y.

$x + y = 80$

$y = 80 - x$

Replace y with $80 - x$ in the 2nd equation.

$25x + 5(80 - x) = 1460$

$25x + 400 - 5x = 1460$

$20x + 400 = 1460$

$20x = 1460 - 400$

$20x = 1060$

$x = \dfrac{1060}{20} = 53$

To find y, use $x = 53$ in the equation.

$y = 80 - x$

$y = 80 - 53$

$y = 27$

53 quarters; 27 nickels

17. Let $x =$ speed in still water

   $y =$ speed of the current

   $\begin{cases} 2(x+y) = 18 \\ 4.5(x-y) = 18 \end{cases}$

   $\dfrac{2(x+y)}{2} = \dfrac{18}{2}$ Simplify the equations.

   $\dfrac{4.5(x-y)}{4.5} = \dfrac{18}{4.5}$

   Add the equations.

   $\begin{array}{r} x + y = 9 \\ x - y = 4 \\ \hline 2x \quad = 13 \end{array}$

   $x \quad = \dfrac{13}{2} = 6.5$

   To find y, use $x = 6.5$ in the equation

   $x + y = 9$

   $6.5 + y = 9$

   $y = 9 - 6.5$

   $y = 2.5$

   rowing 6.5 mph; current 2.5 mph

19. let $x =$ speed of plane

   $y =$ speed of wind

   $\begin{cases} 2(x-y) = 780 \\ 1.5(x+y) = 780 \end{cases}$

   $\dfrac{2(x-y)}{2} = \dfrac{780}{2}$ Simplify the equation.

   $\dfrac{1.5(x+y)}{1.5} = \dfrac{780}{1.5}$

   Add the equations.

   $\begin{array}{r} x - y = 390 \\ x + y = 520 \\ \hline 2x \quad = 910 \end{array}$

   $x \quad = \dfrac{910}{2} = 455$

   To find y, use $x = 455$ in the equation

   $x + y = 520$

   $455 + y = 520$

   $y = 520 - 455$

   $y = 65$

   plane speed 455 mph; wind 65 mph

21. Let $x =$ ounces of 4%

   $y =$ ounces of 12%

   $\begin{cases} x + y = 12 \\ 0.04x + 0.12y = 0.09(12) \end{cases}$

   Solve the first equation for x.

   $x = 12 - y$

   Replace x with $12 - y$ in the second equation.

   $0.04(12 - y) + 0.12y = 0.09(12)$

   $0.48 - 0.04y + 0.12y = 1.08$

   $0.48 + 0.08y = 1.08$

   $0.08y = 1.08 - 0.48$

   $0.08y = 0.60$

   $y = \dfrac{0.60}{0.08} = 7.5$

   To find x, use $y = 7.5$ in the equation

   $x = 12 - y$

   $x = 12 - 7.5$

   $x = 4.5$

   7.5 oz. of 12%; 4.5 oz. of 4%

23. Let $x =$ pounds of $4.95 coffee

   $y =$ pounds of $2.65 coffee

   $\begin{cases} x + y = 200 \\ 4.95x + 2.65y = 3.95(200) \end{cases}$

   Solve the 1st equation for x.

   $x = 200 - y$

   Replace x with $200 - y$ in the second equation.

   $4.95(200 - y) + 2.65y = 3.95(200)$

   $990 - 4.95y + 2.65y = 790$

   $990 - 2.30y = 790$

   $-2.30y = 790 - 990$

   $-2.30y = -200$

   $y = \dfrac{-200}{-2.30} \approx 87$

   To find x, use $y = 87$ in the equation

   $x = 200 - y$

   $x = 200 - 87$

   $x = 113$

   113 lbs. of $4.95 coffee; 87 lbs. of $2.65 coffee

25. Let $x =$ first number

   $y =$ second number

   $\begin{cases} x + 2y = 8 \\ 2x + y = 25 \end{cases}$

   Multiply the 1st equation by -2.

   $-2(x + 2y) = -2(8)$

   $-2x - 4y = -16$

   Add the equations.

   $\begin{array}{r} -2x - 4y = -16 \\ 2x + y = 25 \\ \hline -3y = 9 \end{array}$

   $y = \dfrac{9}{-3} = -3$

To find x, use $y = -3$ in the equation
$x + 2y = 8$
$x + 2(-3) = 8$
$x - 6 = 8$
$x = 8 + 6$
$x = 14$
The numbers are 14 and -3.

27. Let $x =$ pieces sold at $9.50
     $y =$ pieces sold at $7.50
$$\begin{cases} x + y = 90 \\ 9.50x + 7.50y = 721 \end{cases}$$
Solve the 1st equation for x.
$x = 90 - y$
Replace x with $90 - y$ in the second equation.
$9.50(90 - y) + 7.50y = 721$
$855 - 9.50y + 7.50y = 721$
$855 - 2.00y = 721$
$-2.00y = 721 - 855$
$-2.00y = -134$
$y = \dfrac{-134}{-2.00} = 67$
To find x, use $y = 67$ in the equation
$x = 90 - y$
$x = 90 - 67$
$x = 23$
23 pieces were sold at the original price.

29. Let $x =$ width
     $y =$ length
$$\begin{cases} y = 2x - 3 \\ 2x + y = 33 \end{cases}$$
Replace y with $2x - 3$ in the second equation.
$2x + 2x - 3 = 33$
$4x - 3 = 33$
$4x = 33 + 3$
$4x = 36$
$x = \dfrac{36}{4} = 9$
To find y, use $x = 9$ in the equation
$y = 2x - 3$
$y = 2(9) - 3$
$y = 18 - 3$
$y = 15$
The dimensions are 9 ft. wide by 15 ft. long.

31. Let $x =$ time bicycling
     $y =$ time walking
$$\begin{cases} x + y = 6 \\ 40x + 4y = 186 \end{cases}$$
Solve the 1st equation for x.
$x = 6 - y$
Replace x with $6 - y$ in the second equation.
$40(6 - y) + 4y = 186$
$240 - 40y + 4y = 186$
$240 - 36y = 186$
$-36y = 186 - 240$
$-36y = -54$
$y = \dfrac{-54}{-36} = 1.5$
To find x, use $y = 1.5$ in the equation
$x = 6 - y$
$x = 6 - 1.5$
$x = 4.5$
The time bicycling is 4.5 hours.

33. Let $x =$ daily car rental fee
     $y =$ cost per mile driven
$$\begin{cases} 5x + 300y = 178 \\ 4x + 500y = 197 \end{cases}$$
Multiply 1st equation by 4.
Multiply 2nd equation by -5.
$4(5x + 300y) = 4(178)$
$-5(4x + 500y) = -5(197)$
Add the equations.
$$\begin{array}{r} 20x + 1200y = 712 \\ -20x - 2500y = -985 \\ \hline -1300y = -985 \end{array}$$
$y = \dfrac{-273}{-1300} = 0.21$
To find x, use $y = 0.21$ in the equation
$5x + 300y = 178$
$5x + 300(0.21) = 178$
$5x + 63 = 178$
$5x = 178 - 63$
$5x = 115$
$x = \dfrac{115}{5} = 23$
The daily rental fee is $23. The additional cost per mile driven is $0.21.

35. $y < 3 - x$
Find two points.
Let $x = 0$	$x = 4$
$y = 3 - 0$	$y = 3 - 4$
$y = 3$	$y = -1$
$(0, 3)$	$(4, -1)$

The boundary line is dashed.

Choose $(0, 0)$ as a test point.
$y < 3 - x$
$0 \; ? \; 3 - 0$
$0 \; ? \; 3$
Since $0 < 3$, the side containing $(0, 0)$ is shaded.

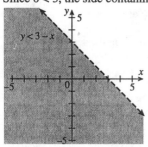

37. $2x - y \geq 6$
Find two points.

Let $x = 0$	$x = 3$
$2(0) - y = 6$	$2(3) - y \geq 6$
$-y = 6$	$6 - y \geq 6$
$y = -6$	$y = 0$
$(0, -6)$	$(3, 0)$

The boundary line is solid. Choose $(0, 0)$ as a test point.
$2(0) - 0 \; ? \; 6$
$0 \geq 6$ False
The side containing $(0, 0)$ is not shaded.

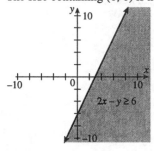

Exercise Set 8.5

1. $\begin{cases} y \geq x + 1 \\ y \geq 3 - x \end{cases}$

Graph $y = x + 1$.
The boundary line is solid.
Choose $(0, 0)$ as a test point.
$y \geq x + 1$
$0 \; ? \; 0 + 1$
$0 \; ? \; 1$
$0 \not\geq 1$, so shade the side *not* containing $(0, 0)$.
Graph $y = 3 - x$.

The boundary line is solid. Choose $(0, 0)$ as a test point.
$y \geq 3 - x$
$0 \; ? \; 3 - 0$
$0 \; ? \; 3$
$0 \not\geq 3$, so shade the side *not* containing $(0, 0)$.

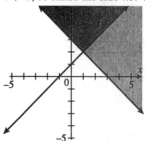

3. $\begin{cases} y < 3x - 4 \\ y \leq x + 2 \end{cases}$

Graph $y = 3x - 4$.
The boundary line is dashed.
Choose $(0, 0)$ as a test point.
$y < 3x - 4$
$0 \; ? \; 3(0) - 4$
$0 \; ? \; -4$
$0 \not< -4$, so shade the side *not* containing $(0, 0)$.
Graph $y = x + 2$.
The boundary line is solid.
Choose $(0, 0)$ as a test point.
$y \leq x + 2$
$0 \; ? \; 0 + 2$
$0 \; ? \; 2$
$0 \leq 2$, so shade the side containing $(0, 0)$.

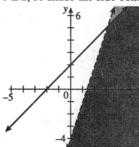

5. $\begin{cases} y \leq -2x - 2 \\ y \geq x + 4 \end{cases}$

Graph $y = -2x - 2$.
The boundary line is solid.
Choose $(0, 0)$ as a test point.
$y \leq -2x - 2$
$0 \; ? \; -2(0) - 2$
$0 \; ? \; -2$
$0 \not\leq -2$, so shade the side *not* containing $(0, 0)$.

Graph $y = x + 4$.
The boundary line is solid.
Choose $(0, 0)$ as a test point.
$y \geq x + 4$
$0 ? 0 + 4$
$0 ? 4$
$0 \ngeq 4$, so shade the side *not* containing $(0, 0)$.

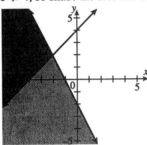

7. $\begin{cases} y \geq -x + 2 \\ y \leq 2x + 5 \end{cases}$

Graph $y = -x + 2$.
The boundary line is solid.
Choose $(0, 0)$ as a test point.
$y \geq -x + 2$
$0 ? -0 + 2$
$0 ? 2$
$0 \ngeq 2$, so shade the side *not* containing $(0, 0)$.

Graph $y = 2x + 5$.
The boundary line is solid.
Choose $(0, 0)$ as a test point.
$y \leq 2x + 5$
$0 ? 2(0) + 5$
$0 ? 5$
$0 \leq 5$, so shade the side containing $(0, 0)$.

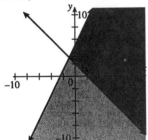

9. $\begin{cases} x \geq 3y \\ x + 3y \leq 6 \end{cases}$

Graph $x = 3y$.
The boundary line is solid.
Choose $(4, 0)$ as a test point.
$x \geq 3y$
$4 ? 3(0)$
$4 ? 0$
$4 \geq 0$, so shade the side containing $(4, 0)$.
Graph $x + 3y = 6$
The boundary line is solid.
Choose $(0, 0)$ as a test point.
$x + 3y \leq 6$
$0 + 3(0) ? 6$
$0 ? 6$
$0 \leq 6$, so shade the side containing $(0, 0)$.

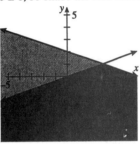

11. $\begin{cases} y + 2x \geq 0 \\ 5x - 3y \leq 12 \end{cases}$

Graph $y + 2x = 0$.
The boundary line is solid.
Choose $(4, 0)$ as a test point.
$y + 2x \geq 0$
$0 + 2(4) ? 0$
$8 ? 0$
$8 \geq 0$, so shade the side containing $(4, 0)$.

Graph $5x - 3y = 12$.
The boundary line is solid.
Choose $(0, 0)$ as a test point.
$5x - 3y \leq 12$
$5(0) - 3(0) ? 12$
$0 ? 12$
$0 \leq 12$, so shade the side containing $(0, 0)$.

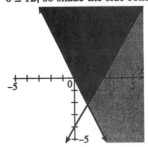

13. $\begin{cases} 3x - 4y \ge -6 \\ 2x + y \le 7 \end{cases}$

Graph $3x - 4y = -6$.
The boundary line is solid.
Choose $(0, 0)$ for a test point.
$3x - 4y \ge -6$
$3(0) - 4(0) \; ? \; -6$
$0 \; ? \; -6$
$0 \ge -6$, so the side containing $(0, 0)$ is shaded.

Graph $2x + y = 7$.
The boundary line is solid.
Choose $(0, 0)$ for a test point.
$2x + y \le 7$
$2(0) + 0 \; ? \; 7$
$0 \; ? \; 7$
$0 \le 7$, so the side containing $(0, 0)$ is shaded.

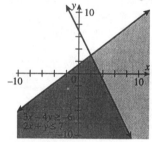

15. $\begin{cases} x \le 2 \\ y \ge -3 \end{cases}$

Graph $x = 2$.
The boundary line is solid.
Since $(0, 0)$ satisfies $x \le 2$, shade the side containing $(0, 0)$.

Graph $y = -3$.
The boundary line is solid.
Since $(0, 0)$ satisfies $y \ge -3$, shade the side containing $(0, 0)$.

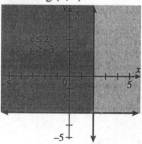

17. $\begin{cases} y \ge 1 \\ x < -3 \end{cases}$

Graph $y = 1$.
The boundary line is solid.
Since $(0, 0)$ does *not* satisfy $y \ge 1$, shade the side *not* containing $(0, 0)$.

Graph $x = -3$
The boundary line is dashed.
Since $(0, 0)$ does *not* satisfy $x < -3$, shade the side *not* containing $(0, 0)$.

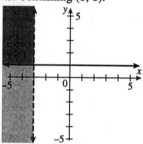

19. $\begin{cases} 2x + 3y < -8 \\ x \ge -4 \end{cases}$

Graph $2x + 3y = -8$.
The boundary line is dashed.
Choose $(0, 0)$ for a test point.
$2x + 3y < -8$
$2(0) + 3(0) \; ? \; -8$
$0 \; ? \; -8$
$0 \not< -8$, so the side *not* containing $(0, 0)$ is shaded.

Graph $x = -4$.
The boundary line is solid.
Since $(0, 0)$ does satisfy $x \ge -4$, shade the side containing $(0, 0)$.

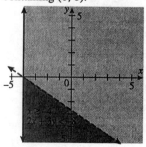

21. $\begin{cases} 2x - 5y \leq 9 \\ \quad y \leq -3 \end{cases}$

Graph $2x - 5y = 9$.
The boundary line is solid.
Choose $(0, 0)$ for a test point.
$2x - 5y \leq 9$
$2(0) - 5(0)\ ?\ 9$
$0\ ?\ 9$
$0 \leq 9$, so the side containing $(0, 0)$ is shaded.

Graph $y = -3$.
The boundary line is solid.
Since $(0, 0)$ does *not* satisfy $y \leq -3$, shade the side not containing $(0, 0)$.

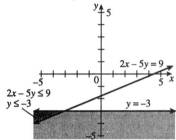

23. $\begin{cases} y \geq \dfrac{1}{2}x + 2 \\ y \leq \dfrac{1}{2}x - 3 \end{cases}$

Graph $y = \dfrac{1}{2}x + 2$.
The boundary line is solid.
Choose $(0, 0)$ for a test point.

$y \geq \dfrac{1}{2}x + 2$

$0\ ?\ \dfrac{1}{2}(0) + 2$

$0\ ?\ 2$

$0 \not\geq 2$, so the side *not* containing $(0, 0)$ is shaded.

Graph $y = \dfrac{1}{2}x - 3$.
The boundary line is solid.
Choose $(0, 0)$ for a test point.

$y \leq \dfrac{1}{2}x - 3$

$0\ ?\ \dfrac{1}{2}(0) - 3$

$0\ ?\ -3$

$0 \not\leq -3$, so shade the side *not* containing $(0, 0)$.

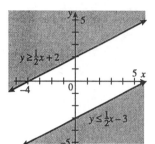

25. C

27. D

29. a. Let x = hours for writing bills
y = hours for writing purchase orders
$$\begin{cases} x + y \leq 8 \\ \quad x < 3 \end{cases}$$

b. Graph $x + y = 8$.
The boundary line is solid.
Choose $(0, 0)$ for a test point.
$x + y \leq 8$
$0 + 0\ ?\ 8$
$0\ ?\ 8$
$0 \leq 8$, so shade the side containing $(0, 0)$.

Graph $x = 3$.
Since $(0, 0)$ does satisfy $x < 3$, shade the side containing $(0, 0)$.

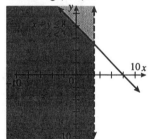

31. Answers may vary.

33. $(-3)^2 = -3 \cdot -3 = 9$

35. $\left(\dfrac{2}{3}\right)^2 = \dfrac{2}{3} \cdot \dfrac{2}{3} = \dfrac{4}{9}$

37. $-2 - (-3) = -2 + 3 = 1$

39. $8 + (-13) = -5$

Chapter 8 - Review

1. $\begin{cases} 2x - 3y = 12 \\ 3x + 4y = 1 \end{cases}$

   a. $(12, 4)$

   $2x - 3y = 12$
   $2(12) - 3(4) \ ? \ 12$
   $24 - 12 \ ? \ 12$
   $12 = 12$

   $3x + 4y = 1$
   $3(12) + 4(4) \ ? \ 1$
   $36 + 16 \ ? \ 1$
   $52 \ ? \ 1$

   no

   b. $(3, -2)$

   $2x - 3y = 12$
   $2(3) - 3(-2) \ ? \ 12$
   $6 + 6 \ ? \ 12$
   $12 = 12$

   $3x + 4y = 1$
   $3(3) + 4(-2) \ ? \ 1$
   $9 - 8 \ ? \ 1$
   $1 = 1$

   yes

   c. $(-3, 6)$

   $2x - 3y = 12$
   $2(-3) - 3(6) \ ? \ 12$
   $-6 - 18 \ ? \ 12$
   $-24 \ ? \ 12$

   $3x + 4y = 1$
   $3(-3) + 4(6) \ ? \ 1$
   $-9 + 24 \ ? \ 1$
   $15 \ ? \ 1$

   no

2. $\begin{cases} 4x + y = 0 \\ -8x - 5y = 9 \end{cases}$

   a. $\left(\dfrac{3}{4}, -3 \right):$

   $4\left(\dfrac{3}{4} \right) + (-3) = 0$
   $3 + (-3) = 0$
   $0 = 0$

   True

   $-8\left(\dfrac{3}{4} \right) - 5(-3) = 9$
   $-6 + 15 = 9$
   $9 = 9$
   True

   b. $(-2, 8):$
   $4(-2) + 8 = 0$
   $-8 + 8 = 0$
   $0 = 0$

   True

   $-8(-2) - 5(8) = -9$
   $16 - 40 = 9$
   $-24 = 9$

   False

   c. $\left(\dfrac{1}{2}, -2 \right):$

   $4\left(\dfrac{1}{2} \right) + (-2) = 0$
   $2 + (-2) = 0$
   $0 = 0$

   True

   $-8\left(\dfrac{1}{2} \right) - 5(-2) = 9$
   $-4 + 10 = 9$
   $6 = 9$

   False

   $\left(\dfrac{3}{4}, -3 \right)$ is a solution.
   $(-2, 8)$ is not a solution.
   $\left(\dfrac{1}{2}, -2 \right)$ is not a solution.

3. $\begin{cases} 5x - 6y = 18 \\ 2y - x = -4 \end{cases}$

   a. $(-6, -8)$

      $5x - 6y = 18$
      $5(-6) - 6(-8) \ ? \ 18$
      $-30 + 48 \ ? \ 18$
      $18 = 18$

      $2y - x = -4$
      $2(-8) - (-6) \ ? \ -4$
      $-16 + 6 \ ? \ -4$
      $-10 \ ? \ -4$

      no

   b. $\left(3, \dfrac{5}{2}\right)$

      $5x - 6y = 18$
      $5(3) - 6\left(\dfrac{5}{2}\right) \ ? \ 18$
      $15 - 15 \ ? \ 18$
      $0 \ ? \ 18$

      $2y - x = -4$
      $2\left(\dfrac{5}{2}\right) - 3 \ ? \ -4$
      $5 - 3 \ ? \ -4$
      $2 \ ? \ -4$

      no

   c. $\left(3, -\dfrac{1}{2}\right)$

      $5x - 6y = 18$
      $5(3) - 6\left(-\dfrac{1}{2}\right) \ ? \ 18$
      $15 + 3 \ ? \ 18$
      $18 = 18$

      $2y - x = -4$
      $2\left(-\dfrac{1}{2}\right) - 3 \ ? \ -4$
      $-1 - 3 \ ? \ -4$
      $-4 = -4$

      yes

4. $\begin{cases} 2x + 3y = 1 \\ 3y - x = 4 \end{cases}$

   a. $(2, 2)$:
      $2(2) + 3(2) = 1$
      $4 + 6 = 1$
      $10 = 1$
      False

   b. $(-1, 1)$:
      $2(-1) + 3(1) = 1$
      $-2 + 3 = 1$
      $1 = 1$
      True

      $3(1) - (-1) = 4$
      $3 + 1 = 4$
      $4 = 4$
      True

   c. $(2, -1)$:
      $2(2) + 3(-1) = 1$
      $4 - 3 = 1$
      $1 = 1$
      True

      $3(-1) - 2 = 4$
      $-3 - 2 = 4$
      $-5 = 4$
      False

      $(2, 2)$ is not a solution.
      $(-1, 1)$ is a solution.
      $(2, -1)$ is not a solution.

5. $\begin{cases} 2x + y = 5 \\ 3y = -x \end{cases}$

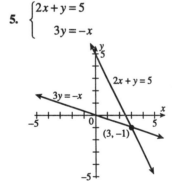

6.

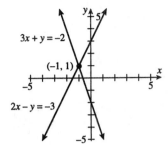

$3x + y = -2$

$(-1, 1)$

$2x - y = -3$

7. $\begin{cases} y - 2x = 4 \\ x + y = -5 \end{cases}$

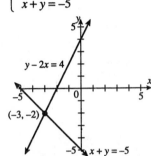

$y - 2x = 4$

$(-3, -2)$

$x + y = -5$

8. $y - 3x = 0$

$\underline{2y - 3 = 6x}$

$y \quad = 3x$

$2(3x) - 3 = 6x$

$6x - 3 = 6x$

$-3 = 0$

no solution, parallel lines

9. $\begin{cases} 3x + y = 2 \\ 3x - 6 = -9y \end{cases}$

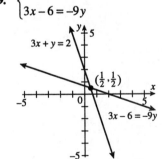

$3x + y = 2$

$(\tfrac{1}{2}, \tfrac{1}{2})$

$3x - 6 = -9y$

10. $2y + x = 2$

$x - y = 5$

$x + 2y = 2 \qquad x - (-1) = 5$

$-x + y = -5 \qquad x + 1 = 5$

$3y = -3 \qquad\quad x = 4$

$y = -1$

$(4, -1)$

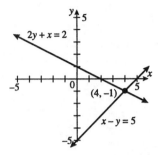

$2y + x = 2$

$(4, -1)$

$x - y = 5$

11. $\begin{cases} 2x - y = 3 \\ \quad y = 3x + 1 \end{cases}$

$2x - y = 3 \qquad\qquad y = 3x + 1$

$-y = -2x + 3 \qquad\quad m = 3,\ b = 1$

$\dfrac{-y}{-1} = \dfrac{-2}{-1}x + \dfrac{3}{-1}$

$y = 2x - 3$

$m = 2,\ b = -3$

Lines intersecting at a single point; one solution

12. $3x + y = 4$

$y = -3x + 1$

$3x + y = 4$

$y = -3x + 4$

parallel lines; no solutions

13. $\begin{cases} \dfrac{2}{3}x + \dfrac{1}{6}y = 0 \\ \qquad\quad y = -4x \end{cases}$

$\dfrac{2}{3}x + \dfrac{1}{6}y = 0 \qquad y = -4x$

$\dfrac{1}{6}y = -\dfrac{2}{3}x$

$6\left(\dfrac{1}{6}y\right) = 6\left(-\dfrac{2}{3}x\right)$

$y = -4x$

identical; infinite number of solutions

14. $\dfrac{1}{4}x + \dfrac{1}{8}y = 0$

$y = -6x$

$\dfrac{1}{4}x + \dfrac{1}{8}y = 0$

$2x + y = 0$

$y = -2x$

intersect at a single point; one solution

15. $\begin{cases} y = 2x + 6 \\ 3x - 2y = -11 \end{cases}$

Replace y with $2x + 6$ in the second equation.

$3x - 2(2x + 6) = -11$

$3x - 4x - 12 = -11$

$-x - 12 = -11$

$-x = -11 + 12$

$-x = 1$

$\dfrac{-x}{-1} = \dfrac{1}{-1}$

$x = -1$

To find y, use $x = -1$ in the equation

$y = 2x + 6$

$y = 2(-1) + 6$

$y = -2 + 6$

$y = 4$

The solution is $(-1, 4)$.

16. $\begin{cases} y = 3x - 7 \\ 2x - 3y = 7 \end{cases}$

$2x - 3(3x - 7) = 7$

$2x - 9x + 21 = 7$

$\dfrac{-7x}{-7} = \dfrac{-14}{-7}$

$x = 2$

$y = 3(2) - 7$

$y = 6 - 7$

$y = -1$

$(2, -1)$

17. $\begin{cases} x + 3y = -3 \\ 2x + y = 4 \end{cases}$

Solve the 1st equation for x.

$x = -3y - 3$

Replace x with $-3y - 3$ in the second equation.

$2(-3y - 3) + y = 4$

$-6y - 6 + y = 4$

$-5y - 6 = 4$

$-5y = 4 + 6$

$-5y = 10$

$y = \dfrac{10}{-5} = -2$

To find x, use $y = -2$ in the equation

$x = -3y - 3$

$x = -3(-2) - 3$

$x = 6 - 3$

$x = 3$

The solution is $(3, -2)$.

18. $\begin{cases} 3x + y = 11 \\ x + 2y = 12 \end{cases}$

$3x + y = 11$

$y = -3x + 11$

$x + 2(-3x + 11) = 12$

$x - 6x + 22 = 12$

$\dfrac{-5x}{-5} = \dfrac{-10}{-5} \Rightarrow x = 2$

$y = -3(2) + 11$

$y = -6 + 11$

$y = 5$

$(2, 5)$

19. $\begin{cases} 4y = 2x - 3 \\ x - 2y = 4 \end{cases}$

Solve the 2nd equation for x.

$x = 2y + 4$

Replace x with $2y + 4$ in the 1st equation.

$4y = 2(2y + 4) - 3$

$4y = 4y + 8 - 3$

$4y = 4y + 5$

$4y - 4y = 5$

$0 = 5$

This is a contradiction, $0 \neq 5$.

There is no solution.

The system is inconsistent.

20. $\begin{cases} 2x = 3y - 18 \\ x + 4y = 2 \end{cases}$

$x = 2 - 4y$

$2(2 - 4y) = 3y - 18$

$4 - 8y = 3y - 18$

$\dfrac{22}{11} = \dfrac{11y}{11}$

$2 = y$

$x = 2 - 4(2)$

$x = 2 - 8$

$x = -6$

$(-6, 2)$

21. $\begin{cases} 2(3x - y) = 7x - 5 \\ 3(x - y) = 4x - 6 \end{cases}$

Simplify the equations.

$2(3x - y) = 7x - 5$

$6x - 2y = 7x - 5$

$6x - 7x - 2y = -5$

$-x - 2y = -5$

$3(x - y) = 4x - 6$

$3x - 3y = 4x - 6$

$3x - 4x - 3y = -6$

$-x - 3y = -6$

Solve the 1st equation for x.

$-x - 2y = -5$

$-x = 2y - 5$

$\dfrac{-x}{-1} = \dfrac{2y}{-1} - \dfrac{5}{-1}$

$x = -2y + 5$

Replace x with $-2y + 5$ in the 2nd equation.

$-(-2y + 5) - 3y = -6$

$2y - 5 - 3y = -6$

$-y - 5 = -6$

$-y = -6 + 5$

$-y = -1$

$\dfrac{-y}{-1} = \dfrac{-1}{-1}$

$y = 1$

To find x, use $y = 1$ in the equation

$x = -2y + 5$

$x = -2(1) + 5$

$x = -2 + 5$

$x = 3$

The solution is (3, 1).

22. $\begin{cases} 4(x - 3y) = 3x - 1 \\ 3(4y - 3x) = 1 - 8x \end{cases}$

$4x - 12y = 3x - 1$

$x = 12y - 1$

$3[4y - 3(12y - 1)] = 1 - 8(12y - 1)$

$3[4y - 36y + 3] = 1 - 96y + 8$

$12y - 108y + 9 = 1 - 96y + 8$

$-96y + 9 = -96y + 9$

infinite solutions; dependent system

23. $\begin{cases} \dfrac{3}{4}x + \dfrac{2}{3}y = 2 \\ 3x + y = 18 \end{cases}$

Solve the 2nd equation for y.

$y = -3x + 18$

Replace y with $-3x + 18$ in the 1st equation.

$\dfrac{3}{4}x + \dfrac{2}{3}(-3x + 18) = 2$

$\dfrac{3}{4}x - 2x + 12 = 2$

$\dfrac{3}{4}x - \dfrac{8}{4}x + 12 = 2$

$-\dfrac{5}{4}x + 12 = 2$

$-\dfrac{5}{4}x = 2 - 12$

$-\dfrac{5}{4}x = -10$

$-\dfrac{4}{5}\left(-\dfrac{5}{4}x\right) = -\dfrac{4}{5}(-10)$

$x = 8$

To find y, use $x = 8$ in the equation

$y = -3x + 18$

$y = -3(8) + 18$

$y = -24 + 18$

$y = -6$

The solution is (8, –6).

24. $\begin{cases} \dfrac{2}{5}x + \dfrac{3}{4}y = 1 \\ x + 3y = -2 \end{cases}$

$x = -3y - 2$

$20\left[\dfrac{2}{5}(-3y - 2) + \dfrac{3}{4}y = 1\right]$

$8(-3y - 2) + 15y = 20$

$-24y - 16 + 15y = 20$

$\dfrac{-9y}{-9} = \dfrac{36}{-9}$

$y = -4$

$x = -3(-4) - 2$

$x = 12 - 2$

$x = 10$

(10, –4)

25. $\begin{cases} 2x + 3y = -6 \\ x - 3y = -12 \end{cases}$

Add the equations.

$\begin{array}{rcl} 2x + 3y &=& -6 \\ x - 3y &=& -12 \\ \hline 3x &=& -18 \end{array}$

$x = \dfrac{-18}{3} = -6$

To find y, use $x = -6$ in the equation

$x - 3y = -12$

$-6 - 3y = -12$

$-3y = -12 + 6$

$-3y = -6$

$y = \dfrac{-6}{-3} = 2$

The solution is (–6, 2).

26. $\begin{cases} 4x + y = 15 \\ -4x + 3y = -19 \end{cases}$

$\dfrac{4y}{4} = \dfrac{-4}{4}$

$y = -1$

$4x + (-1) = 15$

$\dfrac{4x}{4} = \dfrac{16}{4}$

$x = 4$

$(4, -1)$

27. $\begin{cases} 2x - 3y = -15 \\ x + 4y = 31 \end{cases}$

Multiply the 2nd equation by -2.

$-2(x + 4y) = -2(31)$

$-2x - 8y = -62$

Add the equations.

$\begin{array}{r} -2x - 8y = -62 \\ 2x - 3y = -15 \\ \hline -11y = -77 \end{array}$

$y = \dfrac{-77}{-11} = 7$

To find x, use $y = 7$ in the equation

$x + 4y = 31$

$x + 4(7) = 31$

$x + 28 = 31$

$x = 31 - 28$

$x = 3$

The solution is $(3, 7)$.

28. $\begin{cases} -4(x - 5y = -22) \\ 4x + 3y = 4 \end{cases}$

$\begin{array}{r} -4x + 20y = 88 \\ 4x + 3y = 4 \\ \hline 23y = 92 \end{array}$

$\dfrac{23y}{23} = \dfrac{92}{23} = 4$

$x - 5(4) = -22$

$x - 20 = -22$

$x = -2$

$(-2, 4)$

29. $\begin{cases} 2x = 6y - 1 \\ \dfrac{1}{3}x - y = -\dfrac{1}{6} \end{cases}$

Multiply the 2nd equation by 6.

$6\left(\dfrac{1}{3}x - y\right) = 6\left(-\dfrac{1}{6}\right)$

$2x - 6y = -1$

Multiply 1st equation by -1.

$-1(2x) = -(6y - 1)$

$-2x = -6y + 1$

$-2x + 6y = 1$

Add the equations.

$\begin{array}{r} 2x - 6y = -1 \\ -2x + 6y = 1 \\ \hline 0 = 0 \end{array}$

This is an identity.

The equations are dependent.

There are an infinite number of solutions.

30. $8x = 3y - 2$

$-14\left(\dfrac{4}{7}x - y = -\dfrac{5}{2}\right)$

$\begin{array}{r} 8x - 3y = -2 \\ -8x + 14y = 35 \\ \hline 11y = 33 \end{array}$

$\dfrac{11y}{11} = \dfrac{33}{11} = 3$

$8x = 3(3) - 2$

$8x = 9 - 2$

$\dfrac{8x}{8} = \dfrac{7}{8}$

$x = \dfrac{7}{8}$

$\left(\dfrac{7}{8}, 3\right)$

31. $\begin{cases} 5x = 6y + 25 \\ -2y = 7x - 9 \end{cases}$

Rearrange the equations.

$5x - 6y = 25$

$-7x - 2y = -9$

Multiply the 2nd equation by -3.

$-3(-7x - 2y) = -3(-9)$

$21x + 6y = 27$

Add the equations.

$\begin{array}{r} 21x + 6y = 27 \\ 5x - 6y = 25 \\ \hline 26x = 52 \end{array}$

$x = \dfrac{52}{26} = 2$

To find y, use $x = 2$ in the equation
$$5x - 6y = 25$$
$$5(2) - 6y = 25$$
$$10 - 6y = 25$$
$$-6y = 25 - 10$$
$$-6y = 15$$
$$y = \frac{15}{-6} = -\frac{5}{2}$$
The solution is $\left(2, \, -\frac{5}{2}\right)$.

32.
$$-4x = 8 + 6y$$
$$\underline{-3y = 2x - 3}$$
$$-4x - 6y = 8$$
$$\underline{-2(-2x - 3y = -3)}$$
$$-4x - 6y = 8$$
$$\underline{4x + 6y = 6}$$
$$0 = 2$$

no solution; inconsistent system

33. $\begin{cases} 3(x - 4) = -2y \\ 2x = 3(y - 19) \end{cases}$
Simplify each equation.
$$3(x - 4) = -2y$$
$$3x - 12 = -2y$$
$$3x + 2y - 12 = 0$$
$$3x + 2y = 12$$

$$2x = 3(y - 19)$$
$$2x = 3y - 57$$
$$2x - 3y = -57$$
Multiply the 1st equation by –2.
$$-2(3x + 2y) = -2(12)$$
$$-6x - 4y = -24$$
Multiply the 2nd equation by 3.
$$3(2x - 3y) = 3(-57)$$
$$6x - 9y = -171$$
Add the equations.
$$-6x - 4y = -24$$
$$\underline{6x - 9y = -171}$$
$$-13y = -195$$
$$y = \frac{-195}{-13} = 15$$
To find x, use $y = 15$ in the equation
$$2x - 3y = -57$$
$$2x - 3(15) = -57$$
$$2x - 45 = -57$$
$$2x = -57 + 45$$
$$2x = -12$$
$$x = \frac{-12}{2} = -6$$
The solution is $(-6, 15)$.

34.
$$4(x + 5) = -3y$$
$$\underline{3x = 2(y + 18)}$$
$$4x + 20 = -3y$$
$$\underline{3x = 2y + 36}$$
$$2(4x + 3y = -20)$$
$$\underline{3(3x - 2y = 36)}$$
$$8x + 6y = -40$$
$$\underline{9x - 6y = 108}$$
$$\frac{17x}{17} = \frac{68}{17} = 4$$

$$4(4 + 5) = -3y$$
$$\frac{36}{-3} = \frac{-3y}{-3}$$
$$-12 = y$$
$$(4, -12)$$

35. $\begin{cases} \dfrac{2x + 9}{3} = \dfrac{y + 1}{2} \\ \dfrac{x}{3} = \dfrac{y - 7}{6} \end{cases}$
Simplify the equations.
$$6\left(\frac{2x + 9}{3}\right) = 6\left(\frac{y + 1}{2}\right)$$
$$2(2x + 9) = 3(y + 1)$$
$$4x + 18 = 3y + 3$$
$$4x - 3y + 18 = 3$$
$$4x - 3y = 3 - 18$$
$$4x - 3y = -15$$
$$6\left(\frac{x}{3}\right) = 6\left(\frac{y - 7}{6}\right)$$
$$2x = y - 7$$
$$2x - y = -7$$
Multiply the 2nd equation by –3.
$$-3(2x - y) = -3(-7)$$
$$-6x + 3y = 21$$
Add the equations.
$$-6x + 3y = 21$$
$$\underline{4x - 3y = -15}$$
$$-2x = 6$$
$$x = \frac{6}{-2} = -3$$
To find y, use $x = -3$ in the equation
$$2x - y = -7$$
$$2(-3) - y = -7$$
$$-6 - y = -7$$
$$-y = -7 + 6$$
$$-y = -1$$
$$y = \frac{-1}{-1} = 1$$
The solution is $(-3, 1)$.

36. $4\left(\dfrac{2-5x}{4} = \dfrac{2y-4}{2}\right)$

$15\left(\dfrac{x+5}{3} = \dfrac{y}{5}\right)$

$2-5x = 2(2y-4)$
$\underline{5(x+5) = 3y}$
$2-5x = 4y-8$
$\underline{5x+25 = 3y}$
$-5x-4y = -10$
$\underline{5x-3y = -25}$
$\dfrac{-7y}{7} = \dfrac{-35}{-7} = 5$

$5x+25 = 3(5)$
$5x+25 = 15$
$\dfrac{5x}{5} = \dfrac{-10}{5} \Rightarrow x = -2$
$(-2, 5)$

37. Let x = smaller number
     y = larger number
$\begin{cases} x+y = 16 \\ 3y-x = 72 \end{cases}$
Solve the 1st equation for x.
$x = 16-y$
Replace x with $16-y$ in the 2nd equation.
$3y-(16-y) = 72$
$3y-16+y = 72$
$4y-16 = 72$
$4y = 72+16$
$4y = 88$
$y = \dfrac{88}{4} = 22$
To find x, use $y = 22$ in the equation
$x = 16-y$
$x = 16-22$
$x = -6$
The numbers are 22 and −6.

38. orchestra seats = x
balcony seats = y
$-45(x+y = 360)$
$\underline{45x+35y = 15150}$
$-45x-45y = -16200$
$\underline{45x+35y = 15150}$
$\dfrac{-10y}{10} = \dfrac{-1050}{-10} = 105$
$x+105 = 360$
$x = 255$
255 people

39. Let x = speed of boat in still water
     y = speed of current

$\begin{cases} 19(x-y) = 340 \\ 14(x+y) = 340 \end{cases}$
Simplify the equation.
$\dfrac{19(x-y)}{19} = \dfrac{340}{19}$
$x-y = 17.9$
$\dfrac{14(x+y)}{14} = \dfrac{340}{14}$
$x+y = 24.3$
Add the equations.
$x-y = 17.9$
$\underline{x+y = 24.3}$
$2x \quad = 42.2$
$x \quad = \dfrac{42.2}{2} = 21.1$
To find y, use $x = 21.1$ in the equation
$x+y = 24.3$
$21.1+y = 24.3$
$y = 24.3-21.1$
$y = 3.2$
Boat's speed = 21.1 mph
Current's speed = 3.2 mph

40. 6% investment = x
10% investment = y
$-6(x+y = 9000)$
$\underline{.06x+.10y = 652.80}$
$-6x-6y = -54000$
$\underline{6x+10y = 65280}$
$\dfrac{4y}{4} = \dfrac{11280}{4} = 2820$
$x+2820 = 9000$
$x = \$6180$
$\$6180$ at 6%
$\$2820$ at 10%

41. Let x = width
     y = length
$\begin{cases} 2x+2y = 6 \\ 1.6x = y \end{cases}$
Replace $y = 1.6x$ in the 1st equation.
$2x+2(1.6x) = 6$
$2x+3.2x = 6$
$5.2x = 6$
$x = \dfrac{6}{5.2} = 1.15$
To find y, use $x = 1.15$ in the equation
$y = 1.6x$
$y = 1.6(1.15)$
$y = 1.84$
width ≈ 1.15 ft.
length ≈ 1.84 ft.

42. 6% solution = x
14% solution = y

$$-6(x + y) = 50$$
$$\underline{.06x + .14y = .12(50)}$$
$$\overline{-6x - 6y = -300}$$
$$\underline{6x + 14y = \ \ \ 600}$$
$$\frac{8y}{8} = \frac{300}{8} = 37.5$$

$x + y = 50$
$x + 37.5 = 50$
$x = 12.5$
12.5 cc of 6% solution;
37.5 cc of 14% solution

43. Let x = cost of an egg
$\quad\quad y$ = cost of a strip of bacon

$$\begin{cases} 3x + 4y = 3.80 \\ 2x + 3y = 2.75 \end{cases}$$

Multiply the 1st equation by -2.
$-2(3x + 4y) = -2(3.80)$
$-6x - 8y = -7.60$
Multiply the 2nd equation by 3.
$3(2x + 3y) = 3(2.75)$
$6x + 9y = 8.25$
Add the equations.
$$-6x - 8y = -7.60$$
$$\underline{6x + 9y = \ \ 8.25}$$
$$\quad\quad\quad y = \ \ 0.65$$

To find x, use $y = 0.65$ in the equation
$2x + 3y = 2.75$
$2x + 3(0.65) = 2.75$
$2x + 1.95 = 2.75$
$2x = 2.75 - 1.95$
$2x = 0.80$
$$x = \frac{0.80}{2} = 0.40$$

One egg is $0.40 and a strip of bacon is $0.65.

44. jogging = x
walking = y

$$7.5x + 4y = 15$$
$$\underline{-4(x + y = 3)}$$
$$\frac{15}{2}x + 4y = \ \ \ 15$$
$$\underline{-4x - 4y = -12}$$
$$\frac{7}{2}x \quad\quad = 3$$
$$7x \quad\quad = 6$$
$$x \quad\quad = \frac{6}{7}$$

$$x = \frac{6}{7} = 0.86\,\text{hr.}$$

45. $\begin{cases} y \ge 2x - 3 \\ y \le -2x + 1 \end{cases}$

Graph $y = 2x + 3$
The boundary line is solid.
Choose (0, 0) as a test point.
$y \ge 2x - 3$
$0 \ ? \ 2(0) - 3$
$0 \ ? -3$
$0 \ge -3$, so shade the side containing (0, 0).

Graph $y = -2x + 1$.
The boundary line is solid.
Choose (0, 0) for a test point.
$y \le -2x + 1$
$0 \ ? -2(0) + 1$
$0 \ ? \ 1$
$0 \le 1$, so shade the side containing (0, 0).

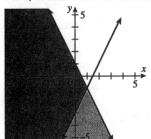

46. $y \le -3x - 3$
$\quad y \le 2x + 7$

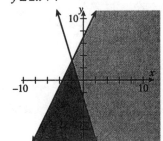

47. $\begin{cases} x + 2y > 0 \\ \quad x - y \le 6 \end{cases}$

Graph $x + 2y = 0$
The boundary line is dashed.
Choose (4, 0) as a test point.
$x + 2y > 0$
$4 + 2(0) \ ? \ 0$
$4 \ ? \ 0$
$4 > 0$, so shade the side containing (4, 0).

Graph $x - y \leq 6$.
The boundary line is solid.
Choose $(0, 0)$ as a test point.
$x - y \leq 6$
$0 - 0 ? 6$
$0 ? 6$
$0 \leq 6$, so shade the side containing $(0, 0)$.

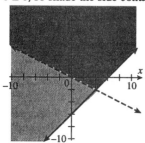

48. $x - 2y \geq 7$
$x + y \leq -5$

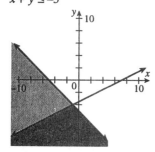

49. $\begin{cases} 3x - 2y \leq 4 \\ 2x + y \geq 5 \end{cases}$

Graph $3x - 2y = 4$.
The boundary line is solid.
Choose $(0, 0)$ as a test point.
$3x - 2y \leq 4$
$3(0) - 2(0) ? 4$
$0 ? 4$
$0 \leq 4$, so shade the side containing $(0, 0)$.

Graph $2x + y = 5$
The boundary line is solid.
Choose $(0, 0)$ as a test point.
$2x + y \geq 5$
$2(0) + 0 ? 5$
$0 ? 5$
$0 \not\geq 5$, so shade the side *not* containing $(0, 0)$.

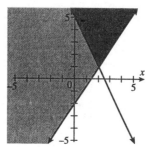

50. $4x - y \leq 0$
$3x - 2y \geq -5$
$3(1) - 2y \geq -5$
$-2y = -8$
$y = 4$
$x = -1$
$-2y = -2$
$y = 1$

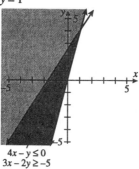

$4x - y \leq 0$
$3x - 2y \geq -5$

51. $\begin{cases} -3x + 2y > -1 \\ y < -2 \end{cases}$

Graph $-3x + 2y = -1$
The boundary line is dashed.
Choose $(0, 0)$ as a test point.
$-3x + 2y > -1$
$-3(0) + 2(0) ? -1$
$0 ? -1$
$0 > -1$, so shade the side containing $(0, 0)$.

Graph $y = -2$.
The boundary line is dashed.
Since $(0, 0)$ does not satisfy $y > -2$, shade the side not containing $(0, 0)$.

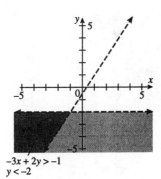

$-3x + 2y > -1$
$y < -2$

52. $-2x + 3y > -7$
$x \geq -2$
$-2(2) + 3y > -7$
$3y > -3$
$y > -1$ $\qquad (-2, -1)$
$-2(-4) + 3y > -7$
$+8 + 3y > -7$
$\dfrac{3y}{3} = \dfrac{-15}{3}$
$y = -5$ $\qquad (-4, -5)$

Chapter 8 - Test

1. $\begin{cases} 2x - 3y = 5 \\ 6x + y = 1 \end{cases}$
$(1, -1)$
$2x - 3y = 5$
$2(1) - 3(-1) ? 5$
$2 + 3 ? 5$
$5 ? 5$

$6x + y = 1$
$6(1) - 1 ? 1$
$6 - 1 ? 1$
$5 ? 1$
no

2. $\begin{cases} 4x - 3y = 24 \\ 4x + 5y = -8 \end{cases}$
$(3, -4)$

$4x - 3y = 24$
$4(3) - 3(-4) ? 24$
$12 + 12 ? 24$
$24 ? 24$

$4x + 5y = -8$
$4(3) + 5(-4) ? -8$
$12 - 20 ? -8$
$-8 ? -8$
yes

3.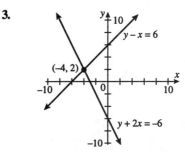

$y - x = 6$
$(-4, 2)$
$y + 2x = -6$

4. $\begin{cases} 3x - 2y = -14 \\ x + 3y = -1 \end{cases}$
Solve the 2nd equation for x.
$x = -3y - 1$
Replace x with $-3y - 1$ in the 1st equation.
$3(-3y - 1) - 2y = -14$
$-9y - 3 - 2y = -14$
$-11y - 3 = -14$
$-11y = -14 + 3$
$-11y = -11$
$y = \dfrac{-11}{-11} = 1$
To find x, use $y = 1$ in the equation
$x = -3y - 1$
$x = -3(1) - 1$
$x = -3 - 1$
$x = -4$
The solution is $(-4, 1)$.

5. $\begin{cases} \dfrac{1}{2}x + 2y = -\dfrac{15}{4} \\ 4x = -y \end{cases}$
Solve the 2nd equation for y.
$4x = -y$
$\dfrac{4x}{-1} = y$
$-4x = y$

Replace y with $-4x$ in the 1st equation.

$$\frac{1}{2}x + 2(-4x) = -\frac{15}{4}$$

$$\frac{1}{2}x - 8x = -\frac{15}{4}$$

$$-\frac{15}{2}x = -\frac{15}{4}$$

$$-\frac{2}{15}\left(-\frac{15}{2}x\right) = -\frac{2}{15}\left(-\frac{15}{4}\right)$$

$$x = \frac{2}{4} = \frac{1}{2}$$

To find y, use $x = \frac{1}{2}$ in the equation

$y = -4x$

$$y = -4\left(\frac{1}{2}\right)$$

$y = -2$

The solution is $\left(\frac{1}{2}, -2\right)$.

6. $\begin{cases} 3x + 5y = 2 \\ 2x - 3y = 14 \end{cases}$

Multiply the 1st equation by 3.

$3(3x + 5y) = 3(2)$

$9x + 15y = 6$

Multiply the 2nd equation by 5.

$5(2x - 3y) = 5(14)$

$10x - 15y = 70$

Add the equations.

$\begin{array}{r} 9x + 15y = \ \ 6 \\ 10x + 15y = 70 \\ \hline 19x \qquad\ = 76 \end{array}$

$$x \qquad = \frac{76}{19} = 4$$

To find y, use $x = 4$ in the equation

$2x - 3y = 14$

$2(4) - 3y = 14$

$8 - 3y = 14$

$-3y = 14 - 8$

$-3y = 6$

$$y = \frac{6}{-3} = -2$$

The solution is $(4, -2)$.

7. $\begin{cases} 5x - 6y = 7 \\ 7x - 4y = 12 \end{cases}$

Multiply the 1st equation by -7.

$-7(5x - 6y) = -7(7)$

$-35x + 42y = -49$

Multiply the 2nd equation by 5.

$5(7x - 4y) = 5(12)$

$35x - 20y = 60$

Add the equations.

$\begin{array}{r} -35x + 42y = -49 \\ 35x - 20y = \ \ 60 \\ \hline 22y = \ \ 11 \end{array}$

$$y = \frac{11}{22} = \frac{1}{2}$$

To find x, use $y = \frac{1}{2}$ in the equation

$5x - 6y = 7$

$$5x - 6\left(\frac{1}{2}\right) = 7$$

$5x - 3 = 7$

$5x = 7 + 3$

$5x = 10$

$$x = \frac{10}{5} = 2$$

The solution is $\left(2, \frac{1}{2}\right)$.

8. $\begin{cases} 3x + y = 7 \\ 4x + 3y = 1 \end{cases}$

Multiply the 1st equation by -3.

$-3(3x + y) = -3(7)$

$-9x - 3y = -21$

Add the equations.

$\begin{array}{r} -9x - 3y = -21 \\ 4x + 3y = \ \ \ 1 \\ \hline -5x \qquad = -20 \end{array}$

$$x \qquad = \frac{-20}{-5} = 4$$

To find y, use $x = 4$ in the equation

$3x + y = 7$

$3(4) + y = 7$

$12 + y = 7$

$y = 7 - 12$

$y = -5$

The solution is $(4, -5)$.

9. $\begin{cases} 3(2x+y) = 4x+20 \\ \quad x - 2y = 3 \end{cases}$

Simplify the 1st equation.

$6x + 3y = 4x + 20$

$6x - 4x + 3y = 20$

$2x + 3y = 20$

Multiply the 2nd equation by –2.

$-2(x - 2y) = -2(3)$

$-2x + 4y = -6$

Add the equations.

$$\begin{array}{r} 2x + 3y = 20 \\ -2x + 4y = -6 \\ \hline 7y = 14 \end{array}$$

$$y = \frac{14}{7} = 2$$

To find x, use $y = 2$ in the equation

$x - 2y = 3$

$x - 2(2) = 3$

$x - 4 = 3$

$x = 3 + 4$

$x = 7$

The solution is (7, 2).

10. $\begin{cases} \dfrac{x-3}{2} = \dfrac{2-y}{4} \\ \dfrac{7-2x}{3} = \dfrac{y}{2} \end{cases}$

Simplify each equation.

$$4\left(\frac{x-3}{2}\right) = 4\left(\frac{2-y}{4}\right)$$

$2(x - 3) = 2 - y$

$2x - 6 = 2 - y$

$2x + y - 6 = 2$

$2x + y = 8$

$$6\left(\frac{7-2x}{3}\right) = 6\left(\frac{y}{2}\right)$$

$2(7 - 2x) = 3y$

$14 - 4x = 3y$

$14 - 4x - 3y = 0$

$-4x - 3y = -14$

Multiply the 1st equation by 3.

$3(2x + y) = 3(8)$

$6x + 3y = 24$

Add the equations.

$$\begin{array}{r} -4x - 3y = -14 \\ 6x + 3y = 24 \\ \hline 2x = 10 \end{array}$$

$$x = \frac{10}{2} = 5$$

To find y, use $x = 5$ in the equation

$2x + y = 8$

$2(5) + y = 8$

$10 + y = 8$

$y = 8 - 10$

$y = -2$

The solution is (5, –2).

11. Let x = number of \$1 bills

  y = number of \$5 bills

$\begin{cases} x + y = 62 \\ 1x + 5y = 230 \end{cases}$

Multiply the 1st equation by –1.

$-1(x + y) = -1(62)$

$-x - y = -62$

Add the equations.

$$\begin{array}{r} x + 5y = 230 \\ -x - y = -62 \\ \hline 4y = 168 \end{array}$$

$$y = \frac{168}{4} = 42$$

To find x, use $y = 42$ in the equation

$x + y = 62$

$x + 42 = 62$

$x = 62 - 42$

$x = 20$

$20 \rightarrow \$1$ bills

$42 \rightarrow \$5$ bills

12. Let x = money at 5%

  y = money at 9%

$\begin{cases} x + y = 4000 \\ 0.05x + 0.09y = 311 \end{cases}$

Solve the 1st equation for x.

$x = 4000 - y$

Replace x with $4000 - y$ in the second equation.

$0.05(4000 - y) + 0.09y = 311$

$200 - 0.05y + 0.09y = 311$

$200 + 0.04y = 311$

$0.04y = 311 - 200$

$0.04y = 111$

$$y = \frac{111}{0.04} = 2775$$

To find x, use $y = 2775$ in the equation

$x = 4000 - y$

$x = 4000 - 2775$

$x = 1225$

\$1225 at 5%

\$2775 at 9%

13. $\begin{cases} y + 2x \le 4 \\ \quad\quad y \ge 2 \end{cases}$

Graph $y + 2x = 4$
The boundary line is solid.
Choose $(0, 0)$ as a test point.
$y + 2x \le 4$
$0 + 2(0) \,?\, 4$
$0 \,?\, 4$
$0 \le 4$, so shade the side containing $(0, 0)$.

Graph $y = 2$.
The boundary line is solid.
Since $(0, 0)$, does *not* satisfy $y \ge 2$, shade the side *not* containing $(0, 0)$.

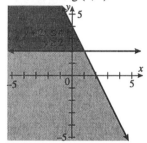

14. $\begin{cases} 2y - x \ge 1 \\ x + y \ge -4 \end{cases}$

Graph $2y - x = 1$
The boundary line is solid.
Choose $(0, 0)$ as a test point.
$2y - x \ge 1$
$2(0) - 0 \,?\, 1$
$0 \,?\, 1$
$0 \not\ge 1$, so shade the side *not* containing $(0, 0)$.

Graph $x + y = -4$
The boundary line is solid.
Choose $(0, 0)$ as a test point.
$x + y \ge -4$
$0 + 0 \,?\, -4$
$0 \,?\, -4$
$0 \ge -4$, so shade the side containing $(0, 0)$.

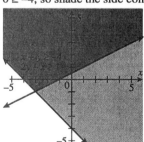

Chapter 8 - Cumulative Review

1. a. $\dfrac{4}{5} \div \dfrac{5}{16} = \dfrac{4}{5} \cdot \dfrac{16}{5} = \dfrac{64}{25}$

   b. $\dfrac{7}{10} \div 14 = \dfrac{7}{10} \cdot \dfrac{1}{14} = \dfrac{1}{20}$

   c. $\dfrac{3}{8} \div \dfrac{3}{10} = \dfrac{3}{8} \cdot \dfrac{10}{3} = \dfrac{5}{4}$

2. $2 - 1 - 3 = -2$; $2 loss

3. a. $\dfrac{1}{22}$ since $22 \cdot \dfrac{1}{22} = 1$

   b. $\dfrac{16}{3}$ since $\dfrac{3}{16} \cdot \dfrac{16}{3} = 1$

   c. $-\dfrac{1}{10}$ since $-10 \cdot \left(-\dfrac{1}{10}\right) = 1$

   d. $-\dfrac{13}{9}$ since $-\dfrac{9}{13} \cdot \left(-\dfrac{13}{9}\right) = 1$

4. a. $x + 3 = 8$; $x = 5$

   b. $8 - x$

5. $-2(x - 5) + 10 = -3(x + 2) + x$
   $-2x + 10 + 10 = -3x - 6 + x$
   $-2x + 20 = -2x - 6$
   $-2x + 2x + 20 = -2x + 2x - 6$
   $20 = -6$
   False
   No solution

6. a. $0.73 = 73\%$

   b. $1.39 = 139\%$

   c. $\dfrac{1}{4} = 0.25 = 25\%$

7. $2x + 7 \le x - 11$
   $2x - x + 7 \le x - x - 11$
   $x + 7 \le -11$
   $x + 7 - 7 \le -11 - 7$
   $x \le -18$

8. $x - 3y = 6$

If $x = 0$, then   If $x = 6$, then
$0 - 3y = 6$ $6 - 3y = 6$
$-3y = 6$ $-3y = 0$
$y = -2$ $y = 0$
$(0, -2)$ $(6, 0)$

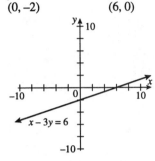

9. $x > 2y$

Points: $(0, 0)$; $(2, 1)$
Line is dashed
Test point: $(2, 0)$
Since $2 > 2(0)$, shade the side with point $(2, 0)$.

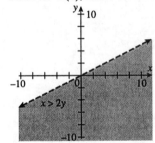

10. a. $\left(\dfrac{-5x^2}{y^3}\right)^2 = \dfrac{(-5)^2 x^{2 \cdot 2}}{y^{3 \cdot 2}} = \dfrac{25x^4}{y^6}$

b. $\dfrac{(x^3)^4 x}{x^7} = \dfrac{x^{12} x}{x^7} = x^{12+1-7} = x^6$

c. $\dfrac{(2x)^5}{x^3} = \dfrac{2^5 x^5}{x^3} = 32x^{5-3} = 32x^2$

d. $\dfrac{(a^2 b)^3}{a^3 b^2} = \dfrac{a^6 b^3}{a^3 b^2} = a^{6-3} b^{3-2} = a^3 b$

11.
$-2x^2 + 5x - 1$
$\underline{+-2x^2 + x + 3}$
$-4x^2 + 6x + 2$

12.

$$2x+3\overline{)8x^3 + 4x^2 + 0x + 7}$$

with quotient $4x^2 - 4x + 6$

$\underline{8x^3 + 12x^2}$
$-8x^2 + 0x$
$\underline{-8x^2 - 12x}$
$12x + 7$
$\underline{12x + 18}$
-11

$4x^2 - 4x + 6 - \dfrac{11}{2x+3}$

13. $x(2x - 7) = 4$
$2x^2 - 7x = 4$
$2x^2 - 7x - 4 = 0$
$(2x + 1)(x - 4) = 0$
$2x + 1 = 0 \qquad$ or $\quad x - 4 = 0$
$x = -\dfrac{1}{2} \qquad$ or $\quad x = 4$

14. Let x = short side
$x + 2$ = medium side
$x + 4$ = long side
$x^2 + (x+2)^2 = (x+4)^2$
$x^2 + x^2 + 4x + 4 = x^2 + 8x + 16$
$2x^2 + 4x + 4 = x^2 + 8x + 16$
$x^2 - 4x - 12 = 0$
$(x - 6)(x + 2) = 0$
$x = 6$
$x + 2 = 8$
$x + 4 = 10$

15. $\dfrac{2y}{2y-7} - \dfrac{7}{2y-7} = \dfrac{2y-7}{2y-7} = 1$

16. $m = \dfrac{3}{4}$

17. $\begin{cases} 2x - 3y = 6 \\ x = 2y \end{cases}$

a. $(12, 6)$
$2(12) - 3(6) = 6$
$24 - 18 = 6$
$6 = 6$

$12 = 2(6)$
$12 = 12$

yes

b. (0, –2)
$$2(0) - 3(-2) = 6$$
$$0 + 6 = 6$$
$$6 = 6$$

$$0 \neq 2(-2)$$
$$0 \neq -4$$

no

18. $\frac{1}{2}x - y = 2$ $x = 2y + 5$

$-y = -\frac{1}{2}x + 2$ $x - 5 = 2y$

$y = \frac{1}{2}x - 2$ $\frac{x}{2} - \frac{5}{2} = y$

$$y = \frac{1}{2}x - \frac{5}{2}$$

slope is same: $m = \frac{1}{2}$

y-intercepts are different
parallel lines
no solutions

19. $x + 2y = 7$
$2x + 2y = 13$

$$-2(x + 2y) = -2(7)$$
$$2x + 2y = 13$$

$$-2x - 4y = -14$$
$$2x + 2y = \ \ 13$$
$$\overline{-2y = -1}$$
$$y = \ \frac{1}{2}$$

$x + 2y = 7$

$x + 2\left(\frac{1}{2}\right) = 7$

$x + 1 = 7$
$x = 6$

$\left(6, \frac{1}{2}\right)$

20. $-x - \frac{y}{2} = \frac{5}{2}$

$-\frac{x}{2} + \frac{y}{4} = 0$

$-2\left(-x - \frac{y}{2}\right) = -2\left(\frac{5}{2}\right)$

$4\left(-\frac{x}{2} + \frac{y}{4}\right) = 4(0)$

$$2x + y = -5$$
$$-2x + y = \ \ 0$$
$$\overline{2y = -5}$$
$$y = -\frac{5}{2}$$

$-x - \frac{y}{2} = \frac{5}{2}$

$-x - \frac{\left(-\frac{5}{2}\right)}{2} = \frac{5}{2}$

$-x + \frac{5}{4} = \frac{5}{2}$

$-x = \frac{5}{4}$

$x = -\frac{5}{4}$

$\left(-\frac{5}{4}, -\frac{5}{2}\right)$

21. Let x = one number
then $37 - x$ = other number
$(37 - x) - x = 21$
$37 - 2x = 21$
$-2x = -16$
$x = 8$
$37 - 8 = 29$
29 and 8

22.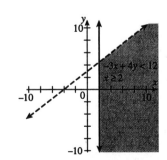

Chapter 9

Exercise Set 9.1

1. $\sqrt{16} = 4$

3. $\sqrt{81} = 9$

5. $\sqrt{\dfrac{1}{25}} = \dfrac{1}{5}$

7. $-\sqrt{100} = -(10) = -10$

9. $\sqrt[3]{64} = 4$

11. $-\sqrt[3]{27} = -(3) = -3$

13. $\sqrt[3]{\dfrac{1}{8}} = \dfrac{1}{2}$

15. $\sqrt[3]{-125} = -5$

17. $\sqrt[5]{32} = 2$

19. $\sqrt[4]{81} = 3$

21. $\sqrt{-4}$; not a real number

23. $\sqrt[3]{\dfrac{1}{27}} = \dfrac{1}{3}$

25. $\sqrt{\dfrac{9}{25}} = \dfrac{3}{5}$

27. $-\sqrt{49} = -(7) = -7$

29. Answers may vary.

31. $\sqrt{z^2} = z$

33. $\sqrt{x^4} = x^2$

35. $\sqrt{9x^8} = 3x^4$

37. $\sqrt{x^2 y^6} = xy^3$

39. $\sqrt[3]{x^{15}} = x^5$

41. $\sqrt{x^{12}} = x^6$

43. $\sqrt{81x^2} = 9x$

45. $-\sqrt{144y^{14}} = -(12y^7) = -12y^7$

47. $\sqrt{x^2 y^2} = xy$

49. $\sqrt{16x^{16}} = 4x^8$

51. $\sqrt{0} = 0$

53. $-\sqrt[5]{\dfrac{1}{32}} = -\left(\dfrac{1}{2}\right) = -\dfrac{1}{2}$

55. $\sqrt{-64}$; not a real number

57. $-\sqrt{64} = -(8) = -8$

59. $-\sqrt{169} = -(13) = -13$

61. $\sqrt{1} = 1$

63. $\sqrt{\dfrac{25}{64}} = \dfrac{5}{8}$

65. $-\sqrt[3]{-8} = -(-2) = 2$

67. $\sqrt{\sqrt{81}} = \sqrt{9} = 3$

69. $\sqrt{9}$; rational, 3

71. $\sqrt{37}$; irrational, 6.083

73. $\sqrt{169}$; rational, 13

75. $\sqrt{4}$; rational, 2

77. $\sqrt{324} = 18$

79. $90 \cdot \sqrt{2} = 90 \cdot (1.41) = 126.90$ feet

81.

x	y
-8	-2
-2	-1.3 (approx.)
-1	-1
0	0
1	1
2	1.3 (approx.)
8	2

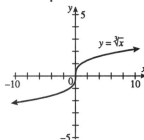

$y = \sqrt[3]{x}$

83. $y = \sqrt{x+3}$
$(-3, 0)$

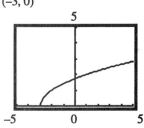

85. $y = \sqrt{x-5}$
$(5, 0)$

87. $8 = 4 \cdot 2$

89. $75 = 25 \cdot 3$

91. $44 = 4 \cdot 11$

93. $90 = 9 \cdot 10$

Section 9.2 Mental Math

1. $\sqrt{4 \cdot 9} = 2 \cdot 3 = 6$

3. $\sqrt{x^2} = x$

5. $\sqrt{0} = 0$

7. $\sqrt{25x^4} = \sqrt{25 \cdot x^2 \cdot x^2} = 5 \cdot x \cdot x = 5x^2$

Exercise Set 9.2

1. $\sqrt{20} = \sqrt{4}\sqrt{5} = 2\sqrt{5}$

3. $\sqrt{18} = \sqrt{9}\sqrt{2} = 3\sqrt{2}$

5. $\sqrt{50} = \sqrt{25}\sqrt{2} = 5\sqrt{2}$

7. $\sqrt{33}$ cannot be simplified.

9. $\sqrt[3]{24} = \sqrt[3]{8}\sqrt[3]{3} = 2\sqrt[3]{3}$

11. $\sqrt[3]{250} = \sqrt[3]{125}\sqrt[3]{2} = 5\sqrt[3]{2}$

13. Answers may vary.

15. $\sqrt{\dfrac{8}{25}} = \dfrac{\sqrt{8}}{\sqrt{25}} = \dfrac{\sqrt{4}\sqrt{2}}{5} = \dfrac{2\sqrt{2}}{5}$

17. $\sqrt{\dfrac{27}{121}} = \dfrac{\sqrt{27}}{\sqrt{121}} = \dfrac{\sqrt{9}\sqrt{3}}{\sqrt{121}} = \dfrac{3\sqrt{3}}{11}$

19. $\sqrt{\dfrac{9}{4}} = \dfrac{\sqrt{9}}{\sqrt{4}} = \dfrac{3}{2}$

21. $\sqrt{\dfrac{125}{9}} = \dfrac{\sqrt{25}\sqrt{5}}{\sqrt{9}} = \dfrac{5\sqrt{5}}{3}$

23. $\sqrt[3]{\dfrac{5}{64}} = \dfrac{\sqrt[3]{5}}{\sqrt[3]{64}} = \dfrac{\sqrt[3]{5}}{4}$

25. $\sqrt[3]{\dfrac{7}{8}} = \dfrac{\sqrt[3]{7}}{\sqrt[3]{8}} = \dfrac{\sqrt[3]{7}}{2}$

27. $\sqrt{x^7} = \sqrt{x^6}\sqrt{x} = x^3\sqrt{x}$

29. $\sqrt{\dfrac{88}{x^4}} = \dfrac{\sqrt{88}}{\sqrt{x^4}} = \dfrac{\sqrt{4}\sqrt{22}}{\sqrt{x^4}} = \dfrac{2\sqrt{22}}{x^2}$

31. $\sqrt[3]{x^{16}} = \sqrt[3]{x^{15}}\sqrt[3]{x} = x^5\sqrt[3]{x}$

33. $\sqrt[3]{\dfrac{2}{x^9}} = \dfrac{\sqrt[3]{2}}{\sqrt[3]{x^9}} = \dfrac{\sqrt[3]{2}}{x^3}$

35. $\sqrt{60} = \sqrt{4}\sqrt{15} = 2\sqrt{15}$

37. $\sqrt{180} = \sqrt{36}\sqrt{5} = 6\sqrt{5}$

39. $\sqrt{52} = \sqrt{4}\sqrt{13} = 2\sqrt{13}$

41. $\sqrt{\dfrac{11}{36}} = \dfrac{\sqrt{11}}{\sqrt{36}} = \dfrac{\sqrt{11}}{6}$

43. $-\sqrt{\dfrac{27}{144}} = -\dfrac{\sqrt{27}}{\sqrt{144}} = -\dfrac{\sqrt{9}\sqrt{3}}{12} = -\dfrac{3\sqrt{3}}{12} = -\dfrac{\sqrt{3}}{4}$

45. $\sqrt[3]{\dfrac{15}{64}} = \dfrac{\sqrt[3]{15}}{\sqrt[3]{64}} = \dfrac{\sqrt[3]{15}}{4}$

47. $\sqrt[3]{80} = \sqrt[3]{8}\sqrt[3]{10} = 2\sqrt[3]{10}$

49. $\sqrt[4]{48} = \sqrt[4]{16}\sqrt[4]{3} = 2\sqrt[4]{3}$

51. $\sqrt{x^{13}} = \sqrt{x^{12}}\sqrt{x} = x^6\sqrt{x}$

53. $\sqrt{75x^2} = \sqrt{25x^2}\sqrt{3} = 5x\sqrt{3}$

55. $\sqrt{96x^4y^2} = \sqrt{16x^4y^2}\sqrt{6} = 4x^2y\sqrt{6}$

57. $\sqrt{\dfrac{12}{y^2}} = \dfrac{\sqrt{12}}{\sqrt{y^2}} = \dfrac{\sqrt{4}\sqrt{3}}{y} = \dfrac{2\sqrt{3}}{y}$

59. $\sqrt{\dfrac{9x}{y^2}} = \dfrac{\sqrt{9x}}{\sqrt{y^2}} = \dfrac{\sqrt{9}\sqrt{x}}{y} = \dfrac{3\sqrt{x}}{y}$

61. $\sqrt[3]{-8x^6} = -2x^2$

63. $\sqrt[3]{80} = \sqrt[3]{8}\sqrt[3]{10} = 2\sqrt[3]{10}$

65. $C = 100\sqrt[3]{n} + 700 = 100\sqrt[3]{1000} + 700$
$= 100 \cdot 10 + 700 = 1000 + 700 = \$1700$

67. $6x + 8x = 14x$

69. $(2x + 3)(x - 5)$
$= 2x^2 - 10x + 3x - 15$
$= 2x^2 - 7x - 15$

71. $9y^2 - 9y^2 = 0$

73. $\dfrac{x}{12} = \dfrac{6}{9}$
$9x = 12 \cdot 6$
$\dfrac{9x}{9} = \dfrac{72}{9}$
$x = 8 \text{ cm}$

Section 9.3 Mental Math

1. $3\sqrt{2} + 5\sqrt{2} = 8\sqrt{2}$

3. $5\sqrt{x} + 2\sqrt{x} = 7\sqrt{x}$

5. $5\sqrt{7} - 2\sqrt{7} = 3\sqrt{7}$

Exercise Set 9.3

1. $4\sqrt{3} - 8\sqrt{3} = -4\sqrt{3}$

3. $3\sqrt{6} + 8\sqrt{6} - 2\sqrt{6} - 5 = 9\sqrt{6} - 5$

5. $6\sqrt{5} - 5\sqrt{5} + \sqrt{2} = \sqrt{5} + \sqrt{2}$

7. $2\sqrt[3]{3} + 5\sqrt[3]{3} - \sqrt{3} = 7\sqrt[3]{3} - \sqrt{3}$

9. $2\sqrt[3]{2} - 7\sqrt[3]{2} - 6 = -5\sqrt[3]{2} - 6$

11. $3\sqrt{5} + 3\sqrt{5} + \sqrt{5} + \sqrt{5} = 8\sqrt{5}$ inches

13. Answers may vary.

15. $\sqrt{12} + \sqrt{27} = \sqrt{4 \cdot 3} + \sqrt{9 \cdot 3} = 2\sqrt{3} + 3\sqrt{3} = 5\sqrt{3}$

17. $\sqrt{45} + 3\sqrt{20} = \sqrt{9 \cdot 5} + 3\sqrt{4 \cdot 5} = 3\sqrt{5} + 3(2)\sqrt{5}$
$= 3\sqrt{5} + 6\sqrt{5} = 9\sqrt{5}$

19. $2\sqrt{8} - \sqrt{128} + \sqrt{48} + \sqrt{18}$
$= 2\sqrt{4 \cdot 2} - \sqrt{64 \cdot 2} + \sqrt{16 \cdot 3} + \sqrt{9 \cdot 2}$
$= 4\sqrt{2} - 8\sqrt{2} + 4\sqrt{3} + 3\sqrt{2} = -4\sqrt{2} + 4\sqrt{3} + 3\sqrt{2}$
$= -1\sqrt{2} + 4\sqrt{3}$

21. $\sqrt[3]{32} - \sqrt[3]{4} = \sqrt[3]{8 \cdot 4} - \sqrt[3]{4} = 2\sqrt[3]{4} - \sqrt[3]{4} = \sqrt[3]{4}$

23. $4x - 3\sqrt{x^2} + \sqrt{x} = 4x - 3x + \sqrt{x} = x + \sqrt{x}$

25. $\sqrt{25x} + \sqrt{36x} - 11\sqrt{x} = 5\sqrt{x} + 6\sqrt{x} - 11\sqrt{x}$
$= 11\sqrt{x} - 11\sqrt{x} = 0$

27. $\sqrt{16x} - \sqrt{x^3} = 4\sqrt{x} - \sqrt{x^2 \cdot x} = 4\sqrt{x} - x\sqrt{x}$

29. $12\sqrt{5} - \sqrt{5} - 4\sqrt{5} = 11\sqrt{5} - 4\sqrt{5} = 7\sqrt{5}$

31. $\sqrt{5} + \sqrt{5} = 2\sqrt{5}$

33. $6 - 2\sqrt{3} - \sqrt{3} = 6 - 3\sqrt{3}$

35. $\sqrt{75} + \sqrt{48} = \sqrt{25 \cdot 3} + \sqrt{16 \cdot 3} = 5\sqrt{3} + 4\sqrt{3} = 9\sqrt{3}$

37. $2\sqrt{80} - \sqrt{45} = 2\sqrt{16 \cdot 5} - \sqrt{9 \cdot 5}$
$= 8\sqrt{5} - 3\sqrt{5} = 5\sqrt{5}$

39. $\sqrt{6} + \sqrt{16} + \sqrt{24} + \sqrt{25} = \sqrt{6} + 4 + \sqrt{4 \cdot 6} + 5$
$= \sqrt{6} + 9 + 2\sqrt{6} = 3\sqrt{6} + 9$

41. $\sqrt{\dfrac{3}{64}} + \sqrt{\dfrac{3}{16}} = \dfrac{\sqrt{3}}{\sqrt{64}} + \dfrac{\sqrt{3}}{\sqrt{16}} = \dfrac{\sqrt{3}}{8} + \dfrac{\sqrt{3}}{4}$
$\dfrac{\sqrt{3} + 2\sqrt{3}}{8} = \dfrac{3\sqrt{3}}{8}$

43. $2\sqrt[3]{8} + 2\sqrt[3]{16} = 2(2) + 2\sqrt[3]{8 \cdot 2} = 4 + 4\sqrt[3]{2}$

45. $\sqrt[3]{8} + \sqrt[3]{54} - 5 = 2 + \sqrt[3]{27 \cdot 2} - 5 = -3 + 3\sqrt[3]{2}$

47. $5\sqrt{18} + 2\sqrt{32} = 5\sqrt{9 \cdot 2} + 2\sqrt{16 \cdot 2} = 15\sqrt{2} + 8\sqrt{2}$
$= 23\sqrt{2}$

49. $5\sqrt{2xz^2} + z\sqrt{98x} = 5z\sqrt{2x} + z\sqrt{49 \cdot 2x}$
$= 5z\sqrt{2x} + 7z\sqrt{2x} = 12z\sqrt{2x}$

51. $5\sqrt{x} + 4\sqrt{4x} = 5\sqrt{x} + 8\sqrt{x} = 13\sqrt{x}$

53. $\sqrt{3x^3} + 3x\sqrt{x} = \sqrt{3x \cdot x^2} + 3x\sqrt{x}$
$= x\sqrt{3x} + 3x\sqrt{x}$

55. $\sqrt{32x^2} + \sqrt[3]{32x^2} + \sqrt{4x^2}$
$= \sqrt{16 \cdot 2x^2} + \sqrt[3]{8 \cdot 4x^2} + 2x$
$= 4x\sqrt{2} + 2\sqrt[3]{4x^2} + 2x$

57. $\sqrt{40x} + \sqrt[3]{40x^4} - 2\sqrt{10x} - \sqrt[3]{5x^4}$
$= \sqrt{4 \cdot 10x} + \sqrt[3]{8 \cdot 5x^3 \cdot x} - 2\sqrt{10x} - \sqrt[3]{5x^3 \cdot x}$
$= 2\sqrt{10x} + 2x\sqrt[3]{5x} - 2\sqrt{10x} - x\sqrt[3]{5x} = x\sqrt[3]{5x}$

59. $2(3 \cdot 8) + 2\left(\dfrac{3\sqrt{27}}{4}\right) = 2(24) + \dfrac{3\sqrt{27}}{2}$
$= 48 + 3\dfrac{\sqrt{9 \cdot 3}}{2} = 48 + \dfrac{9\sqrt{3}}{2} = 48 + \dfrac{9\sqrt{3}}{2}$ sq. ft.

61. $(x+6)^2 = (x)^2 + 2(x)(6) + (6)^2 = x^2 + 12x + 36$

63. $(2x-1)^2 = (2x)^2 + 2(2x)(-1) + (1)^2$
$= 4x^2 - 4x + 1$

65. $\begin{cases} x = 2y \\ x + 5y = 14 \end{cases}$
Replace x with $2y$ in the 2nd equation.
$2y + 5y = 14$
$7y = 14$
$y = \dfrac{14}{7} = 2$
To find x, use $y = 2$ in the equation,
$x = 2y$
$x = 2(2)$
$x = 4$
The solution is $(4, 2)$.

Section 9.4 Mental Math

1. $\sqrt{2} \cdot \sqrt{3} = \sqrt{2 \cdot 3} = \sqrt{6}$

3. $\sqrt{1} \cdot \sqrt{6} = \sqrt{1 \cdot 6} = \sqrt{6}$

5. $\sqrt{10} \cdot \sqrt{y} = \sqrt{10y}$

Exercise Set 9.4

1. $\sqrt{8}\sqrt{2} = \sqrt{16} = 4$

3. $\sqrt{10}\sqrt{5} = \sqrt{50} = \sqrt{25}\sqrt{2} = 5\sqrt{2}$

5. $\sqrt[3]{12}\sqrt[3]{4} = \sqrt[3]{48} = \sqrt[3]{8}\sqrt[3]{6} = 2\sqrt[3]{6}$

7. $\sqrt{10}(\sqrt{2} + \sqrt{5}) = \sqrt{20} + \sqrt{50} = \sqrt{4}\sqrt{5} + \sqrt{25}\sqrt{2}$
$= 2\sqrt{5} + 5\sqrt{2}$

9. $(3\sqrt{5} - \sqrt{10})(\sqrt{5} - 4\sqrt{3})$
$= 3\sqrt{25} - 12\sqrt{15} - \sqrt{50} + 4\sqrt{30}$
$= 3(5) - 12\sqrt{15} - \sqrt{25}\sqrt{2} + 4\sqrt{30}$
$= 15 - 12\sqrt{15} - 5\sqrt{2} + 4\sqrt{30}$

11. $(\sqrt{x} + 6)(\sqrt{x} - 6) = (\sqrt{x})^2 - (6)^2 = x - 36$

13. $(\sqrt{3} + 8)^2 = (\sqrt{3})^2 + 2(\sqrt{3})(8) + (8)^2$
$= 3 + 16\sqrt{3} + 64 = 67 + 16\sqrt{3}$

15. $A = lw$
$A = (13\sqrt{2})(5\sqrt{6})$
$A = 65\sqrt{12}$
$A = 65\sqrt{4}\sqrt{3}$
$A = 65(2)\sqrt{3}$
$A = 130\sqrt{3}$ sq. meters

17. $\dfrac{\sqrt{32}}{\sqrt{2}} = \sqrt{\dfrac{32}{2}} = \sqrt{16} = 4$

19. $\dfrac{\sqrt{90}}{\sqrt{5}} = \sqrt{\dfrac{90}{5}} = \sqrt{18} = \sqrt{9}\sqrt{2} = 3\sqrt{2}$

21. $\dfrac{\sqrt{75y^5}}{\sqrt{3y}} = \sqrt{\dfrac{75y^5}{3y}} = \sqrt{25y^4} = 5y^2$

23. $\sqrt{\dfrac{3}{5}} = \dfrac{\sqrt{3}}{\sqrt{5}} \cdot \dfrac{\sqrt{5}}{\sqrt{5}} = \dfrac{\sqrt{15}}{5}$

25. $\dfrac{1}{\sqrt{6y}} \cdot \dfrac{\sqrt{6y}}{\sqrt{6y}} = \dfrac{\sqrt{6y}}{6y}$

27. $\sqrt{\dfrac{5}{18}} = \dfrac{\sqrt{5}}{\sqrt{18}} = \dfrac{\sqrt{5}}{\sqrt{9}\sqrt{2}} = \dfrac{\sqrt{5}}{3\sqrt{2}} \cdot \dfrac{\sqrt{2}}{\sqrt{2}} = \dfrac{\sqrt{10}}{3(2)} = \dfrac{\sqrt{10}}{6}$

29. $\dfrac{6}{\sqrt[3]{2}} = \dfrac{6}{\sqrt[3]{2}} \cdot \dfrac{\sqrt[3]{4}}{\sqrt[3]{4}} = \dfrac{6\sqrt[3]{4}}{\sqrt[3]{8}} = \dfrac{6\sqrt[3]{4}}{2} = 3\sqrt[3]{4}$

31. $\sqrt[3]{\dfrac{1}{9}} = \dfrac{\sqrt[3]{1}}{\sqrt[3]{9}} = \dfrac{1}{\sqrt[3]{9}} \cdot \dfrac{\sqrt[3]{3}}{\sqrt[3]{3}} = \dfrac{\sqrt[3]{3}}{\sqrt[3]{27}} = \dfrac{\sqrt[3]{3}}{3}$

33. $\sqrt[3]{\dfrac{2}{9x^2}} = \dfrac{\sqrt[3]{2}}{\sqrt[3]{9x^2}} \cdot \dfrac{\sqrt[3]{3x}}{\sqrt[3]{3x}} = \dfrac{\sqrt[3]{6x}}{\sqrt[3]{27x^3}} = \dfrac{\sqrt[3]{6x}}{3x}$

35. $r = \sqrt{\dfrac{A}{\pi}} = \dfrac{\sqrt{A}}{\sqrt{\pi}} \cdot \dfrac{\sqrt{\pi}}{\sqrt{\pi}} = \dfrac{\sqrt{A\pi}}{\sqrt{\pi^2}} = \dfrac{\sqrt{A\pi}}{\pi}$

37. Answers may vary.

39. $\dfrac{3}{\sqrt{2} + 1} = \dfrac{3(\sqrt{2} - 1)}{(\sqrt{2} + 1)(\sqrt{2} - 1)} = \dfrac{3(\sqrt{2} - 1)}{2 - 1}$
$= \dfrac{3\sqrt{2} - 3}{1} = 3\sqrt{2} - 3$

41. $\dfrac{2}{\sqrt{10} - 3} = \dfrac{2(\sqrt{10} + 3)}{(\sqrt{10} - 3)(\sqrt{10} + 3)} = \dfrac{2(\sqrt{10} + 3)}{10 - 9}$
$= \dfrac{2\sqrt{10} + 6}{1} = 2\sqrt{10} + 6$

43. $\dfrac{\sqrt{5} + 1}{\sqrt{6} - \sqrt{5}} = \dfrac{(\sqrt{5} + 1)(\sqrt{6} + \sqrt{5})}{(\sqrt{6} - \sqrt{5})(\sqrt{6} + \sqrt{5})}$
$= \dfrac{\sqrt{30} + \sqrt{25} + \sqrt{6} + \sqrt{5}}{6 - 5} = \dfrac{\sqrt{30} + 5 + \sqrt{6} + \sqrt{5}}{1}$
$= \sqrt{30} + 5 + \sqrt{6} + \sqrt{5}$

45. $\dfrac{6 + 2\sqrt{3}}{2} = \dfrac{2(3 + \sqrt{3})}{2} = 3 + \sqrt{3}$

47. $\dfrac{18 - 12\sqrt{5}}{6} = \dfrac{6(3 - 2\sqrt{5})}{6} = 3 - 2\sqrt{5}$

49. $\dfrac{15\sqrt{3} + 5}{5} = \dfrac{5(3\sqrt{3} + 1)}{5} = 3\sqrt{3} + 1$

51. $2\sqrt{3} \cdot 4\sqrt{15} = 8\sqrt{45} = 8\sqrt{9}\sqrt{5} = 8(3)\sqrt{5} = 24\sqrt{5}$

53. $(2\sqrt{5})^2 = 4(5) = 20$

55. $(6\sqrt{x})^2 = 36x$

57. $\sqrt{6}(\sqrt{5} + \sqrt{7}) = \sqrt{30} + \sqrt{42}$

59. $4\sqrt{5x}(\sqrt{x} - 3\sqrt{5}) = 4\sqrt{5x^2} - 12\sqrt{25x}$
$= 4\sqrt{x^2}\sqrt{5} - 12\sqrt{25}\sqrt{x} = 4x\sqrt{5} - 12(5)\sqrt{x}$
$= 4x\sqrt{5} - 60\sqrt{x}$

61. $(\sqrt{3} + \sqrt{5})(\sqrt{2} - \sqrt{5}) = \sqrt{6} - \sqrt{15} + \sqrt{10} - \sqrt{25}$
$= \sqrt{6} - \sqrt{15} + \sqrt{10} - 5$

63. $(\sqrt{7} - 2\sqrt{3})(\sqrt{7} + 2\sqrt{3}) = (\sqrt{7})^2 - (2\sqrt{3})^2$
$= 7 - 4(3) = 7 - 12 = -5$

65. $(\sqrt{x} - 3)(\sqrt{x} + 3) = (\sqrt{x})^2 - (3)^2 = x - 9$

67. $(\sqrt{6} + 3)^2 = (\sqrt{6})^2 + 2(\sqrt{6})(3) + (3)^2$
$= 6 + 6\sqrt{6} + 9 = 15 + 6\sqrt{6}$

69. $(3\sqrt{x} - 5)^2 = (3\sqrt{x})^2 + 2(3\sqrt{x})(-5) + (5)^2$
$= 9x - 30\sqrt{x} + 25$

71. $\dfrac{\sqrt{150}}{\sqrt{2}} = \sqrt{\dfrac{150}{2}} = \sqrt{75} = \sqrt{25}\sqrt{3} = 5\sqrt{3}$

73. $\dfrac{\sqrt{72y^5}}{\sqrt{3y^3}} = \sqrt{\dfrac{72y^5}{3y^3}} = \sqrt{24y^2} = \sqrt{4y^2}\sqrt{6} = 2y\sqrt{6}$

75. $\dfrac{\sqrt{24x^3y^4}}{\sqrt{2xy}} = \sqrt{\dfrac{24x^3y^4}{2xy}} = \sqrt{12x^2y^3}$
$= \sqrt{4x^2y^2}\sqrt{3y} = 2xy\sqrt{3y}$

77. $2\sqrt[3]{5} \cdot 6\sqrt[3]{2} = 12\sqrt[3]{10}$

79. $\sqrt[3]{15}\sqrt[3]{25} = \sqrt[3]{375} = \sqrt[3]{125}\sqrt[3]{3} = 5\sqrt[3]{3}$

81. $\dfrac{\sqrt[3]{54x^2}}{\sqrt[3]{2}} = \sqrt[3]{\dfrac{54x^2}{2}} = \sqrt[3]{27} \cdot \sqrt[3]{x^2} = 3\sqrt[3]{x^2}$

83. $\sqrt{\dfrac{2}{15}} = \dfrac{\sqrt{2}}{\sqrt{15}} \cdot \dfrac{\sqrt{15}}{\sqrt{15}} = \dfrac{\sqrt{30}}{15}$

85. $\sqrt{\dfrac{3}{20}} = \dfrac{\sqrt{3}}{\sqrt{20}} = \dfrac{\sqrt{3}}{\sqrt{4}\sqrt{5}} = \dfrac{\sqrt{3}}{2\sqrt{5}} \cdot \dfrac{\sqrt{5}}{\sqrt{5}}$
$= \dfrac{\sqrt{15}}{2(5)} = \dfrac{\sqrt{15}}{10}$

87. $\dfrac{3x}{\sqrt{2x}} = \dfrac{3x}{\sqrt{2x}} \cdot \dfrac{\sqrt{2x}}{\sqrt{2x}} = \dfrac{3x\sqrt{2x}}{2x} = \dfrac{3\sqrt{2x}}{2}$

89. $\sqrt[3]{\dfrac{5}{4}} = \dfrac{\sqrt[3]{5}}{\sqrt[3]{4}} \cdot \dfrac{\sqrt[3]{2}}{\sqrt[3]{2}} = \dfrac{\sqrt[3]{10}}{\sqrt[3]{8}} = \dfrac{\sqrt[3]{10}}{2}$

91. $\dfrac{4}{2 - \sqrt{5}} = \dfrac{4(2 + \sqrt{5})}{(2 - \sqrt{5})(2 + \sqrt{5})} = \dfrac{4(2 + \sqrt{5})}{4 - 5}$
$= \dfrac{8 + 4\sqrt{5}}{-1} = \dfrac{-1(-8 - 4\sqrt{5})}{-1} = -8 - 4\sqrt{5}$

93. $\dfrac{5}{3 + \sqrt{10}} = \dfrac{5(3 - \sqrt{10})}{(3 + \sqrt{10})(3 - \sqrt{10})} = \dfrac{5(3 - \sqrt{10})}{9 - 10}$
$= \dfrac{15 - 5\sqrt{10}}{-1} = \dfrac{-1(-15 + 5\sqrt{10})}{-1} = -15 + 5\sqrt{10}$

95. $\dfrac{2\sqrt{3}}{\sqrt{15} + 2} = \dfrac{2\sqrt{3}(\sqrt{15} - 2)}{(\sqrt{15} + 2)(\sqrt{15} - 2)} = \dfrac{2\sqrt{3}(\sqrt{15} - 2)}{15 - 4}$
$= \dfrac{2\sqrt{45} - 4\sqrt{3}}{11} = \dfrac{2\sqrt{9}\sqrt{5} - 4\sqrt{3}}{11} = \dfrac{2(3)\sqrt{5} - 4\sqrt{3}}{11}$
$= \dfrac{6\sqrt{5} - 4\sqrt{3}}{11}$

97. $\dfrac{\sqrt{3} + 1}{\sqrt{2} - 1} = \dfrac{(\sqrt{3} + 1)(\sqrt{2} + 1)}{(\sqrt{2} - 1)(\sqrt{2} + 1)} = \dfrac{(\sqrt{3} + 1)(\sqrt{2} + 1)}{2 - 1}$
$= \dfrac{\sqrt{6} + \sqrt{3} + \sqrt{2} + 1}{1} = \sqrt{6} + \sqrt{3} + \sqrt{2} + 1$

99. $\dfrac{\sqrt{3} + 1}{\sqrt{2} - 1} = \dfrac{(\sqrt{3} + 1)}{(\sqrt{2} - 1)} \cdot \dfrac{(\sqrt{3} - 1)}{(\sqrt{3} - 1)} = \dfrac{3 - 1}{(\sqrt{2} - 1)(\sqrt{3} - 1)}$
$= \dfrac{2}{\sqrt{6} - \sqrt{2} - \sqrt{3} + 1}$

101. $\dfrac{3x + 12}{3} = \dfrac{3(x + 4)}{3} = x + 4$

103. $\dfrac{6x^2 - 3x}{3x} = \dfrac{3x(2x - 1)}{3x} = 2x - 1$

105. $x + 5 = 7^2$
$x + 5 = 49$
$x = 44$

107. $4z^2 + 6z - 12 = (2z)^2$
$4z^2 + 6z - 12 = 4z^2$
$6z - 12 = 0$
$6z = 12$
$z = 2$

Exercise Set 9.5

1. $\sqrt{x} = 9$
   $(\sqrt{x})^2 = 9^2$
   $x = 81$

3. $\sqrt{x+5} = 2$
   $(\sqrt{x+5})^2 = 2^2$
   $x + 5 = 4$
   $x = 4 - 5$
   $x = -1$

5. $\sqrt{x} + 3 = 7$
   $\sqrt{x} = 7 - 3$
   $\sqrt{x} = 4$
   $(\sqrt{x})^2 = 4^2$
   $x = 16$

7. $\sqrt{4x-3} = \sqrt{x+3}$
   $(\sqrt{4x-3})^2 = (\sqrt{x+3})^2$
   $4x - 3 = x + 3$
   $4x - x - 3 = 3$
   $3x - 3 = 3$
   $3x = 3 + 3$
   $3x = 6$
   $x = \dfrac{6}{3} = 2$
   Check
   $\sqrt{4(2)-3} = \sqrt{2+3}$
   $\sqrt{5} = \sqrt{5}$
   True
   The solution is 2.

9. $\sqrt{x+7} = x + 5$
   $(\sqrt{x+7})^2 = (x+5)^2$
   $x + 7 = (x)^2 + 2(x)(5) + (5)^2$
   $x + 7 = x^2 + 10x + 25$
   $0 = x^2 + 10x - x + 25 - 7$
   $0 = x^2 + 9x + 18$
   $0 = (x+6)(x+3)$
   $x + 6 = 0$ or $x + 3 = 0$
   $x = -6$ or $x = -3$
   Check $x = -6$
   $\sqrt{-6+7} = -6 + 5$
   $\sqrt{1} = -1$ False
   Check $x = -3$
   $\sqrt{-3+7} = -3 + 5$

$\sqrt{4} = 2$
$2 = 2$ True
The solution is –3.

11. $\sqrt{9x^2+2x-4} = 3x$
    $\left(\sqrt{9x^2+2x-4}\right)^2 = (3x)^2$
    $9x^2 + 2x - 4 = 9x^2$
    $9x^2 - 9x^2 + 2x - 4 = 0$
    $2x - 4 = 0$
    $2x = 4$
    $x = \dfrac{4}{2} = 2$
    Check
    $\sqrt{9(2)^2 + 2(2) - 4} = 3(2)$
    $\sqrt{9(4) + 4 - 4} = 6$
    $\sqrt{36 + 4 - 4} = 6$
    $\sqrt{36} = 6$
    $6 = 6$ True
    The solution is 2.

13. $\sqrt{x+6} + 5 = 3$
    $\sqrt{x+6} = 3 - 5$
    $\sqrt{x+6} = -2$
    $(\sqrt{x+6})^2 = (-2)^2$
    $x + 6 = 4$
    $x = 4 - 6$
    $x = -2$
    Check
    $\sqrt{-2+6} + 5 = 3$
    $\sqrt{4} + 5 = 3$
    $2 + 5 = 3$
    $7 = 3$ False
    There is no solution.

15. $\sqrt{2x+6} = 4$
    $(\sqrt{2x+6})^2 = 4^2$
    $2x + 6 = 16$
    $2x = 16 - 6$
    $2x = 10$
    $x = \dfrac{10}{2} = 5$
    Check
    $\sqrt{2(5)+6} = 4$
    $\sqrt{10+6} = 4$
    $\sqrt{16} = 4$
    $4 = 4$ True
    The solution is 5.

17. $\sqrt{x} - 2 = 5$

$\sqrt{x} = 5 + 2$

$\sqrt{x} = 7$

$(\sqrt{x})^2 = 7^2$

$x = 49$

Check

$\sqrt{49} - 2 = 5$

$7 - 2 = 5$

$5 = 5$ True

The solution is 49.

19. $3\sqrt{x} + 5 = 2$

$3\sqrt{x} = 2 - 5$

$3\sqrt{x} = -3$

$(3\sqrt{x})^2 = (-3)^2$

$9x = 9$

$x = \dfrac{9}{9} = 1$

Check

$3\sqrt{1} + 5 = 2$

$3 + 5 = 2$

$8 = 2$ False

There is no solution.

21. $\sqrt{x+6} + 1 = 3$

$\sqrt{x+6} = 3 - 1$

$\sqrt{x+6} = 2$

$(\sqrt{x+6})^2 = 2^2$

$x + 6 = 4$

$x = 4 - 6$

$x = -2$

Check

$\sqrt{-2+6} + 1 = 3$

$\sqrt{4} + 1 = 3$

$2 + 1 = 3$

$3 = 3$ True

The solution is -2.

23. $\sqrt{2x+1} + 3 = 5$

$\sqrt{2x+1} = 5 - 3$

$\sqrt{2x+1} = 2$

$(\sqrt{2x+1})^2 = 2^2$

$2x + 1 = 4$

$2x = 4 - 1$

$2x = 3$

$x = \dfrac{3}{2}$

Check

$\sqrt{2\left(\dfrac{3}{2}\right) + 1} + 3 = 5$

$\sqrt{3+1} + 3 = 5$

$\sqrt{4} + 3 = 5$

$2 + 3 = 5$

$5 = 5$ True

The solution is $\dfrac{3}{2}$.

25. $\sqrt{x} = \sqrt{3x - 8}$

$(\sqrt{x})^2 = (\sqrt{3x-8})^2$

$x = 3x - 8$

$x - 3x = -8$

$-2x = -8$

$x = \dfrac{-8}{-2} = 4$

Check

$\sqrt{4} = \sqrt{3(4) - 8}$

$\sqrt{4} = \sqrt{12 - 8}$

$\sqrt{4} = \sqrt{4}$ True

The solution is 4.

27. $\sqrt{4x} - \sqrt{2x+6} = 0$

$\sqrt{4x} = \sqrt{2x+6}$

$(\sqrt{4x})^2 = (\sqrt{2x+6})^2$

$4x = 2x + 6$

$4x - 2x = 6$

$2x = 6$

$x = \dfrac{6}{2} = 3$

Check

$\sqrt{4(3)} - \sqrt{2(3) + 6} = 0$

$\sqrt{12} - \sqrt{12} = 0$

$0 = 0$ True

The solution is 3.

29. $\sqrt{x} = x - 6$

$(\sqrt{x})^2 = (x-6)^2$

$x = (x)^2 + 2(x)(-6) + (6)^2$

$x = x^2 - 12x + 36$

$0 = x^2 - 12x - x + 36$

$0 = x^2 - 13x + 36$

$0 = (x-9)(x-4)$

$x - 9 = 0$ or $x - 4 = 0$

$x = 9$ or $x = 4$

Check $x = 9$

$\sqrt{9} = 9 - 6$

$3 = 3$ True

Check $x = 4$

$\sqrt{4} = 4 - 6$

$2 = -2$ False

The solution is 9.

31. $\sqrt{2x+1} = x - 7$

$(\sqrt{2x+1})^2 = (x-7)^2$

$2x + 1 = (x)^2 + 2(x)(-7) + (7)^2$

$2x + 1 = x^2 - 14x + 49$

$0 = x^2 - 14x - 2x + 49 - 1$

$0 = x^2 - 16x + 48$

$0 = (x - 12)(x - 4)$

$x - 12 = 0$ or $x - 4 = 0$

$x = 12$ or $x = 4$

Check $x = 12$

$\sqrt{2(12)+1} = 12 - 7$

$\sqrt{24+1} = 5$

$\sqrt{25} = 5$

$5 = 5$ True

Check $x = 4$

$\sqrt{2(4)+1} = 4 - 7$

$\sqrt{8+1} = -3$

$\sqrt{9} = -3$

$3 = -3$ False

The solution is 12.

33. $x = \sqrt{2x-2} + 1$

$x - 1 = \sqrt{2x-2}$

$(x-1)^2 = (\sqrt{2x-2})^2$

$(x)^2 + 2(x)(-1) + (1)^2 = 2x - 2$

$x^2 - 2x + 1 = 2x - 2$

$x^2 - 2x - 2x + 1 + 2 = 0$

$x^2 - 4x + 3 = 0$

$(x - 3)(x - 1) = 0$

$x - 3 = 0$ or $x - 1 = 0$

$x = 3$ or $x = 1$

Check $x = 3$

$3 = \sqrt{2(3)-2} + 1$

$3 = \sqrt{6-2} + 1$

$3 = \sqrt{4} + 1$

$3 = 2 + 1$

$3 = 3$ True

Check $x = 1$

$1 = \sqrt{2(1)-2} + 1$

$1 = \sqrt{2-2} + 1$

$1 = \sqrt{0} + 1$

$1 = 1$ True

The solutions are 3 and 1.

35. $\sqrt{1-8x} - x = 4$

$\sqrt{1-8x} = 4 + x$

$(\sqrt{1-8x})^2 = (4+x)^2$

$1 - 8x = (4)^2 + 2(4)(x) + (x)^2$

$1 - 8x = 16 + 8x + x^2$

$0 = 16 - 1 + 8x + 8x + x^2$

$0 = 15 + 16x + x^2$

$0 = x^2 + 16x + 15$

$0 = (x + 15)(x + 1)$

$x + 15 = 0$ or $x + 1 = 0$

$x = -15$ or $x = -1$

Check $x = -15$

$\sqrt{1-8(-15)} - (-15) = 4$

$\sqrt{1+120} + 15 = 4$

$\sqrt{121} + 15 = 4$

$11 + 15 = 4$

$26 = 4$ False

Check $x = -1$

$\sqrt{1-8(-1)} - (-1) = 4$

$\sqrt{1+8} + 1 = 4$

$\sqrt{9} + 1 = 4$

$3 + 1 = 4$

$4 = 4$ True

The solution is -1.

37. $\sqrt{2x+5} - 1 = x$

$\sqrt{2x+5} = x + 1$

$(\sqrt{2x+5})^2 = (x+1)^2$

$2x + 5 = (x)^2 + 2(x)(1) + (1)^2$

$2x + 5 = x^2 + 2x + 1$

$0 = x^2 + 2x - 2x + 1 - 5$

$0 = x^2 - 4$

$0 = (x - 2)(x + 2)$

$x - 2 = 0$ or $x + 2 = 0$

$x = 2$ or $x = -2$

Check $x = 2$

$\sqrt{2(2)+5} - 1 = 2$

$\sqrt{4+5} - 1 = 2$

$\sqrt{9} - 1 = 2$

$3 - 1 = 2$

$2 = 2$ True

Check $x = -2$
$\sqrt{2(-2)+5} - 1 = -2$
$\sqrt{-4+5} - 1 = -2$
$\sqrt{1} - 1 = -2$
$1 - 1 = -2$
$0 = -2$ False
The solution is 2.

39. $\sqrt{16x^2 - 3x + 6} = 4x$

$\left(\sqrt{16x^2 - 3x + 6}\right)^2 = (4x)^2$

$16x^2 - 3x + 6 = 16x^2$
$16x^2 - 16x^2 - 3x + 6 = 0$
$-3x + 6 = 0$
$-3x = -6$
$x = \dfrac{-6}{-3} = 2$

Check
$\sqrt{16(2)^2 - 3(2) + 6} = 4(2)$
$\sqrt{16(4) - 6 + 6} = 8$
$\sqrt{64 - 6 + 6} = 8$
$\sqrt{64} = 8$
$8 = 8$ True
The solution is 2.

41. $\sqrt{16x^2 + 2x + 2} = 4x$

$\left(\sqrt{16x^2 + 2x + 2}\right)^2 = (4x)^2$

$16x^2 + 2x + 2 = 16x^2$
$16x^2 - 16x^2 + 2x + 2 = 0$
$2x + 2 = 0$
$2x = -2$
$x = \dfrac{-2}{2} = -1$

Check
$\sqrt{16(-1)^2 + 2(-1) + 2} = 4(-1)$
$\sqrt{16(1) - 2 + 2} = -4$
$\sqrt{16 - 2 + 2} = -4$
$\sqrt{16} = -4$
$4 = -4$ False
There is no solution.

43. $\sqrt{2x^2 + 6x + 9} = 3$

$\left(\sqrt{2x^2 + 6x + 9}\right)^2 = (3)^2$

$2x^2 + 6x + 9 = 9$
$2x^2 + 6x + 9 - 9 = 0$
$2x^2 + 6x = 0$
$2x(x + 3) = 0$
$2x = 0$ or $x + 3 = 0$
$x = 0$ or $x = -3$
Check $x = 0$
$\sqrt{2(0)^2 + 6(0) + 9} = 3$
$\sqrt{2(0) + 0 + 9} = 3$
$\sqrt{9} = 3$
$3 = 3$ True
Check $x = -3$
$\sqrt{2(-3)^2 + 6(-3) + 9} = 3$
$\sqrt{2(9) - 18 + 9} = 3$
$\sqrt{18 - 18 + 9} = 3$
$\sqrt{9} = 3$
$3 = 3$ True
The solutions are 0 or -3.

45. Let $x =$ number
$x = 6 + \sqrt{x}$
$x - 6 = \sqrt{x}$
$(x - 6)^2 = (\sqrt{x})^2$
$(x)^2 + 2(x)(-6) + (-6)^2 = x$
$x^2 - 12x + 36 = x$
$x^2 - 12x - x + 36 = 0$
$x^2 - 13x + 36 = 0$
$(x - 9)(x - 4) = 0$
$x - 9 = 0$ or $x - 4 = 0$
$x = 9$ or $x = 4$
Check $x = 9$
$9 = 6 + \sqrt{9}$
$9 = 6 + 3$
$9 = 9$ True
Check $x = 4$
$4 = 6 + \sqrt{4}$
$4 = 6 + 2$
$4 = 8$ False
The solution is 9.

47. a. $b = \sqrt{\dfrac{V}{2}}$

V	20	200	2000
b	3.2	10	31.6

b. no

49. Answers may vary.

51. $\sqrt{x+1} = 2x - 3$
$x = 2.43$

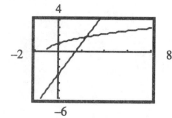

53. $-\sqrt{x+5} = -7x + 1$
$x = 0.48$

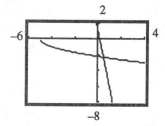

55. $2x - (3 + x) = 11$
$2x - 3 - x = 11$
$x - 3 = 11$
$x = 14$

57. length $= x + 2$
width $= x$
$2l + 2w = p$
$2(x + 2) + 2x = 24$
$2x + 4 + 2x = 24$
$4x + 4 = 24$
$\dfrac{4x}{4} = \dfrac{20}{4}$
$x = 5$
length $= x + 2$
length $= 7$ inches

59. 1st number $= x$
2nd number $= 3x$
$3x - x = 58$
$\dfrac{2x}{2} = \dfrac{58}{2}$
$x = 29$
$3x = 87$
$29 = $ 1st number
$87 = $ 2nd number

Exercise Set 9.6

1.

$a^2 + b^2 = c^2$
$2^2 + 3^2 = c^2$
$4 + 9 = c^2$
$13 = c^2$
$\sqrt{13} = \sqrt{c^2}$
$\sqrt{13} = c$

3.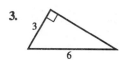

$a^2 + b^2 = c^2$
$3^2 + b^2 = 6^2$
$9 + b^2 = 36$
$b^2 = 36 - 9$
$b^2 = 27$
$\sqrt{b^2} = \sqrt{27}$
$b = \sqrt{27} = 3\sqrt{3}$

5.

$a^2 + b^2 = c^2$
$7^2 + 24^2 = c^2$
$49 + 576 = c^2$
$625 = c^2$
$\sqrt{625} = \sqrt{c^2}$
$25 = c$

7.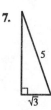

$a^2 + b^2 = c^2$

$a^2 + (\sqrt{3})^2 = 5^2$

$a^2 + 3 = 25$

$a^2 = 25 - 3$

$a^2 = 22$

$\sqrt{a^2} = \sqrt{22}$

$a = \sqrt{22}$

9.

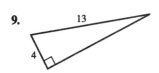

$a^2 + b^2 = c^2$

$4^2 + b^2 = 13^2$

$16 + b^2 = 169$

$b^2 = 169 - 16$

$b^2 = 153$

$\sqrt{b^2} = \sqrt{153}$

$b = \sqrt{153} = 3\sqrt{17}$

11. $a = 4,\ b = 5$

$a^2 + b^2 = c^2$

$4^2 + 5^2 = c^2$

$16 + 25 = c^2$

$41 = c^2$

$\sqrt{41} = \sqrt{c^2}$

$\sqrt{41} = c$

13. $b = 2,\ c = 6$

$a^2 + b^2 = c^2$

$a^2 + 2^2 = 6^2$

$a^2 + 4 = 36$

$a^2 = 36 - 4$

$a^2 = 32$

$\sqrt{a^2} = \sqrt{32}$

$a = \sqrt{16}\sqrt{2}$

$a = 4\sqrt{2}$

15. $a = \sqrt{10},\ c = 10$

$a^2 + b^2 = c^2$

$(\sqrt{10})^2 + b^2 = 10^2$

$10 + b^2 = 100$

$b^2 = 100 - 10$

$b^2 = 90$

$\sqrt{b^2} = \sqrt{90}$

$b = \sqrt{9}\sqrt{10}$

$b = 3\sqrt{10}$

17. $\sqrt{7^2 - 3^2} = \sqrt{49 - 9} = \sqrt{40} = 2\sqrt{10}$

$\sqrt{5^2 - 3^2} = \sqrt{25 - 9} = \sqrt{16} = 4$

$x = -4 + 2\sqrt{10}$

19. $(3,\ 6),\ (5,\ 11)$

$d = \sqrt{(x_2 - x_1)^2 + (y_2 - y_1)^2}$

$d = \sqrt{(5 - 3)^2 + (11 - 6)^2}$

$d = \sqrt{2^2 + 5^2}$

$d = \sqrt{4 + 25}$

$d = \sqrt{29}$

21. $(-3,\ 1),\ (5,\ -2)$

$d = \sqrt{(x_2 - x_1)^2 + (y_2 - y_1)^2}$

$d = \sqrt{[5 - (-3)]^2 + (-2 - 1)^2}$

$d = \sqrt{(5 + 3)^2 + (-2 - 1)^2}$

$d = \sqrt{8^2 + (-3)^2}$

$d = \sqrt{64 + 9}$

$d = \sqrt{73}$

23. $(3,\ -2),\ (1,\ -8)$

$d = \sqrt{(x_2 - x_1)^2 + (y_2 - y_1)^2}$

$d = \sqrt{(1 - 3)^2 + [-8 - (-2)]^2}$

$d = \sqrt{(-2)^2 + (-8 + 2)^2}$

$d = \sqrt{(-2)^2 + (-6)^2}$

$d = \sqrt{4 + 36}$

$d = \sqrt{40}$

$d = \sqrt{4}\sqrt{10}$

$d = 2\sqrt{10}$

25. $\left(\frac{1}{2},\ 2\right),\ (2,\ -1)$

$$d = \sqrt{(x_2 - x_1)^2 + (y_2 - y_1)^2}$$

$$d = \sqrt{\left(2 - \frac{1}{2}\right)^2 + (-1 - 2)^2}$$

$$d = \sqrt{\left(\frac{3}{2}\right)^2 + (-3)^2}$$

$$d = \sqrt{\frac{9}{4} + 9}$$

$$d = \sqrt{\frac{9}{4} + \frac{36}{4}}$$

$$d = \sqrt{\frac{45}{4}} = \frac{\sqrt{45}}{\sqrt{4}} = \frac{\sqrt{45}}{2}$$

$$d = \frac{\sqrt{45}}{2} = \frac{\sqrt{9}\sqrt{5}}{2} = \frac{3\sqrt{5}}{2}$$

27. $\sqrt{(5-3)^2 + (7+2)^2} = \sqrt{2^2 + 9^2} = \sqrt{4 + 81} = \sqrt{85}$

$\sqrt{85}$ units

29.

20 | hypotenuse

5

$$a^2 + b^2 = c^2$$
$$20^2 + 5^2 = c^2$$
$$400 + 25 = c^2$$
$$425 = c^2$$
$$\sqrt{425} = \sqrt{c^2}$$
$$20.6 \text{ ft} \approx c$$

31. $b = \sqrt{\dfrac{3V}{H}}$

$$6 = \sqrt{\frac{3V}{2}}$$

$$(6)^2 = \left(\sqrt{\frac{3V}{2}}\right)^2$$

$$36 = \frac{3V}{2}$$

$$2 \cdot 36 = 3V$$

$$72 = 3V$$

$$24 \text{ cu. ft.} = \frac{72}{3} = V$$

33.

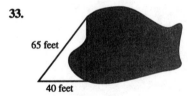

65 feet

40 feet

$$a^2 + b^2 = c^2$$
$$40^2 + b^2 = 65^2$$
$$1600 + b^2 = 4225$$
$$b^2 = 4225 - 1600$$
$$b^2 = 2625$$
$$\sqrt{b^2} = \sqrt{2625}$$
$$b \approx 51.2 \text{ ft.}$$

35.

6

10

$$a^2 + b^2 = c^2$$
$$6^2 + 10^2 = c^2$$
$$36 + 100 = c^2$$
$$136 = c^2$$
$$\sqrt{136} = \sqrt{c^2}$$
$$11.7 \text{ ft} \approx c$$

37. $b = \sqrt{\dfrac{3V}{H}}$

$$b = \sqrt{\frac{3(18)}{12}}$$

$$b = \sqrt{\frac{54}{12}}$$

$$b = \sqrt{\frac{9}{2}}$$

$$b = \frac{\sqrt{9}}{\sqrt{2}}$$

$$b = \frac{3}{\sqrt{2}} \cdot \frac{\sqrt{2}}{\sqrt{2}}$$

$$b = \frac{3\sqrt{2}}{2} \text{ in.}$$

39. $s = \sqrt{30\,fd}$

$60 = \sqrt{30(0.95)d}$

$60 = \sqrt{28.5d}$

$(60)^2 = (\sqrt{28.5d})^2$

$3600 = 28.5d$

$\dfrac{3600}{28.5} = d$

$126 \text{ ft.} \approx d$

41. $v = \sqrt{2.5r}$

$30 = \sqrt{2.5r}$

$(30)^2 = (\sqrt{2.5r})^2$

$900 = 2.5r$

$\dfrac{900}{2.5} = r$

$360 \text{ ft.} = r$

43. $d = 3.5\sqrt{h}$

$40 = 3.5\sqrt{h}$

$\dfrac{40}{3.5} = \sqrt{h}$

$11.43 = \sqrt{h}$

$(11.43)^2 = (\sqrt{h})^2$

$130.6 \text{ m} \approx h$

45. Answers may vary.

47. $(-3)^3 = (-3)(-3)(-3) = -27$

49. $\left(\dfrac{2}{7}\right)^3 = \left(\dfrac{2}{7}\right)\left(\dfrac{2}{7}\right)\left(\dfrac{2}{7}\right) = \dfrac{8}{343}$

51. $x^4 \cdot x^2 = x^6$

53. $x \cdot x^7 = x^8$

Exercise Set 9.7

1. $8^{1/3} = \sqrt[3]{8} = 2$

3. $9^{1/2} = \sqrt{9} = 3$

5. $16^{3/4} = (\sqrt[4]{16})^3 = 2^3 = 8$

7. $32^{2/5} = (\sqrt[5]{32})^2 = 2^2 = 4$

9. $-16^{-1/4} = -\dfrac{1}{16^{1/4}} = -\dfrac{1}{\sqrt[4]{16}} = -\dfrac{1}{2}$

11. $16^{-3/2} = \dfrac{1}{16^{3/2}} = \dfrac{1}{(\sqrt{16})^3} = \dfrac{1}{4^3} = \dfrac{1}{64}$

13. $81^{-3/2} = \dfrac{1}{81^{3/2}} = \dfrac{1}{(\sqrt{81})^3} = \dfrac{1}{9^3} = \dfrac{1}{729}$

15. $\left(\dfrac{4}{25}\right)^{-1/2} = \dfrac{4^{-1/2}}{25^{-1/2}} = \dfrac{25^{1/2}}{4^{1/2}} = \dfrac{\sqrt{25}}{\sqrt{4}} = \dfrac{5}{2}$

17. Answers may vary.

19. $2^{1/3} \cdot 2^{2/3} = 2^{3/3} = 2^1 = 2$

21. $\dfrac{4^{3/4}}{4^{1/4}} = 4^{3/4 - 1/4} = 4^{2/4} = 4^{1/2} = \sqrt{4} = 2$

23. $\dfrac{x^{1/6}}{x^{5/6}} = x^{\frac{1}{6} - \frac{5}{6}} = x^{-4/6} = x^{-2/3} = \dfrac{1}{x^{2/3}}$

25. $(x^{1/2})^6 = x^{6/2} = x^3$

27. Answers may vary.

29. $81^{1/2} = \sqrt{81} = 9$

31. $(-8)^{1/3} = \sqrt[3]{-8} = -2$

33. $-81^{1/4} = -(81^{1/4}) = -(\sqrt[4]{81}) = -(3) = -3$

35. $\left(\dfrac{1}{81}\right)^{1/2} = \dfrac{1^{1/2}}{81^{1/2}} = \dfrac{\sqrt{1}}{\sqrt{81}} = \dfrac{1}{9}$

37. $\left(\dfrac{27}{64}\right)^{1/3} = \dfrac{27^{1/3}}{64^{1/3}} = \dfrac{\sqrt[3]{27}}{\sqrt[3]{64}} = \dfrac{3}{4}$

39. $9^{3/2} = (\sqrt{9})^3 = (3)^3 = 27$

41. $64^{3/2} = (\sqrt{64})^3 = (8)^3 = 512$

43. $-8^{2/3} = -(8^{2/3}) = -(\sqrt[3]{8})^2 = -(2)^2 = -(4) = -4$

45. $4^{5/2} = (\sqrt{4})^5 = (2)^5 = 32$

47. $\left(\dfrac{4}{9}\right)^{3/2} = \dfrac{4^{3/2}}{9^{3/2}} = \dfrac{(\sqrt{4})^3}{(\sqrt{9})^3} = \dfrac{2^3}{3^3} = \dfrac{8}{27}$

49. $\left(\dfrac{1}{81}\right)^{3/4} = \dfrac{1^{3/4}}{81^{3/4}} = \dfrac{(\sqrt[4]{1})^3}{(\sqrt[4]{81})^3} = \dfrac{1^3}{3^3} = \dfrac{1}{27}$

51. $4^{-1/2} = \dfrac{1}{4^{1/2}} = \dfrac{1}{\sqrt{4}} = \dfrac{1}{2}$

53. $125^{-1/3} = \dfrac{1}{125^{1/3}} = \dfrac{1}{\sqrt[3]{125}} = \dfrac{1}{5}$

55. $625^{-3/4} = \dfrac{1}{625^{3/4}} = \dfrac{1}{(\sqrt[4]{625})^3} = \dfrac{1}{5^3} = \dfrac{1}{125}$

57. $3^{4/3} \cdot 3^{2/3} = 3^{4/3+2/3} = 3^{6/3} = 3^2 = 9$

59. $\dfrac{6^{2/3}}{6^{1/3}} = 6^{2/3-1/3} = 6^{1/3}$

61. $\left(x^{2/3}\right)^9 = x^{\frac{2}{3}\bullet 9} = x^6$

63. $\dfrac{6^{1/3}}{6^{-5/3}} = 6^{1/3-(-5/3)} = 6^{\frac{1}{3}+\frac{5}{3}} = 6^{6/3} = 6^2 = 36$

65. $\dfrac{3^{-3/5}}{3^{2/5}} = 3^{-3/5-2/5} = 3^{-5/5} = 3^{-1} = \dfrac{1}{3}$

67. $\left(\dfrac{x^{1/3}}{y^{3/4}}\right)^2 = \dfrac{x^{\frac{1}{3}\bullet 2}}{x^{\frac{3}{4}\bullet 2}} = \dfrac{x^{2/3}}{y^{3/2}}$

69. $\left(\dfrac{x^{2/5}}{y^{3/4}}\right)^8 = \dfrac{x^{\frac{2}{5}\bullet 8}}{x^{\frac{3}{4}\bullet 8}} = \dfrac{x^{16/5}}{y^6}$

71. $P = P_0(1.08)^N$

$P = 10,000(1.08)^{3/2}$

$P = 10,000(\sqrt{1.08})^3$

$P = 10,000(1.0392)^3$

$P = 10,000(1.1224)$

$P = 11,224$

73. $5^{3/4} = 3.344$

75. $18^{3/5} = 5.665$

77. $x + y < 6$
    $y \geq 2x$

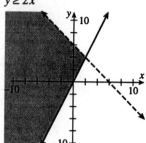

79. $x^2 - 4 = 3x$

  $x^2 - 3x - 4 = 0$

  $(x - 4)(x + 1) = 0$

  $x - 4 = 0$ or $x + 1 = 0$

  $x = 4$ or $x = -1$

  $x = -1, 4$

81. $2x^2 - 5x - 3 = 0$

  $(2x + 1)(x - 3) = 0$

  $2x + 1 = 0$ or $x - 3 = 0$

  $2x = -1$ or $x = 3$

  $x = -\dfrac{1}{2}$

Chapter 9 - Review

1. $\sqrt{81} = 9$

2. $-\sqrt{49} = -7$

3. $\sqrt[3]{27} = 3$

4. $\sqrt[4]{16} = 2$

5. $-\sqrt{\dfrac{9}{64}} = -\dfrac{3}{8}$

6. $\sqrt{\dfrac{36}{81}} = \dfrac{6}{9} = \dfrac{2}{3}$

7. $\sqrt[4]{-\dfrac{16}{81}}$; not a real number

8. $\sqrt[3]{-\dfrac{27}{64}} = -\dfrac{3}{4}$

9. $\sqrt{76}$; irrational, 8.718

10. $\sqrt{576}$; rational, 24

11. $\sqrt{x^{12}} = x^6$

12. $\sqrt{x^8} = x^4$

13. $\sqrt{9x^6 y^2} = 3x^3 y$

14. $\sqrt{25x^4 y^{10}} = 5x^2 y^5$

15. $-\sqrt[3]{8x^6} = -(2x^2) = -2x^2$

16. $-\sqrt[4]{16x^8} = -2x^2$

17. $\sqrt{54} = \sqrt{9}\sqrt{6} = 3\sqrt{6}$

18. $\sqrt{88} = \sqrt{4 \cdot 22} = 2\sqrt{22}$

19. $\sqrt{150x^3 y^6} = \sqrt{25x^2 y^6}\sqrt{6x} = 5xy^3\sqrt{6x}$

20. $\sqrt{92x^8 y^5} = \sqrt{4 \cdot 23 \cdot x^8 \cdot y^4 \cdot y} = 2x^4 y^2\sqrt{23y}$

21. $\sqrt[3]{54} = \sqrt[3]{27}\sqrt[3]{2} = 3\sqrt[3]{2}$

22. $\sqrt[3]{88} = \sqrt[3]{8 \cdot 11} = 2\sqrt[3]{11}$

23. $\sqrt[4]{48x^3 y^6} = \sqrt[4]{16y^4}\sqrt[4]{3x^3 y^2} = 2y\sqrt[4]{3x^3 y^2}$

24. $\sqrt[4]{162x^8 y^5} = \sqrt[4]{81 \cdot 2 \cdot x^8 \cdot y^4 \cdot y} = 3x^2 y\sqrt[4]{2y}$

25. $\sqrt{\dfrac{18}{25}} = \dfrac{\sqrt{18}}{\sqrt{25}} = \dfrac{\sqrt{9}\sqrt{2}}{5} = \dfrac{3\sqrt{2}}{5}$

26. $\sqrt{\dfrac{75}{64}} = \dfrac{\sqrt{75}}{\sqrt{64}} = \dfrac{\sqrt{25 \cdot 3}}{8} = \dfrac{5\sqrt{3}}{8}$

27. $\sqrt{\dfrac{45x^2 y^2}{4x^6}} = \sqrt{\dfrac{45y^2}{4x^4}} = \dfrac{\sqrt{45y^2}}{\sqrt{4x^4}} = \dfrac{\sqrt{9y^2}\sqrt{5}}{2x^2}$

$= \dfrac{3y\sqrt{5}}{2x^2}$

28. $\sqrt{\dfrac{20x^5}{9x^2}} = \sqrt{\dfrac{20x^3}{9}} = \dfrac{\sqrt{4 \cdot 5 \cdot x^2 \cdot x}}{\sqrt{9}} = \dfrac{2x\sqrt{5x}}{3}$

29. $\sqrt[4]{\dfrac{9}{16}} = \dfrac{\sqrt[4]{9}}{\sqrt[4]{16}} = \dfrac{\sqrt[4]{9}}{2}$

30. $\sqrt[3]{\dfrac{40}{27}} = \dfrac{\sqrt[3]{8 \cdot 5}}{\sqrt[3]{27}} = \dfrac{2\sqrt[3]{5}}{3}$

31. $\sqrt[3]{\dfrac{3y^6}{8x^3}} = \dfrac{\sqrt[3]{3y^6}}{\sqrt[3]{8x^3}} = \dfrac{\sqrt[3]{y^6}\sqrt[3]{3}}{2x} = \dfrac{y^2\sqrt[3]{3}}{2x}$

32. $\sqrt[4]{\dfrac{5x^2}{81x^8}} = \dfrac{\sqrt[4]{5x^2}}{\sqrt[4]{81x^8}} = \dfrac{4\sqrt{5x^2}}{3x^2}$

33. $3\sqrt[3]{2} + 2\sqrt[3]{3} - 4\sqrt[3]{2} = -\sqrt[3]{2} + 2\sqrt[3]{3}$

34. $5\sqrt{2} + 2\sqrt[3]{2} - 8\sqrt{2} = -3\sqrt{2} + 2\sqrt[3]{2}$

35. $\sqrt{6} + 2\sqrt[3]{6} - 4\sqrt[3]{6} + 5\sqrt{6} = 6\sqrt{6} - 2\sqrt[3]{6}$

36. $3\sqrt{5} - \sqrt[3]{5} - 2\sqrt{5} + 3\sqrt[3]{5} = \sqrt{5} + 2\sqrt[3]{5}$

37. $\sqrt{28} + \sqrt{63} + \sqrt[3]{56} = \sqrt{4}\sqrt{7} + \sqrt{9}\sqrt{7} + \sqrt[3]{8}\sqrt[3]{7}$
$= 2\sqrt{7} + 3\sqrt{7} + 2\sqrt[3]{7} = 5\sqrt{7} + 2\sqrt[3]{7}$

38. $\sqrt{75} + \sqrt{48} - \sqrt[4]{16} = \sqrt{25 \cdot 3} + \sqrt{16 \cdot 3} - 2$
$= 5\sqrt{3} + 4\sqrt{3} - 2 = 9\sqrt{3} - 2$

39. $\sqrt{\dfrac{5}{9}} - \sqrt{\dfrac{5}{36}} = \dfrac{\sqrt{5}}{\sqrt{9}} - \dfrac{\sqrt{5}}{\sqrt{36}} = \dfrac{\sqrt{5}}{3} - \dfrac{\sqrt{5}}{6}$

$= \dfrac{\sqrt{5}}{3} \cdot \dfrac{2}{2} - \dfrac{\sqrt{5}}{6} = \dfrac{2\sqrt{5}}{6} - \dfrac{\sqrt{5}}{6} = \dfrac{\sqrt{5}}{6}$

40. $\sqrt{\dfrac{11}{25}} + \sqrt{\dfrac{11}{16}} = \dfrac{\sqrt{11}}{5} + \dfrac{\sqrt{11}}{4} = \dfrac{4\sqrt{11} + 5\sqrt{11}}{20}$

$= \dfrac{9\sqrt{11}}{20}$

41. $2\sqrt[3]{125x^3} - 5x\sqrt[3]{8} = 2(5x) - 5x(2) = 10x - 10x = 0$

42. $3\sqrt[3]{16x^4} - 2x\sqrt[3]{2x} = 3\sqrt[3]{8 \cdot 2x^3 \cdot x} - 2x\sqrt[3]{2x}$
$= 6x\sqrt[3]{2x} - 2x\sqrt[3]{2x} = 4x\sqrt[3]{2x}$

43. $3\sqrt{10} \cdot 2\sqrt{5} = 6\sqrt{50} = 6\sqrt{25}\sqrt{2} = 6(5)\sqrt{2} = 30\sqrt{2}$

44. $2\sqrt[3]{4} \cdot 5\sqrt[3]{6} = 10\sqrt[3]{24} = 10\sqrt[3]{8 \cdot 3} = 20\sqrt[3]{3}$

45. $\sqrt{3}(2\sqrt{6} - 3\sqrt{12}) = 2\sqrt{18} - 3\sqrt{36}$
$= 2\sqrt{9}\sqrt{2} - 3(6) = 2(3)\sqrt{2} - 18 = 6\sqrt{2} - 18$

46. $4\sqrt{5}(2\sqrt{10}-5\sqrt{5})=8\sqrt{50}-20\sqrt{25}$
$=8\sqrt{25\cdot2}-20(5)=40\sqrt{2}-100$

47. $(\sqrt{3}+2)(\sqrt{6}-5)=\sqrt{18}-5\sqrt{3}+2\sqrt{6}-10$
$=\sqrt{9}\sqrt{2}-5\sqrt{3}+2\sqrt{6}-10$
$=3\sqrt{2}-5\sqrt{3}+2\sqrt{6}-10$

48. $(2\sqrt{5}+1)(4\sqrt{5}-3)=8\sqrt{25}-6\sqrt{5}+4\sqrt{5}-3$
$=40-3-2\sqrt{5}=37-2\sqrt{5}$

49. $\dfrac{\sqrt{96}}{\sqrt{3}}=\sqrt{\dfrac{96}{3}}=\sqrt{32}=\sqrt{16}\sqrt{2}=4\sqrt{2}$

50. $\dfrac{\sqrt{160}}{\sqrt{8}}=\sqrt{\dfrac{160}{8}}=\sqrt{20}=\sqrt{4\cdot5}=2\sqrt{5}$

51. $\dfrac{\sqrt{15x^6y}}{\sqrt{12x^3y^9}}=\sqrt{\dfrac{15x^6y}{12x^3y^9}}=\sqrt{\dfrac{5x^3}{4y^8}}=\dfrac{\sqrt{5x^3}}{\sqrt{4y^8}}$
$=\dfrac{\sqrt{x^2}\sqrt{5x}}{2y^4}=\dfrac{x\sqrt{5x}}{2y^4}$

52. $\dfrac{\sqrt{50xy^8}}{\sqrt{72x^7y^3}}=\dfrac{\sqrt{25\cdot2xy^8}}{\sqrt{36\cdot2\cdot x^6\cdot x\cdot y^2\cdot y}}=\dfrac{5y^4\sqrt{2x}}{6x^3y\sqrt{2xy}}$
$=\dfrac{5y^3}{6x^3}\sqrt{\dfrac{2x}{2xy}}=\dfrac{5y^3}{6x^3}\sqrt{\dfrac{1}{y}}=\dfrac{5y^3}{6x^3\sqrt{y}}=\dfrac{5y^2\sqrt{y}}{6x^3}$

53. $\sqrt{\dfrac{5}{6}}=\dfrac{\sqrt{5}}{\sqrt{6}}\cdot\dfrac{\sqrt{6}}{\sqrt{6}}=\dfrac{\sqrt{30}}{6}$

54. $\sqrt{\dfrac{7}{10}}=\dfrac{\sqrt{7}}{\sqrt{10}}\cdot\dfrac{\sqrt{10}}{\sqrt{10}}=\dfrac{\sqrt{70}}{\sqrt{100}}=\dfrac{\sqrt{70}}{10}$

55. $\sqrt{\dfrac{3}{2x}}=\dfrac{\sqrt{3}}{\sqrt{2x}}\cdot\dfrac{\sqrt{2x}}{\sqrt{2x}}=\dfrac{\sqrt{6x}}{2x}$

56. $\sqrt{\dfrac{6}{5y}}=\dfrac{\sqrt{6}}{\sqrt{5y}}\cdot\dfrac{\sqrt{5y}}{\sqrt{5y}}=\dfrac{\sqrt{30y}}{\sqrt{25y^2}}=\dfrac{\sqrt{30y}}{5y}$

57. $\sqrt{\dfrac{7}{20y^2}}=\dfrac{\sqrt{7}}{\sqrt{20y^2}}=\dfrac{\sqrt{7}}{\sqrt{4y^2}\sqrt{5}}=\dfrac{\sqrt{7}}{2y\sqrt{5}}\cdot\dfrac{\sqrt{5}}{\sqrt{5}}$
$=\dfrac{\sqrt{35}}{2y(5)}=\dfrac{\sqrt{35}}{10y}$

58. $\sqrt{\dfrac{5z}{12x^2}}=\dfrac{\sqrt{5z}}{\sqrt{12x^2}}\cdot\dfrac{\sqrt{3}}{\sqrt{3}}=\dfrac{\sqrt{15z}}{\sqrt{36x^2}}=\dfrac{\sqrt{15z}}{6x}$

59. $\sqrt[3]{\dfrac{7}{9}}=\dfrac{\sqrt[3]{7}}{\sqrt[3]{9}}\cdot\dfrac{\sqrt[3]{3}}{\sqrt[3]{3}}=\dfrac{\sqrt[3]{21}}{\sqrt[3]{27}}=\dfrac{\sqrt[3]{21}}{3}$

60. $\sqrt[3]{\dfrac{3}{4}}=\dfrac{\sqrt[3]{3}}{\sqrt[3]{4}}\cdot\dfrac{\sqrt[3]{2}}{\sqrt[3]{2}}=\dfrac{\sqrt[3]{6}}{\sqrt[3]{8}}=\dfrac{\sqrt[3]{6}}{2}$

61. $\sqrt[3]{\dfrac{3}{2x^2}}=\dfrac{\sqrt[3]{3}}{\sqrt[3]{2x^2}}\cdot\dfrac{\sqrt[3]{4x}}{\sqrt[3]{4x}}=\dfrac{\sqrt[3]{12x}}{\sqrt[3]{8x^3}}=\dfrac{\sqrt[3]{12x}}{2x}$

62. $\sqrt[3]{\dfrac{5x}{4y}}=\dfrac{\sqrt[3]{5x}}{\sqrt[3]{4y}}\cdot\dfrac{\sqrt[3]{2y^2}}{\sqrt[3]{2y^2}}=\dfrac{\sqrt[3]{10xy^2}}{\sqrt[3]{8y^3}}=\dfrac{\sqrt[3]{10xy^2}}{2y}$

63. $\dfrac{3}{\sqrt{5}-2}=\dfrac{3(\sqrt{5}+2)}{(\sqrt{5}-2)(\sqrt{5}+2)}=\dfrac{3(\sqrt{5}+2)}{5-4}$
$=\dfrac{3\sqrt{5}+6}{1}=3\sqrt{5}+6$

64. $\dfrac{8}{\sqrt{10}-3}\cdot\dfrac{\sqrt{10}+3}{\sqrt{10}+3}=\dfrac{8(\sqrt{10}+3)}{10-9}=\dfrac{8\sqrt{10}+24}{1}$
$=8\sqrt{10}+24$

65. $\dfrac{8}{\sqrt{6}+2}=\dfrac{8(\sqrt{6}-2)}{(\sqrt{6}+2)(\sqrt{6}-2)}=\dfrac{8(\sqrt{6}-2)}{6-4}$
$=\dfrac{8(\sqrt{6}-2)}{2}=4(\sqrt{6}-2)=4\sqrt{6}-8$

66. $\dfrac{12}{\sqrt{15}-3}\cdot\dfrac{\sqrt{15}+3}{\sqrt{15}+3}=\dfrac{12(\sqrt{15}+3)}{\sqrt{225}-9}=\dfrac{12\sqrt{15}+36}{15-9}$
$=\dfrac{12\sqrt{15}+36}{6}=2\sqrt{15}+6$

67. $\dfrac{\sqrt{2}}{4+\sqrt{2}}=\dfrac{\sqrt{2}(4-\sqrt{2})}{(4+\sqrt{2})(4-\sqrt{2})}=\dfrac{\sqrt{2}(4-\sqrt{2})}{16-2}$
$\dfrac{4\sqrt{2}-2}{14}=\dfrac{2(2\sqrt{2}-1)}{14}=\dfrac{2\sqrt{2}-1}{7}$

68. $\dfrac{\sqrt{3}}{5+\sqrt{3}}\cdot\dfrac{5-\sqrt{3}}{5-\sqrt{3}}=\dfrac{\sqrt{3}(5-\sqrt{3})}{25-\sqrt{9}}=\dfrac{5\sqrt{3}-\sqrt{9}}{25-3}$
$=\dfrac{5\sqrt{3}-3}{22}$

68. $\dfrac{\sqrt{3}}{5+\sqrt{3}} \cdot \dfrac{5-\sqrt{3}}{5-\sqrt{3}} = \dfrac{\sqrt{3}(5-\sqrt{3})}{25-\sqrt{9}} = \dfrac{5\sqrt{3}-\sqrt{9}}{25-3}$

$\qquad = \dfrac{5\sqrt{3}-3}{22}$

69. $\dfrac{2\sqrt{3}}{\sqrt{3}-5} = \dfrac{2\sqrt{3}(\sqrt{3}+5)}{(\sqrt{3}-5)(\sqrt{3}+5)} = \dfrac{2\sqrt{3}(\sqrt{3}+5)}{3-25}$

$\qquad = \dfrac{2(3)+10\sqrt{3}}{-22} = \dfrac{6+10\sqrt{3}}{-22} = \dfrac{2(3+5\sqrt{3})}{-22}$

$\qquad = -\dfrac{3+5\sqrt{3}}{11}$

70. $\dfrac{7\sqrt{2}}{\sqrt{2}-4} \cdot \dfrac{\sqrt{2}+4}{\sqrt{2}+4} = \dfrac{7\sqrt{2}(\sqrt{2}+4)}{\sqrt{4}-16} = \dfrac{7\sqrt{4}+28\sqrt{2}}{2-16}$

$\qquad = \dfrac{14+28\sqrt{2}}{-14} = -1-2\sqrt{2}$

71. $\sqrt{2x} = 6$

$\qquad (\sqrt{2x})^2 = 6^2$

$\qquad 2x = 36$

$\qquad x = \dfrac{36}{2} = 18$

Check

$\qquad \sqrt{2(18)} = 6$

$\qquad \sqrt{36} = 6$

$\qquad 6 = 6 \qquad\qquad$ True

The solution is 18.

72. $\sqrt{x+3} = 4$

$\qquad (\sqrt{x}+3)^2 = (4)^2$

$\qquad x+3 = 16$

$\qquad x = 13$

73. $\sqrt{x}+3 = 8$

$\qquad \sqrt{x} = 8-3$

$\qquad \sqrt{x} = 5$

$\qquad (\sqrt{x})^2 = 5^2$

$\qquad x = 25$

Check

$\qquad \sqrt{25}+3 = 8$

$\qquad 5+3 = 8$

$\qquad 8 = 8 \qquad\qquad$ True

The solution is 25.

74. $\sqrt{x}+8 = 3$

$\qquad \sqrt{x} = -5$

No solution because the positive square root of a number cannot equal a negative.

75. $\sqrt{2x+1} = x-7$

$\qquad (\sqrt{2x+1})^2 = (x-7)^2$

$\qquad 2x+1 = (x)^2 + 2(x)(-7)+(7)^2$

$\qquad 2x+1 = x^2 -14x+49$

$\qquad 0 = x^2 -14x-2x+49-1$

$\qquad 0 = x^2 -16x+48$

$\qquad 0 = (x-12)(x-4)$

$\qquad x-12 = 0 \qquad$ or $\qquad x-4 = 0$

$\qquad x = 12 \qquad\quad$ or $\qquad x = 4$

Check $x = 12$

$\qquad \sqrt{2(12)+1} = 12-7$

$\qquad \sqrt{24+1} = 5$

$\qquad \sqrt{25} = 5$

$\qquad 5 = 5 \qquad\qquad$ True

Check $x = 4$

$\qquad \sqrt{2(4)+1} = 4-7$

$\qquad \sqrt{8+1} = -3$

$\qquad \sqrt{9} = -3$

$\qquad 3 = -3 \qquad\qquad$ False

The solution is 12.

76. $\sqrt{3x+1} = x-1$

$\qquad (\sqrt{3x+1})^2 = (x-1)^2$

$\qquad 3x+1 = x^2 -2x+1$

$\qquad 0 = x^2 -5x$

$\qquad 0 = x(x-5)$

$\qquad x = 0 \qquad$ or $\qquad x-5 = 0$

$\qquad$ extraneous $\qquad\qquad x = 5$

77. $\sqrt{x+3}+x = 9$

$\qquad \sqrt{x+3} = 9-x$

$\qquad (\sqrt{x+3})^2 = (9-x)^2$

$\qquad x+3 = (9)^2 + 2(9)(-x)+(x)^2$

$\qquad x+3 = 81-18x+x^2$

$\qquad 0 = 81-3-18x-x+x^2$

$\qquad 0 = 78-19x+x^2$

$\qquad 0 = x^2 -19x+78$

$\qquad 0 = (x-6)(x-13)$

$\qquad x-6 = 0 \qquad$ or $\qquad x-13 = 0$

$\qquad x = 6 \qquad\quad$ or $\qquad x = 13$

Check $x = 6$

$\qquad \sqrt{6+3}+6 = 9$

$\sqrt{9} + 6 = 9$

$3 + 6 = 9$

$9 = 9$ ⠀⠀⠀⠀⠀True

Check $x = 13$

$\sqrt{13 + 3} + 13 = 9$

$\sqrt{16} + 13 = 9$

$4 + 13 = 9$

$17 = 9$ ⠀⠀⠀⠀⠀False

The solution is 6.

78. $\sqrt{2x} + x = 4$

$(\sqrt{2x})^2 = (4 - x)^2$

$2x = 16 - 8x + x^2$

$0 = x^2 - 10x + 16$

$0 = (x - 8)(x - 2)$

$x - 8 = 0$ ⠀⠀or⠀⠀ $x - 2 = 0$

$x = 8$ ⠀⠀⠀or⠀⠀ $x = 2$

$x = 2$; 8 is extraneous.

79.

$a^2 + b^2 = c^2$

$5^2 + b^2 = 9^2$

$25 + b^2 = 81$

$b^2 = 81 - 25$

$b^2 = 56$

$\sqrt{b^2} = \sqrt{56}$

$b = \sqrt{4}\sqrt{14}$

$b = 2\sqrt{14}$

80. $9^2 + 6^2 = c^2$

$81 + 36 = c^2$

$\sqrt{117} = \sqrt{c^2}$

$\sqrt{117}$ units $= c$

81. $a^2 + b^2 = c^2$

$12^2 + 20^2 = c^2$

$144 + 400 = c^2$

$544 = c^2$

$\sqrt{544} = \sqrt{c^2}$

$\sqrt{16}\sqrt{34} = c$

$4\sqrt{34}$ ft $= c$

82. $5^2 + b^2 = 10^2$

$25 + b^2 = 100$

$\sqrt{b^2} = \sqrt{75}$

$b = \sqrt{25 \cdot 3}$

$b = 5\sqrt{3}$ inches

83. $(6, -2)$ and $(-3, 5)$

$d = \sqrt{(x_2 - x_1)^2 + (y_2 - y_1)^2}$

$d = \sqrt{(-3 - 6)^2 + [5 - (-2)]^2}$

$d = \sqrt{(-9)^2 + (5 + 2)^2}$

$d = \sqrt{81 + (7)^2}$

$d = \sqrt{81 + 49}$

$d = \sqrt{130}$

84. $\sqrt{(-6 - 2)^2 + (10 - 8)^2} = \sqrt{64 + 4} = \sqrt{68}$

⠀⠀ $= \sqrt{4 \cdot 17} = 2\sqrt{17}$ units

85. $r = \sqrt{\dfrac{A}{4\pi}}$

$r = \sqrt{\dfrac{72}{4(3.14)}}$

$r = \sqrt{\dfrac{72}{12.56}}$

$r = \sqrt{5.732}$

$r \approx 2.4$ in.

86. $r = \sqrt{\dfrac{A}{4\pi}}$

$6 = \sqrt{\dfrac{A}{4\pi}}$

$36 = \dfrac{A}{4\pi}$

$36(4\pi) = A \implies A = 144\pi$

144π sq. in.

87. $\sqrt{a^5} = a^{5/2}$

88. $\sqrt[5]{a^3} = a^{3/5}$

89. $\sqrt[6]{x^{15}} = x^{15/6} = x^{5/2}$

90. $\sqrt[4]{x^{12}} = x^{12/4} = x^3$

91. $16^{1/2} = \sqrt{16} = 4$

92. $36^{1/2} = \sqrt{36} = 6$

93. $(-8)^{1/3} = \sqrt[3]{-8} = -2$

94. $(-32)^{1/5} = \sqrt[5]{-32} = -2$

95. $-64^{3/2} = -(64^{3/2}) = -(\sqrt{64})^3 = -(8)^3 = -512$

96. $-8^{2/3} = -\sqrt[3]{8^2} = -\sqrt[3]{64} = -4$

97. $\left(\dfrac{16}{81}\right)^{3/4} = \dfrac{16^{3/4}}{81^{3/4}} = \dfrac{(\sqrt[4]{16})^3}{(\sqrt[4]{81})^3} = \dfrac{2^3}{3^3} = \dfrac{8}{27}$

98. $\left(\dfrac{9}{25}\right)^{3/2} = \left(\sqrt{\dfrac{9}{25}}\right)^3 = \left(\dfrac{3}{5}\right)^3 = \dfrac{27}{125}$

99. $25^{-1/2} = \dfrac{1}{25^{1/2}} = \dfrac{1}{\sqrt{25}} = \dfrac{1}{5}$

100. $64^{-2/3} = \dfrac{1}{64^{2/3}} = \dfrac{1}{\left(\sqrt[3]{64}\right)^2} = \dfrac{1}{(4)^2} = \dfrac{1}{16}$

101. $8^{1/3} \cdot 8^{4/3} = 8^{5/3} = (\sqrt[3]{8})^5 = 2^5 = 32$

102. $4^{3/2} \cdot 4^{1/2} = 4^{\frac{3}{2}+\frac{1}{2}} = 4^{4/2} = 4^2 = 16$

103. $\dfrac{3^{1/6}}{3^{5/6}} = 3^{\frac{1}{6}-\frac{5}{6}} = 3^{-4/6} = 3^{-2/3} = \dfrac{1}{3^{2/3}}$

104. $\dfrac{2^{1/4}}{2^{-3/5}} = 2^{\frac{1}{4}-\left(-\frac{3}{5}\right)} = 2^{\frac{1}{4}+\frac{3}{5}} = 2^{\frac{5+12}{20}} = 2^{17/20}$

105. $\left(x^{-1/3}\right)^6 = x^{-\frac{1}{3}\cdot 6} = x^{-2} = \dfrac{1}{x^2}$

106. $\left(\dfrac{x^{1/2}}{y^{1/3}}\right)^2 = \dfrac{x^{2/2}}{y^{2/3}} = \dfrac{x}{y^{2/3}}$

Chapter 9 - Test

1. $\sqrt{16} = 4$

2. $\sqrt[3]{125} = 5$

3. $16^{3/4} = (\sqrt[4]{16})^3 = 2^3 = 8$

4. $\left(\dfrac{9}{16}\right)^{1/2} = \dfrac{9^{1/2}}{16^{1/2}} = \dfrac{\sqrt{9}}{\sqrt{16}} = \dfrac{3}{4}$

5. $\sqrt[4]{-81}$; not a real number

6. $27^{-2/3} = \dfrac{1}{27^{2/3}} = \dfrac{1}{(\sqrt[3]{27})^2} = \dfrac{1}{3^2} = \dfrac{1}{9}$

7. $\sqrt{54} = \sqrt{9}\sqrt{6} = 3\sqrt{6}$

8. $\sqrt{92} = \sqrt{4}\sqrt{23} = 2\sqrt{23}$

9. $\sqrt{3x^6} = \sqrt{x^6}\sqrt{3} = x^3\sqrt{3}$

10. $\sqrt{8x^4y^7} = \sqrt{4x^4y^6}\sqrt{2y} = 2x^2y^3\sqrt{2y}$

11. $\sqrt{9x^9} = \sqrt{9x^8}\sqrt{x} = 3x^4\sqrt{x}$

12. $\sqrt[3]{40} = \sqrt[3]{8}\sqrt[3]{5} = 2\sqrt[3]{5}$

13. $\sqrt[3]{8x^6y^{10}} = \sqrt[3]{8x^6y^9}\sqrt[3]{y} = 2x^2y^3\sqrt[3]{y}$

14. $\sqrt{12} - 2\sqrt{75} = \sqrt{4}\sqrt{3} - 2\sqrt{25}\sqrt{3} = 2\sqrt{3} - 2(5)\sqrt{3}$
$= 2\sqrt{3} - 10\sqrt{3} = -8\sqrt{3}$

15. $\sqrt{2x^2} + \sqrt[3]{54} - x\sqrt{18}$
$= \sqrt{x^2}\sqrt{2} + \sqrt[3]{27}\sqrt[3]{2} - x\sqrt{9}\sqrt{2}$
$= x\sqrt{2} + 3\sqrt[3]{2} - x(3)\sqrt{2} = x\sqrt{2} + 3\sqrt[3]{2} - 3x\sqrt{2}$
$= -2x\sqrt{2} + 3\sqrt[3]{2}$

16. $\sqrt{\dfrac{5}{16}} = \dfrac{\sqrt{5}}{\sqrt{16}} = \dfrac{\sqrt{5}}{4}$

17. $\sqrt[3]{\dfrac{2x^3}{27}} = \dfrac{\sqrt[3]{2x^3}}{\sqrt[3]{27}} = \dfrac{\sqrt[3]{x^3}\sqrt[3]{2}}{\sqrt[3]{27}} = \dfrac{x\sqrt[3]{2}}{3}$

18. $3\sqrt{8x} = 3\sqrt{4}\sqrt{2x} = 3(2)\sqrt{2x} = 6\sqrt{2x}$

19. $\sqrt{\dfrac{2}{3}} = \dfrac{\sqrt{2}}{\sqrt{3}} \cdot \dfrac{\sqrt{3}}{\sqrt{3}} = \dfrac{\sqrt{6}}{3}$

20. $\sqrt[3]{\dfrac{5}{9}} = \dfrac{\sqrt[3]{5}}{\sqrt[3]{9}} \cdot \dfrac{\sqrt[3]{3}}{\sqrt[3]{3}} = \dfrac{\sqrt[3]{15}}{\sqrt[3]{27}} = \dfrac{\sqrt[3]{15}}{3}$

21. $\sqrt[3]{\dfrac{3}{4x^2}} = \dfrac{\sqrt[3]{3}}{\sqrt[3]{4x^2}} \cdot \dfrac{\sqrt[3]{2x}}{\sqrt[3]{2x}} = \dfrac{\sqrt[3]{6x}}{\sqrt[3]{8x^3}} = \dfrac{\sqrt[3]{6x}}{2x}$

22. $\dfrac{8}{\sqrt{6}+2} = \dfrac{8(\sqrt{6}-2)}{(\sqrt{6}+2)(\sqrt{6}-2)} = \dfrac{8(\sqrt{6}-2)}{6-4}$

$= \dfrac{8(\sqrt{6}-2)}{2} = 4(\sqrt{6}-2) = 4\sqrt{6}-8$

23. $\dfrac{2\sqrt{3}}{\sqrt{3}-3} = \dfrac{2\sqrt{3}(\sqrt{3}+3)}{(\sqrt{3}-3)(\sqrt{3}+3)} = \dfrac{2\sqrt{3}(\sqrt{3}+3)}{3-9}$

$= \dfrac{2(3)+6\sqrt{3}}{-6} = \dfrac{6+6\sqrt{3}}{-6} = \dfrac{6(1+\sqrt{3})}{-6}$

$= -1(1+\sqrt{3}) = -1-\sqrt{3}$

24. $\sqrt{x}+8=11$

$\sqrt{x}=11-8$

$\sqrt{x}=3$

$(\sqrt{x})^2=(3)^2$

$x=9$

Check

$\sqrt{9}+8=11$

$3+8=11$

$11=11$ True

The solution is 9.

25. $\sqrt{3x-6}=\sqrt{x+4}$

$(\sqrt{3x-6})^2=(\sqrt{x+4})^2$

$3x-6=x+4$

$3x-x-6=4$

$2x-6=4$

$2x=4+6$

$2x=10$

$x=\dfrac{10}{2}=5$

Check

$\sqrt{3(5)-6}=\sqrt{5+4}$

$\sqrt{15-6}=\sqrt{9}$

$\sqrt{9}=\sqrt{9}$ True

The solution is 5.

26. $\sqrt{2x-2}=x-5$

$(\sqrt{2x-2})^2=(x-5)^2$

$2x-2=(x)^2+2(x)(-5)+(5)^2$

$2x-2=x^2-10x+25$

$0=x^2-10x-2x+25+2$

$0=x^2-12x+27$

$0=(x-9)(x-3)$

$x-9=0$ or $x-3=0$

$x=9$ or $x=3$

Check $x=9$

$\sqrt{2(9)-2}=9-5$

$\sqrt{18-2}=4$

$\sqrt{16}=4$

$4=4$ True

Check $x=3$

$\sqrt{2(3)-2}=3-5$

$\sqrt{6-2}=-2$

$\sqrt{4}=-2$

$2=-2$ False

The solution is 9.

27.

$a^2+b^2=c^2$

$a^2+8^2=12^2$

$a^2+64=144$

$a^2=144-64$

$a^2=80$

$\sqrt{a^2}=\sqrt{80}$

$a=\sqrt{16}\sqrt{5}$

$a=4\sqrt{5}$ in.

28. $(-3, 6)$ and $(-2, 8)$

$d=\sqrt{(x_2-x_1)^2+(y_2-y_1)^2}$

$d=\sqrt{[-2-(-3)]^2+(8-6)^2}$

$d=\sqrt{(-2+3)^2+(2)^2}$

$d=\sqrt{1^2+4}$

$d=\sqrt{1+4}$

$d=\sqrt{5}$

29. $16^{-3/4} \cdot 16^{-1/4} = 16^{-4/4} = 16^{-1} = \dfrac{1}{16}$

30. $\left(x^{2/3}\right)^5 = x^{10/3}$

Chapter 9 - Cumulative Review

1. $3[4(5+2)-10] = 3[4(7)-10] = 3[28-10] = 3[18]$
$= 54$

2. $2x+3x-5+7 = 10x+3-6x-4$
$5x+2 = 4x-1$
$x+2 = -1$
$x = -3$

3. a. 45%

  b. $45+38 = 83\%$

  c. $22,000(0.45) = 9900$

  d. $360(0.16) = 57.6°$

4. $-5x+7 < 2(x-3)$
$-5x+7 < 2x-6$
$7 < 7x-6$
$13 < 7x$
$\dfrac{13}{7} < x$

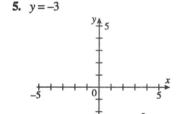

5. $y = -3$

6. $5x+4y \le 20$

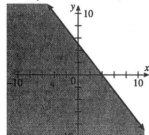

7. $(3x+2)(2x-5) = 6x^2-15x+4x-10$
$= 6x^2-11x-10$

8. a. $1.02 \times 10^5 = 102,000$

  b. $7.358 \times 10^{-3} = 0.007358$

  c. $8.4 \times 10^7 = 84,000,000$

  d. $3.007 \times 10^{-5} = 0.00003007$

9. $-9a^5+18a^2-3a = -3a\left(3a^4-6a+1\right)$

10. a. $x-3 = 0; \ x = 3$

  b. $x^2-3x+2 = 0; \ (x-2)(x-1) = 0; \ x = 2, \ x = 1$

  c. none

  d. none; x^2+1 is never zero.

11. $\dfrac{4-x^2}{3x^2-5x-2} = \dfrac{(2-x)(2+x)}{(3x+1)(x-2)} = -\dfrac{2+x}{3x+1}$

12. a. $\dfrac{a}{4} - \dfrac{2a}{8} = \dfrac{a}{4} \cdot \dfrac{2}{2} - \dfrac{2a}{8} = \dfrac{2a}{8} - \dfrac{2a}{8} = 0$

  b. $\dfrac{3}{10x^2} + \dfrac{7}{25x} = \dfrac{3}{10x^2} \cdot \dfrac{5}{5} + \dfrac{7}{25x} \cdot \dfrac{2x}{2x}$
  $= \dfrac{15}{50x^2} + \dfrac{14x}{50x^2} = \dfrac{15+14x}{50x^2}$

13. $\dfrac{x+1}{x+3} = \dfrac{1}{2x}$
$2x \cdot (x+1) = 1 \cdot (x+3)$
$2x^2+2x = x+3$
$2x^2+x-3 = 0$
$(2x+3)(x-1) = 0$
$2x+3 = 0 \qquad\qquad \text{or} \qquad x-1 = 0$
$x = -\dfrac{3}{2} \qquad\qquad \text{or} \qquad x = 1$

14. $y-5 = -2(x-(-1))$
$y-5 = -2(x+1)$
$y-5 = -2x-2$
$2x+y-5 = -2$
$2x+y = 3$

15. $3x+4y = 13$
$5x-9y = 6$

$-5(3x+4y) = -5 \cdot 13$
$3(5x-9y) = 3 \cdot 6$

$$-15x - 20y = -65$$
$$\underline{15x - 27y = 18}$$
$$\underline{-47y = -47}$$
$$\frac{-47y}{-47} = \frac{-47}{-47}$$
$$y = 1$$

$$3x + 4(1) = 13$$
$$3x + 4 = 13$$
$$3x = 9$$
$$x = 3$$
$$(3, 1)$$

16. $\begin{cases} x - y < 2 \\ x + 2y > -1 \end{cases}$

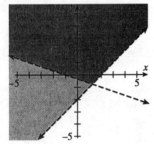

17. a. $\sqrt[3]{1} = 1$

 b. $\sqrt[3]{-27} = -3$

 c. $\sqrt[3]{\dfrac{1}{25}} = \dfrac{\sqrt[3]{1}}{\sqrt[3]{25}} \cdot \dfrac{\sqrt[3]{5}}{\sqrt[3]{5}} = \dfrac{\sqrt[3]{5}}{5}$

18. a. $\sqrt{54} = \sqrt{9 \cdot 6} = 3\sqrt{6}$

 b. $\sqrt{12} = \sqrt{4 \cdot 3} = 2\sqrt{3}$

 c. $\sqrt{200} = \sqrt{100 \cdot 2} = 10\sqrt{2}$

 d. $\sqrt{35}$ cannot be simplified.

19. a. $2\sqrt{x^2} - \sqrt{25x} + \sqrt{x} = 2x - 5\sqrt{x} + \sqrt{x}$
$$= 2x - 4\sqrt{x}$$

 b. $3\sqrt[3]{54x^4} + 5x\sqrt[3]{16x} = 9x\sqrt[3]{2x} + 10x\sqrt[3]{2x}$
$$= 19x\sqrt[3]{2x}$$

20. a. $\dfrac{2}{\sqrt{7}} = \dfrac{2}{\sqrt{7}} \cdot \dfrac{\sqrt{7}}{\sqrt{7}} = \dfrac{2\sqrt{7}}{7}$

 b. $\dfrac{\sqrt{5}}{\sqrt{12}} = \dfrac{\sqrt{5}}{\sqrt{12}} \cdot \dfrac{\sqrt{3}}{\sqrt{3}} = \dfrac{\sqrt{15}}{\sqrt{36}} = \dfrac{\sqrt{15}}{6}$

 c. $\sqrt{\dfrac{1}{18x}} = \dfrac{\sqrt{1}}{\sqrt{18x}} = \dfrac{1}{3\sqrt{2x}} \cdot \dfrac{\sqrt{2x}}{\sqrt{2x}} = \dfrac{\sqrt{2x}}{3(2x)}$
$$= \dfrac{\sqrt{2x}}{6x}$$

21. $\sqrt{x} = \sqrt{5x - 2}$
$$(\sqrt{x})^2 = (\sqrt{5x - 2})^2$$
$$x = 5x - 2$$
$$-4x = -2$$
$$x = \dfrac{1}{2}$$

22. $v = \sqrt{2gh}$
$$v = \sqrt{2(32)(5)}$$
$$v = \sqrt{64 \cdot 5}$$
$$v = 8\sqrt{5} \text{ ft per second}$$

23. a. $25^{1/2} = \sqrt{25} = 5$

 b. $8^{1/3} = \sqrt[3]{8} = 2$

 c. $-16^{1/4} = -\sqrt[4]{16} = -2$

 d. $(-27)^{1/3} = \sqrt[3]{-27} = -3$

 e. $\left(\dfrac{1}{9}\right)^{1/2} = \sqrt{\dfrac{1}{9}} = \dfrac{1}{3}$

Chapter 10

Exercise Set 10.1

1. $k^2 - 9 = 0$
$(k-3)(k+3) = 0$
$k-3 = 0$ or $k+3 = 0$
$k = 3$ or $k = -3$

3. $m^2 + 2m = 15$
$m^2 + 2m - 15 = 0$
$(m+5)(m-3) = 0$
$m+5 = 0$ or $m-3 = 0$
$m = -5$ or $m = 3$

5. $2x^2 - 81 = 0$
$(\sqrt{2}x + 9)(\sqrt{2}x - 9) = 0$
$\sqrt{2}x + 9 = 0$ or $\sqrt{2}x - 9 = 0$
$\sqrt{2}x = -9$ or $\sqrt{2}x = 9$
$x = \dfrac{-9}{\sqrt{2}}$ or $x = \dfrac{9}{\sqrt{2}}$
$x = \dfrac{-9}{\sqrt{2}} \cdot \dfrac{\sqrt{2}}{\sqrt{2}}$ or $x = \dfrac{9}{\sqrt{2}} \cdot \dfrac{\sqrt{2}}{\sqrt{2}}$
$x = \dfrac{-9\sqrt{2}}{2}$ or $x = \dfrac{9\sqrt{2}}{2}$

7. $4a^2 - 36 = 0$
$4(a^2 - 9) = 0$
$4(a+3)(a-3) = 0$
$a+3 = 0$ or $a-3 = 0$
$a = -3$ or $a = 3$

9. $x^2 = 64$
$x = \pm\sqrt{64}$
$x = \pm 8$

11. $p^2 = \dfrac{1}{49}$
$p = \pm\sqrt{\dfrac{1}{49}}$
$p = \pm\dfrac{1}{7}$

13. $y^2 = -36$
$y = \pm\sqrt{-36}$
$\sqrt{-36}$ is not a real number. No real solutions.

15. $2x^2 = 50$
$x^2 = \dfrac{50}{2}$
$x^2 = 25$
$x^2 = \pm\sqrt{25}$
$x = \pm 5$

17. $3x^2 = 4$
$x^2 = \dfrac{4}{3}$
$x = \pm\sqrt{\dfrac{4}{3}}$
$x = \pm\dfrac{\sqrt{4}}{\sqrt{3}}$
$x = \pm\dfrac{2}{\sqrt{3}} \cdot \dfrac{\sqrt{3}}{\sqrt{3}}$
$x = \pm\dfrac{2\sqrt{3}}{3}$

19. $x^2 = 225$
$x = \sqrt{225}$
$x = 15$ ft.
Dimensions are 15 ft. by 15 ft.

21. $(x-5)^2 = 49$
$x - 5 = \pm\sqrt{49}$
$x - 5 = \pm 7$
$x = 5 \pm 7$
$x = 5 + 7$ or $x = 5 - 7$
$x = 12$ or $x = -2$

23. $(x+2)^2 = 7$
$x + 2 = \pm\sqrt{7}$
$x = -2 \pm\sqrt{7}$

25. $\left(m - \dfrac{1}{2}\right)^2 = \dfrac{1}{4}$
$m - \dfrac{1}{2} = \pm\sqrt{\dfrac{1}{4}}$
$m - \dfrac{1}{2} = \pm\dfrac{1}{2}$
$m = \dfrac{1}{2} \pm \dfrac{1}{2}$

$$m = \frac{1}{2} + \frac{1}{2} \qquad \text{or} \qquad m = \frac{1}{2} - \frac{1}{2}$$
$$m = 1 \qquad \text{or} \qquad m = 0$$

27. $(p+2)^2 = 10$
$$p + 2 = \pm\sqrt{10}$$
$$p = -2 \pm \sqrt{10}$$
$$p = -2 + \sqrt{10} \qquad \text{or} \qquad p = -2 - \sqrt{10}$$

29. $(3y+2)^2 = 100$
$$3y + 2 = \pm\sqrt{100}$$
$$3y + 2 = \pm 10$$
$$3y = -2 \pm 10$$
$$y = \frac{-2 \pm 10}{3}$$
$$y = \frac{-2 + 10}{3} \qquad \text{or} \qquad y = \frac{-2 - 10}{3}$$
$$y = \frac{8}{3} \qquad \text{or} \qquad y = \frac{-12}{3} = -4$$

31. $(z-4)^2 = -9$
$$z - 4 = \pm\sqrt{-9}$$
$\sqrt{-9}$ is not a real number. No real solutions.

33. $(2x-11)^2 = 50$
$$2x - 11 = \pm\sqrt{50}$$
$$2x - 11 = \pm\sqrt{25}\sqrt{2}$$
$$2x - 11 = \pm 5\sqrt{2}$$
$$2x = 11 \pm 5\sqrt{2}$$
$$x = \frac{11 \pm 5\sqrt{2}}{2}$$

35.

$$a^2 + b^2 = c^2$$
$$x^2 + x^2 = 20^2$$
$$2x^2 = 400$$
$$x^2 = 200$$
$$x = \sqrt{200}$$
$$x = \sqrt{100}\sqrt{2}$$
$$x = 10\sqrt{2} \text{ cm}$$

37. $q^2 = 100$
$$q = \pm\sqrt{100}$$
$$q = \pm 10$$

39. $(x-13)^2 = 16$
$$x - 13 = \pm\sqrt{16}$$
$$x - 13 = \pm 4$$
$$x = 13 \pm 4$$
$$x = 13 + 4 \qquad \text{or} \qquad x = 13 - 4$$
$$x = 17 \qquad \text{or} \qquad x = 9$$

41. $z^2 = 12$
$$z = \pm\sqrt{12}$$
$$z = \pm\sqrt{4}\sqrt{3}$$
$$z = \pm 2\sqrt{3}$$

43. $(x+5)^2 = 10$
$$x + 5 = \pm\sqrt{10}$$
$$x = -5 \pm \sqrt{10}$$

45. $m^2 = -10$
$$m = \pm\sqrt{-10}$$
$\sqrt{-10}$ is not a real number. No real solutions.

47. $2y^2 = 11$
$$y^2 = \frac{11}{2}$$
$$y = \pm\sqrt{\frac{11}{2}}$$
$$y = \pm\frac{\sqrt{11}}{\sqrt{2}}$$
$$y = \pm\frac{\sqrt{11}}{\sqrt{2}} \cdot \frac{\sqrt{2}}{\sqrt{2}}$$
$$y = \pm\frac{\sqrt{22}}{2}$$

49. $(2p-5)^2 = 121$
$$2p - 5 = \pm\sqrt{121}$$
$$2p - 5 = \pm 11$$
$$2p = 5 \pm 11$$
$$p = \frac{5 \pm 11}{2}$$
$$p = \frac{5 + 11}{2} \qquad \text{or} \qquad p = \frac{5 - 11}{2}$$
$$p = 8 \qquad \text{or} \qquad p = -3$$

51. $(3x-1)^2 = 7$

$3x - 1 = \pm\sqrt{7}$

$3x = 1 \pm \sqrt{7}$

$x = \dfrac{1 \pm \sqrt{7}}{3}$

53. $(3x-7)^2 = 32$

$3x - 7 = \pm\sqrt{32}$

$3x - 7 = \pm\sqrt{16}\sqrt{2}$

$3x - 7 = \pm 4\sqrt{2}$

$3x = 7 \pm 4\sqrt{2}$

$x = \dfrac{7 \pm 4\sqrt{2}}{3}$

55. $x^2 = 1.78$

$x = \pm\sqrt{1.78}$

$x = \pm 1.33$

57. $(x - 1.37)^2 = 5.71$

$x - 1.37 = \pm\sqrt{5.71}$

$x = 1.37 \pm \sqrt{5.71}$

$x = 1.37 \pm 2.39$

$x = -1.02, 3.76$

59. $(2y + 1.58)^2 = 21.11$

$2y + 1.58 = \pm\sqrt{21.11}$

$2y = -1.58 \pm \sqrt{21.11}$

$y = \dfrac{-1.58 \pm 4.59}{2}$

$y = -3.09, 1.51$

61. $x^2 + 4x + 4 = 16$

$(x+2)^2 = 16$

$x + 2 = \pm\sqrt{16}$

$x + 2 = \pm 4$

$x = -2 \pm 4$

$x = -6, 2$

63. $y^2 - 10y + 25 = 11$

$(y - 5)^2 = 11$

$y - 5 = \pm\sqrt{11}$

$y = 5 \pm \sqrt{11}$

65. $A = \pi r^2$

$36\pi = \pi r^2$

$\dfrac{36\pi}{\pi} = \dfrac{\pi r^2}{\pi}$

$36 = r^2$

$\pm\sqrt{36} = r$

$\pm 6 = r$

Disregard the negative.

radius = 6 inches

67. $x^2 + 6x + 9 = (x+3)(x+3) = (x+3)^2$

69. $x^2 - 4x + 4 = (x-2)(x-2) = (x-2)^2$

71. 35 million

73. $x = 9$ years since 1991

$y = 1.7(9) + 29.5 = 44.8$

44.8 million

Section 10.2 Mental Math

1. 16; $p^2 + 8p + 16 = (p+4)^2$

3. 100; $x^2 + 20x + 100 = (x+10)^2$

5. 49; $y^2 + 14y + 49 = (y+7)^2$

Exercise Set 10.2

1. $x^2 + 4x$

$x^2 + 4x + 4 = (x+2)^2$

3. $k^2 - 12k$

$k^2 - 12k + 36 = (k-6)^2$

5. $x^2 - 3x$

$x^2 - 3x + \dfrac{9}{4} = \left(x - \dfrac{3}{2}\right)^2$

7. $m^2 - m$

$m^2 - m + \dfrac{1}{4} = \left(m - \dfrac{1}{2}\right)^2$

9. $x^2 - 6x = 0$

$x^2 - 6x + 9 = 0 + 9$

$(x-3)^2 = 9$

$x - 3 = \pm\sqrt{9}$

$x - 3 = \pm 3$

$x = 3 \pm 3$

$x = 3 + 3$ or $x = 3 - 3$

$x = 6$ or $x = 0$

11. $x^2 + 8x = -12$

$x^2 + 8x + 16 = -12 + 16$

$(x+4)^2 = 4$

$x + 4 = \pm\sqrt{4}$

$x + 4 = \pm 2$

$x = -4 \pm 2$

$x = -4 + 2$ or $x = -4 - 2$

$x = -2$ or $x = -6$

13. $x^2 + 2x - 5 = 0$

$x^2 + 2x = 5$

$x^2 + 2x + 1 = 5 + 1$

$(x+1)^2 = 6$

$x + 1 = \pm\sqrt{6}$

$x = -1 \pm\sqrt{6}$

15. $x^2 + kx + 16$

$\left(\dfrac{k}{2}\right)^2 = 16$

$\dfrac{k}{2} = \pm\sqrt{16}$

$\dfrac{k}{2} = \pm 4$

$k = \pm 8$

17. $4x^2 - 24x = 13$

$\dfrac{4x^2}{4} - \dfrac{24x}{4} = \dfrac{13}{4}$

$x^2 - 6x = \dfrac{13}{4}$

$x^2 - 6x + 9 = \dfrac{13}{4} + 9$

$(x-3)^2 = \dfrac{13}{4} + \dfrac{36}{4}$

$(x-3)^2 = \dfrac{49}{4}$

$x - 3 = \pm\sqrt{\dfrac{49}{4}}$

$x - 3 = \pm\dfrac{7}{2}$

$x = 3 \pm\dfrac{7}{2}$

$x = 3 + \dfrac{7}{2}$ or $x = 3 - \dfrac{7}{2}$

$x = \dfrac{6}{2} + \dfrac{7}{2}$ or $x = \dfrac{6}{2} - \dfrac{7}{2}$

$x = \dfrac{13}{2}$ or $x = -\dfrac{1}{2}$

19. $5x^2 + 10x + 6 = 0$

$5x^2 + 10x = -6$

$\dfrac{5x^2}{5} + \dfrac{10x}{5} = \dfrac{-6}{5}$

$x^2 + 2x = -\dfrac{6}{5}$

$x^2 + 2x + 1 = -\dfrac{6}{5} + 1$

$(x+1)^2 = -\dfrac{6}{5} + \dfrac{5}{5}$

$x + 1 = \pm\sqrt{-\dfrac{1}{5}}$

$\sqrt{-\dfrac{1}{5}}$ is not a real number. No real solutions.

21. $2x^2 = 6x + 5$

$2x^2 - 6x = 5$

$\dfrac{2x^2}{2} - \dfrac{6x}{2} = \dfrac{5}{2}$

$x^2 - 3x = \dfrac{5}{2}$

$x^2 - 3x + \dfrac{9}{4} = \dfrac{5}{2} + \dfrac{9}{4}$

$\left(x - \dfrac{3}{2}\right)^2 = \dfrac{10}{4} + \dfrac{9}{4}$

$\left(x - \dfrac{3}{2}\right)^2 = \dfrac{19}{4}$

$x - \dfrac{3}{2} = \pm\sqrt{\dfrac{19}{4}}$

$x - \dfrac{3}{2} = \pm\dfrac{\sqrt{19}}{2}$

$$x = \frac{3}{2} \pm \frac{\sqrt{19}}{2}$$

$$x = \frac{3 \pm \sqrt{19}}{2}$$

23. $x^2 + 6x - 25 = 0$

$x^2 + 6x = 25$

$x^2 + 6x + 9 = 25 + 9$

$(x + 3)^2 = 34$

$x + 3 = \pm\sqrt{34}$

$x = -3 \pm \sqrt{34}$

25. $z^2 + 5z = 7$

$z^2 + 5z + \frac{25}{4} = 7 + \frac{25}{4}$

$\left(z + \frac{5}{2}\right)^2 = \frac{28}{4} + \frac{25}{4}$

$\left(z + \frac{5}{2}\right)^2 = \frac{53}{4}$

$z + \frac{5}{2} = \pm\sqrt{\frac{53}{4}}$

$z + \frac{5}{2} = \pm\frac{\sqrt{53}}{2}$

$z = -\frac{5}{2} \pm \frac{\sqrt{53}}{2}$

$z = \frac{-5 \pm \sqrt{53}}{2}$

27. $x^2 - 2x - 1 = 0$

$x^2 - 2x = 1$

$x^2 - 2x + 1 = 1 + 1$

$(x - 1)^2 = 2$

$x - 1 = \pm\sqrt{2}$

$x = 1 \pm \sqrt{2}$

29. $y^2 + 5y + 4 = 0$

$y^2 + 5y = -4$

$y^2 + 5y + \frac{25}{4} = -4 + \frac{25}{4}$

$\left(y + \frac{5}{2}\right)^2 = -\frac{16}{4} + \frac{25}{4}$

$\left(y + \frac{5}{2}\right)^2 = \frac{9}{4}$

$y + \frac{5}{2} = \pm\sqrt{\frac{9}{4}}$

$y + \frac{5}{2} = \pm\frac{3}{2}$

$y = -\frac{5}{2} \pm \frac{3}{2}$

$y = -\frac{5}{2} + \frac{3}{2}$ or $y = -\frac{5}{2} - \frac{3}{2}$

$y = -\frac{2}{2} = -1$ or $y = -\frac{8}{2} = -4$

31. $3x^2 - 6x = 24$

$\frac{3x^2}{3} - \frac{6x}{3} = \frac{24}{3}$

$x^2 - 2x = 8$

$x^2 - 2x + 1 = 8 + 1$

$(x - 1)^2 = 9$

$x - 1 = \pm\sqrt{9}$

$x - 1 = \pm 3$

$x = 1 \pm 3$

$x = 1 + 3$ or $x = 1 - 3$

$x = 4$ or $x = -2$

33. $2y^2 + 8y + 5 = 0$

$2y^2 + 8y = -5$

$\frac{2y^2}{2} + \frac{8y}{2} = \frac{-5}{2}$

$y^2 + 4y = -\frac{5}{2}$

$y^2 + 4y + 4 = -\frac{5}{2} + 4$

$(y + 2)^2 = -\frac{5}{2} + \frac{8}{2}$

$(y + 2)^2 = \frac{3}{2}$

$y + 2 = \pm\sqrt{\frac{3}{2}}$

$y + 2 = \pm\frac{\sqrt{3}}{\sqrt{2}}$

$y + 2 = \pm\frac{\sqrt{3}}{\sqrt{2}} \cdot \frac{\sqrt{2}}{\sqrt{2}}$

$y + 2 = \pm\frac{\sqrt{6}}{2}$

$y = -2 \pm \frac{\sqrt{6}}{2}$

$$y = -\frac{4}{2} \pm \frac{\sqrt{6}}{2}$$

$$y = \frac{-4 \pm \sqrt{6}}{2}$$

35. $2y^2 - 3y + 1 = 0$

$$2y^2 - 3y = -1$$

$$\frac{2y^2}{2} - \frac{3y}{2} = \frac{-1}{2}$$

$$y^2 - \frac{3}{2}y = -\frac{1}{2}$$

$$y^2 - \frac{3}{2}y + \frac{9}{16} = -\frac{1}{2} + \frac{9}{16}$$

$$\left(y - \frac{3}{4}\right)^2 = -\frac{8}{16} + \frac{9}{16}$$

$$\left(y - \frac{3}{4}\right)^2 = \frac{1}{16}$$

$$y - \frac{3}{4} = \pm\sqrt{\frac{1}{16}}$$

$$y - \frac{3}{4} = \pm\frac{1}{4}$$

$$y = \frac{3}{4} \pm \frac{1}{4}$$

$$y = \frac{3}{4} + \frac{1}{4} \quad \text{or} \quad y = \frac{3}{4} - \frac{1}{4}$$

$$y = \frac{4}{4} = 1 \quad \text{or} \quad y = \frac{2}{4} = \frac{1}{2}$$

37. $3y^2 - 2y - 4 = 0$

$$3y^2 - 2y = 4$$

$$\frac{3y^2}{3} - \frac{2y}{3} = \frac{4}{3}$$

$$y^2 - \frac{2}{3}y = \frac{4}{3}$$

$$y^2 - \frac{2}{3}y + \frac{1}{9} = \frac{4}{3} + \frac{1}{9}$$

$$\left(y - \frac{1}{3}\right)^2 = \frac{12}{9} + \frac{1}{9}$$

$$\left(y - \frac{1}{3}\right)^2 = \frac{13}{9}$$

$$y - \frac{1}{3} = \pm\sqrt{\frac{13}{9}}$$

$$y - \frac{1}{3} = \pm\frac{\sqrt{13}}{3}$$

$$y = \frac{1}{3} \pm \frac{\sqrt{13}}{3}$$

$$y = \frac{1 \pm \sqrt{13}}{3}$$

39. $y^2 = 5y + 14$

$$y^2 - 5y = 14$$

$$y^2 - 5y + \frac{25}{4} = 14 + \frac{25}{4}$$

$$\left(y - \frac{5}{2}\right)^2 = \frac{56}{4} + \frac{25}{4}$$

$$\left(y - \frac{5}{2}\right)^2 = \frac{81}{4}$$

$$y - \frac{5}{2} = \pm\sqrt{\frac{81}{4}}$$

$$y - \frac{5}{2} = \pm\frac{9}{2}$$

$$y = \frac{5}{2} \pm \frac{9}{2}$$

$$y = \frac{5}{2} + \frac{9}{2} \quad \text{or} \quad y = \frac{5}{2} - \frac{9}{2}$$

$$y = \frac{14}{2} = 7 \quad \text{or} \quad y = -\frac{4}{2} = -2$$

41. $x(x + 3) = 18$

$$x^2 + 3x = 18$$

$$x^2 + 3x + \frac{9}{4} = 18 + \frac{9}{4}$$

$$\left(x + \frac{3}{2}\right)^2 = \frac{72}{4} + \frac{9}{4}$$

$$\left(x + \frac{3}{2}\right)^2 = \frac{81}{4}$$

$$x + \frac{3}{2} = \pm\sqrt{\frac{81}{4}}$$

$$x + \frac{3}{2} = \pm\frac{9}{2}$$

$$x = -\frac{3}{2} \pm \frac{9}{2}$$

$$x = -\frac{3}{2} + \frac{9}{2} \quad \text{or} \quad x = -\frac{3}{2} - \frac{9}{2}$$

$$x = \frac{6}{2} = 3 \quad \text{or} \quad x = \frac{-12}{2} = -6$$

43. Answers may vary.

45. $x^2 + 8x = -12$
$x = -6, -2$

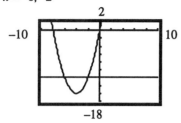

47. $2x^2 = 6x + 5$
$x \approx -0.68, 3.68$

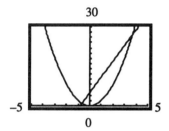

49. $\dfrac{3}{4} - \sqrt{\dfrac{25}{16}} = \dfrac{3}{4} - \dfrac{\sqrt{25}}{\sqrt{16}} = \dfrac{3}{4} - \dfrac{5}{4} = -\dfrac{2}{4} = -\dfrac{1}{2}$

51. $\dfrac{1}{2} - \sqrt{\dfrac{9}{4}} = \dfrac{1}{2} - \dfrac{\sqrt{9}}{\sqrt{4}} = \dfrac{1}{2} - \dfrac{3}{2} = \dfrac{-2}{2} = -1$

53. $\dfrac{6 + 4\sqrt{5}}{2} = 3 + 2\sqrt{5}$

55. $\dfrac{3 - 9\sqrt{2}}{6} = \dfrac{1 - 3\sqrt{2}}{2}$

Section 10.3 Mental Math

1. $2x^2 + 5x + 3 = 0$
$a = 2, b = 5, c = 3$

3. $10x^2 - 13x - 2 = 0$
$a = 10, b = -13, c = -2$

5. $x^2 - 6 = 0$
$a = 1, b = 0, c = -6$

Exercise Set 10.3

1. $\dfrac{-1 \pm \sqrt{1^2 - 4(1)(-2)}}{2(1)} = \dfrac{-1 \pm \sqrt{1 + 8}}{2}$

$= \dfrac{-1 \pm \sqrt{9}}{2} = \dfrac{-1 \pm 3}{2}$

$= \dfrac{-1 + 3}{2}$ or $\dfrac{-1 - 3}{2}$

$\dfrac{2}{2} = 1$ or $\dfrac{-4}{2} = -2$

3. $\dfrac{-5 \pm \sqrt{5^2 - 4(1)(2)}}{2(1)} = \dfrac{-5 \pm \sqrt{25 - 8}}{2} = \dfrac{-5 \pm \sqrt{17}}{2}$

5. $\dfrac{-(-4) \pm \sqrt{(-4)^2 - 4(2)(1)}}{2(2)}$

$= \dfrac{4 \pm \sqrt{16 - 8}}{4} = \dfrac{4 \pm \sqrt{8}}{4} = \dfrac{4 \pm \sqrt{4}\sqrt{2}}{4}$

$= \dfrac{4 \pm 2\sqrt{2}}{4} = \dfrac{2(2 \pm \sqrt{2})}{4} = \dfrac{2 \pm \sqrt{2}}{2}$

7. $x^2 - 3x + 2 = 0$
$a = 1, b = -3, c = 2$

$x = \dfrac{-(-3) \pm \sqrt{(-3)^2 - 4(1)(2)}}{2(1)}$

$x = \dfrac{3 \pm \sqrt{9 - 8}}{2}$

$x = \dfrac{3 \pm \sqrt{1}}{2}$

$x = \dfrac{3 \pm 1}{2}$

$x = \dfrac{3 + 1}{2}$ or $x = \dfrac{3 - 1}{2}$

$x = \dfrac{4}{2} = 2$ or $x = \dfrac{2}{2} = 1$

9. $3k^2 + 7k + 1 = 0$
$a = 3, b = 7, c = 1$

$k = \dfrac{-7 \pm \sqrt{7^2 - 4(3)(1)}}{2(3)}$

$$k = \frac{-7 \pm \sqrt{49 - 12}}{6}$$

$$k = \frac{-7 \pm \sqrt{37}}{6}$$

11. $49x^2 - 4 = 0$
$a = 49, b = 0, c = -4$

$$x = \frac{-0 \pm \sqrt{0^2 - 4(49)(-4)}}{2(49)}$$

$$x = \frac{\pm\sqrt{784}}{98}$$

$$x = \pm\frac{28}{98} = \pm\frac{2}{7}$$

13. $5z^2 - 4z + 3 = 0$
$a = 5, b = -4, c = 3$

$$z = \frac{-(-4) \pm \sqrt{(-4)^2 - 4(5)(3)}}{2(5)}$$

$$z = \frac{4 \pm \sqrt{16 - 60}}{10}$$

$$z = \frac{4 \pm \sqrt{-44}}{10}$$

$\sqrt{-44}$ is not a real number. No real solutions.

15. $y^2 = 7y + 30$
$y^2 - 7y - 30 = 0$
$a = 1, b = -7, c = -30$

$$y = \frac{-(-7) \pm \sqrt{(-7)^2 - 4(1)(-30)}}{2(1)}$$

$$y = \frac{7 \pm \sqrt{49 + 120}}{2}$$

$$y = \frac{7 \pm \sqrt{169}}{2}$$

$$y = \frac{7 \pm 13}{2}$$

$$y = \frac{7 + 13}{2} \quad \text{or} \quad y = \frac{7 - 13}{2}$$

$$y = \frac{20}{2} = 10 \quad \text{or} \quad y = \frac{-6}{2} = -3$$

17. $2x^2 = 10$
$2x^2 - 10 = 0$
$a = 2, b = 0, c = -10$

$$x = \frac{-0 \pm \sqrt{0^2 - 4(2)(-10)}}{2(2)}$$

$$x = \frac{\pm\sqrt{80}}{4}$$

$$x = \frac{\pm\sqrt{16}\sqrt{5}}{4}$$

$$x = \pm\frac{4\sqrt{5}}{4}$$

$$x = \pm\sqrt{5}$$

19. $m^2 - 12 = m$
$m^2 - m - 12 = 0$
$a = 1, b = -1, c = -12$

$$m = \frac{-(-1) \pm \sqrt{(-1)^2 - 4(1)(-12)}}{2(1)}$$

$$m = \frac{1 \pm \sqrt{1 + 48}}{2}$$

$$m = \frac{1 \pm \sqrt{49}}{2}$$

$$m = \frac{1 \pm 7}{2}$$

$$m = \frac{1 + 7}{2} \quad \text{or} \quad m = \frac{1 - 7}{2}$$

$$m = \frac{8}{2} = 4 \quad \text{or} \quad m = \frac{-6}{2} = -3$$

21. $3 - x^2 = 4x$
$-x^2 - 4x + 3 = 0$
$a = -1, b = -4, c = 3$

$$x = \frac{-(-4) \pm \sqrt{(-4)^2 - 4(-1)(3)}}{2(-1)}$$

$$x = \frac{4 \pm \sqrt{16 + 12}}{-2}$$

$$x = \frac{4 \pm \sqrt{28}}{-2}$$

$$x = \frac{4 \pm \sqrt{4}\sqrt{7}}{-2}$$

$$x = \frac{4 \pm 2\sqrt{7}}{-2}$$

$$x = \frac{2(2 \pm \sqrt{7})}{-2}$$

$$x = -1(2 \pm \sqrt{7})$$

$$x = -2 \pm \sqrt{7}$$

23. $3p^2 - \dfrac{2}{3}p + 1 = 0$

$$3\left(3p^2 - \dfrac{2}{3}p + 1\right) = 3(0)$$

$9p^2 - 2p + 3 = 0$

$a = 9,\ b = -2,\ c = 3$

$$p = \dfrac{-(-2) \pm \sqrt{(-2)^2 - 4(9)(3)}}{2(9)}$$

$$p = \dfrac{2 \pm \sqrt{4 - 108}}{18}$$

$$p = \dfrac{2 \pm \sqrt{-104}}{18}$$

$\sqrt{-104}$ is not a real number. No real solutions.

25. $\dfrac{m^2}{2} = m + \dfrac{1}{2}$

$$2\left(\dfrac{m^2}{2}\right) = 2\left(m + \dfrac{1}{2}\right)$$

$m^2 = 2m + 1$

$m^2 - 2m - 1 = 0$

$a = 1,\ b = -2,\ c = -1$

$$m = \dfrac{-(-2) \pm \sqrt{(-2)^2 - 4(1)(-1)}}{2(1)}$$

$$m = \dfrac{2 \pm \sqrt{4 + 4}}{2}$$

$$m = \dfrac{2 \pm \sqrt{8}}{2}$$

$$m = \dfrac{2 \pm \sqrt{4}\sqrt{2}}{2}$$

$$m = \dfrac{2 \pm 2\sqrt{2}}{2}$$

$$m = \dfrac{2(1 \pm \sqrt{2})}{2}$$

$m = 1 \pm \sqrt{2}$

27. $4p^2 + \dfrac{3}{2} = -5p$

$$2\left(4p^2 + \dfrac{3}{2}\right) = 2(-5p)$$

$8p^2 + 3 = -10p$

$8p^2 + 10p + 3 = 0$

$a = 8,\ b = 10,\ c = 3$

$$p = \dfrac{-10 \pm \sqrt{10^2 - 4(8)(3)}}{2(8)}$$

$$p = \dfrac{-10 \pm \sqrt{100 - 96}}{16}$$

$$p = \dfrac{-10 \pm \sqrt{4}}{16}$$

$$p = \dfrac{-10 \pm 2}{16}$$

$p = \dfrac{-10 + 2}{16}$ or $p = \dfrac{-10 - 2}{16}$

$p = \dfrac{-8}{16} = -\dfrac{1}{2}$ or $p = \dfrac{-12}{16} = -\dfrac{3}{4}$

29. $2a^2 - 7a + 3 = 0$

$a = 2,\ b = -7,\ c = 3$

$$a = \dfrac{-(-7) \pm \sqrt{(-7)^2 - 4(2)(3)}}{2(2)}$$

$$a = \dfrac{7 \pm \sqrt{49 - 24}}{4}$$

$$a = \dfrac{7 \pm \sqrt{25}}{4}$$

$$a = \dfrac{7 \pm 5}{4}$$

$a = \dfrac{7 + 5}{4}$ or $a = \dfrac{7 - 5}{4}$

$a = \dfrac{12}{4} = 3$ or $a = \dfrac{2}{4} = \dfrac{1}{2}$

31. $x^2 - 5x - 2 = 0$

$a = 1,\ b = -5,\ c = -2$

$$x = \dfrac{-(-5) \pm \sqrt{(-5)^2 - 4(1)(-2)}}{2(1)}$$

$$x = \dfrac{5 \pm \sqrt{25 + 8}}{2}$$

$$x = \dfrac{5 \pm \sqrt{33}}{2}$$

33. $3x^2 - x - 14 = 0$

$a = 3, b = -1, c = -14$

$$x = \frac{-(-1) \pm \sqrt{(-1)^2 - 4(3)(-14)}}{2(3)}$$

$$x = \frac{1 \pm \sqrt{1 + 168}}{6}$$

$$x = \frac{1 \pm \sqrt{169}}{6}$$

$$x = \frac{1 \pm 13}{6}$$

$$x = \frac{1 + 13}{6} \quad \text{or} \quad x = \frac{1 - 13}{6}$$

$$x = \frac{14}{6} = \frac{7}{3} \quad \text{or} \quad x = \frac{-12}{6} = -2$$

35. $6x^2 + 9x = 2$

$6x^2 + 9x - 2 = 0$

$a = 6, b = 9, c = -2$

$$x = \frac{-9 \pm \sqrt{9^2 - 4(6)(-2)}}{2(6)}$$

$$x = \frac{-9 \pm \sqrt{81 + 48}}{12}$$

$$x = \frac{-9 \pm \sqrt{129}}{12}$$

37. $7p^2 + 2 = 8p$

$7p^2 - 8p + 2 = 0$

$a = 7, b = -8, c = 2$

$$p = \frac{-8(-8) \pm \sqrt{(-8)^2 - 4(7)(2)}}{2(7)}$$

$$p = \frac{8 \pm \sqrt{64 - 56}}{14}$$

$$p = \frac{8 \pm \sqrt{8}}{14}$$

$$p = \frac{8 \pm \sqrt{4}\sqrt{2}}{14}$$

$$p = \frac{8 \pm 2\sqrt{2}}{14}$$

$$p = \frac{2(4 \pm \sqrt{2})}{14}$$

$$p = \frac{4 \pm \sqrt{2}}{7}$$

39. $a^2 - 6a + 2 = 0$

$a = 1, b = -6, c = 2$

$$a = \frac{-(-6) \pm \sqrt{(-6)^2 - 4(1)(2)}}{2(1)}$$

$$a = \frac{6 \pm \sqrt{36 - 8}}{2}$$

$$a = \frac{6 \pm \sqrt{28}}{2}$$

$$a = \frac{6 \pm \sqrt{4}\sqrt{7}}{2}$$

$$a = \frac{6 \pm 2\sqrt{7}}{2}$$

$$a = \frac{2(3 \pm \sqrt{7})}{2}$$

$$a = 3 \pm \sqrt{7}$$

41. $2x^2 - 6x + 3 = 0$

$a = 2, b = -6, c = 3$

$$x = \frac{-(-6) \pm \sqrt{(-6)^2 - 4(2)(3)}}{2(2)}$$

$$x = \frac{6 \pm \sqrt{36 - 24}}{4}$$

$$x = \frac{6 \pm \sqrt{12}}{4}$$

$$x = \frac{6 \pm \sqrt{4}\sqrt{3}}{4}$$

$$x = \frac{6 \pm 2\sqrt{3}}{4}$$

$$x = \frac{2(3 \pm \sqrt{3})}{4}$$

$$x = \frac{3 \pm \sqrt{3}}{2}$$

43. $3x^2 = 1 - 2x$

$3x^2 + 2x - 1 = 0$

$a = 3, b = 2, c = -1$

$$x = \frac{-2 \pm \sqrt{2^2 - 4(3)(-1)}}{2(3)}$$

$$x = \frac{-2 \pm \sqrt{4 + 12}}{6}$$

$$x = \frac{-2 \pm \sqrt{16}}{6}$$

$$x = \frac{-2 \pm 4}{6}$$

$$x = \frac{-2+4}{6} \quad \text{or} \quad x = \frac{-2-4}{6}$$

$$x = \frac{2}{6} = \frac{1}{3} \quad \text{or} \quad x = \frac{-6}{6} = -1$$

45. $20y^2 = 3 - 11y$

$20y^2 + 11y - 3 = 0$

$a = 20,\ b = 11,\ c = -3$

$$y = \frac{-11 \pm \sqrt{11^2 - 4(20)(-3)}}{2(20)}$$

$$y = \frac{-11 \pm \sqrt{121 + 240}}{40}$$

$$y = \frac{-11 \pm \sqrt{361}}{40}$$

$$y = \frac{-11 \pm 19}{40}$$

$$y = \frac{-11+19}{40} \quad \text{or} \quad y = \frac{-11-19}{40}$$

$$y = \frac{8}{40} = \frac{1}{5} \quad \text{or} \quad y = \frac{-30}{40} = -\frac{3}{4}$$

47. $x^2 + x + 1 = 0$

$a = 1,\ b = 1,\ c = 1$

$$x = \frac{-1 \pm \sqrt{1^2 - 4(1)(1)}}{2(1)}$$

$$x = \frac{-1 \pm \sqrt{1-4}}{2}$$

$$x = \frac{-1 \pm \sqrt{-3}}{2}$$

$\sqrt{-3}$ is not a real number. **No real solutions.**

49. $4y^2 = 6y + 1$

$4y^2 - 6y - 1 = 0$

$a = 4,\ b = -6,\ c = -1$

$$y = \frac{-(-6) \pm \sqrt{(-6)^2 - 4(4)(-1)}}{2(4)}$$

$$y = \frac{6 \pm \sqrt{36 + 16}}{8}$$

$$y = \frac{6 \pm \sqrt{52}}{8}$$

$$y = \frac{6 \pm \sqrt{4}\sqrt{13}}{8}$$

$$y = \frac{6 \pm 2\sqrt{13}}{8}$$

$$y = \frac{2(3 \pm \sqrt{13})}{8}$$

$$y = \frac{3 \pm \sqrt{13}}{4}$$

51. $5x^2 = \frac{7}{2}x + 1$

$$2(5x^2) = 2\left(\frac{7}{2}x + 1\right)$$

$10x^2 = 7x + 2$

$10x^2 - 7x - 2 = 0$

$a = 10,\ b = -7,\ c = -2$

$$x = \frac{-(-7) \pm \sqrt{(-7)^2 - 4(10)(-2)}}{2(10)}$$

$$x = \frac{7 \pm \sqrt{49 + 80}}{20}$$

$$x = \frac{7 \pm \sqrt{129}}{20}$$

53. $28x^2 + 5x + \frac{11}{4} = 0$

$$4\left(28x^2 + 5x + \frac{11}{4}\right) = 4(0)$$

$112x^2 + 20x + 11 = 0$

$a = 112,\ b = 20,\ c = 11$

$$x = \frac{-20 \pm \sqrt{20^2 - 4(112)(11)}}{2(112)}$$

$$x = \frac{-20 \pm \sqrt{400 - 4928}}{224}$$

$$x = \frac{-20 \pm \sqrt{-4528}}{224}$$

$\sqrt{-4528}$ is not a real number. **No real solutions.**

55. $5z^2 - 2z = \dfrac{1}{5}$

$5(5z^2 - 2z) = 5\left(\dfrac{1}{5}\right)$

$25z^2 - 10z = 1$

$25z^2 - 10z - 1 = 0$

$a = 25,\ b = -10,\ c = -1$

$z = \dfrac{-(-10) \pm \sqrt{(-10)^2 - 4(25)(-1)}}{2(25)}$

$z = \dfrac{10 \pm \sqrt{100 + 100}}{50}$

$z = \dfrac{10 \pm \sqrt{200}}{50}$

$z = \dfrac{10 \pm \sqrt{100}\sqrt{2}}{50}$

$z = \dfrac{10 \pm 10\sqrt{2}}{50}$

$z = \dfrac{10(1 \pm \sqrt{2})}{50}$

$z = \dfrac{1 \pm \sqrt{2}}{5}$

57. $x^2 + 3\sqrt{2}x - 5 = 0$

$a = 1,\ b = 3\sqrt{2},\ c = -5$

$x = \dfrac{-3\sqrt{2} \pm \sqrt{(3\sqrt{2})^2 - 4(1)(-5)}}{2(1)}$

$x = \dfrac{-3\sqrt{2} \pm \sqrt{18 + 20}}{2}$

$x = \dfrac{-3\sqrt{2} \pm \sqrt{38}}{2}$

59. $x^2 + x = 15$

$x^2 + x - 15 = 0$

$a = 1,\ b = 1,\ c = -15$

$x = \dfrac{-1 \pm \sqrt{1^2 - 4(1)(-15)}}{2(1)}$

$x = \dfrac{-1 \pm \sqrt{61}}{2}$

$x \approx -4.4,\ 3.4$

61. $1.2x^2 - 5.2x - 3.9 = 0$

$a = 1.2,\ b = -5.2,\ c = -3.9$

$x = \dfrac{-(-5.2) \pm \sqrt{(-5.2)^2 - 4(1.2)(-3.9)}}{2(1.2)}$

$x = \dfrac{5.2 \pm \sqrt{45.76}}{2.4}$

$x \approx -0.7,\ 5.0$

63. $h = -16t^2 + 120t + 80$

$30 = -16t^2 + 120t + 80$

$0 = -16t^2 + 120t + 50$

$0 = -8t^2 + 60t + 25$

$a = -8,\ b = 60,\ c = 25$

$t = \dfrac{-60 \pm \sqrt{(60)^2 - 4(-8)(25)}}{2(-8)}$

$t = \dfrac{-60 \pm \sqrt{4400}}{-16}$

$t \approx 7.9,\ -0.4$

Disregard the negative.

7.9 seconds

65. Answers may vary.

67. $\dfrac{5x}{3} = 1$

$\dfrac{3}{5}\left(\dfrac{5x}{3}\right) = \dfrac{3}{5}(1)$

$x = \dfrac{3}{5}$

69. $\dfrac{6}{11}x + \dfrac{1}{5} = 0$

$\dfrac{6}{11}x = -\dfrac{1}{5}$

$\dfrac{11}{6}\left(\dfrac{6}{11}x\right) = \dfrac{11}{6}\left(-\dfrac{1}{5}\right)$

$x = -\dfrac{11}{30}$

71. $\dfrac{5}{2}z + 10 = 0$

$\dfrac{5}{2}z = -10$

$\dfrac{2}{5}\left(\dfrac{5}{2}z\right) = \dfrac{2}{5}(-10)$

$z = -4$

Exercise Set 10.4

1. $5x^2 - 11x + 2 = 0$
$(5x - 1)(x - 2) = 0$
$5x - 1 = 0$ or $x - 2 = 0$
$5x = 1$ or $x = 2$
$x = \dfrac{1}{5}$

3. $x^2 - 1 = 2x$
$x^2 - 2x - 1 = 0$
$a = 1,\ b = -2,\ c = -1$
$x = \dfrac{-(-2) \pm \sqrt{(-2)^2 - 4(1)(-1)}}{2(1)}$
$x = \dfrac{2 \pm \sqrt{4 + 4}}{2}$
$x = \dfrac{2 \pm \sqrt{8}}{2}$
$x = \dfrac{2 \pm \sqrt{4}\sqrt{2}}{2}$
$x = \dfrac{2 \pm 2\sqrt{2}}{2}$
$x = \dfrac{2(1 \pm \sqrt{2})}{2}$
$x = 1 \pm \sqrt{2}$

5. $a^2 = 20$
$a = \pm\sqrt{20}$
$a = \pm\sqrt{4}\sqrt{5}$
$a = \pm 2\sqrt{5}$

7. $x^2 - x + 4 = 0$
$a = 1,\ b = -1,\ c = 4$
$x = \dfrac{-(-1) \pm \sqrt{(-1)^2 - 4(1)(4)}}{2(1)}$
$x = \dfrac{1 \pm \sqrt{1 - 16}}{2}$
$x = \dfrac{1 \pm \sqrt{-15}}{2}$
$\sqrt{-15}$ is not a real number. No real solutions.

9. $3x^2 - 12x + 12 = 0$
$3(x^2 - 4x + 4) = 0$
$3(x - 2)^2 = 0$
$(x - 2)^2 = 0$
$x - 2 = \pm\sqrt{0}$
$x - 2 = 0$
$x = 2$

11. $9 - 6p + p^2 = 0$
$(3 - p)^2 = 0$
$3 - p = \pm\sqrt{0}$
$3 - p = 0$
$3 = p$

13. $4y^2 - 16 = 0$
$4(y^2 - 4) = 0$
$4(y + 2)(y - 2) = 0$
$y + 2 = 0$ or $y - 2 = 0$
$y = -2$ or $y = 2$

15. $x^4 - 3x^3 + 2x^2 = 0$
$x^2(x^2 - 3x + 2) = 0$
$x^2(x - 2)(x - 1) = 0$
$x^2 = 0$ or $x - 2 = 0$ or $x - 1 = 0$
$x = 0$ or $x = 2$ or $x = 1$

17. $(2z + 5)^2 = 25$
$(2z)^2 + 2(2z)(5) + (5)^2 = 25$
$4z^2 + 20z + 25 = 25$
$4z^2 + 20z + 25 - 25 = 0$
$4z^2 + 20z = 0$
$4z(z + 5) = 0$
$4z = 0$ or $z + 5 = 0$
$z = \dfrac{0}{4} = 0$ or $z = -5$

19. $30x = 25x^2 + 2$
$0 = 25x^2 - 30x + 2$
$a = 25,\ b = -30,\ c = 2$
$x = \dfrac{-(-30) \pm \sqrt{(-30)^2 - 4(25)(2)}}{2(25)}$
$x = \dfrac{30 \pm \sqrt{900 - 200}}{50}$
$x = \dfrac{30 \pm \sqrt{700}}{50}$

$$x = \frac{30 \pm \sqrt{100}\sqrt{7}}{50}$$

$$x = \frac{30 \pm 10\sqrt{7}}{50}$$

$$x = \frac{10(3 \pm \sqrt{7})}{50}$$

$$x = \frac{3 \pm \sqrt{7}}{5}$$

21. $\frac{2}{3}m^2 - \frac{1}{3}m - 1 = 0$

$$3\left(\frac{2}{3}m^2 - \frac{1}{3}m - 1\right) = 3(0)$$

$$2m^2 - m - 3 = 0$$
$$(2m - 3)(m + 1) = 0$$
$$2m - 3 = 0 \qquad \text{or} \qquad m + 1 = 0$$
$$2m = 3 \qquad \text{or} \qquad m = -1$$
$$m = \frac{3}{2}$$

23. $x^2 - \frac{1}{2}x - \frac{1}{5} = 0$

$$10\left(x^2 - \frac{1}{2}x - \frac{1}{5}\right) = 10(0)$$

$$10x^2 - 5x - 2 = 0$$
$$a = 10, \, b = -5, \, c = -2$$

$$x = \frac{-(-5) \pm \sqrt{(-5)^2 - 4(10)(-2)}}{2(10)}$$

$$x = \frac{5 \pm \sqrt{25 + 80}}{20}$$

$$x = \frac{5 \pm \sqrt{105}}{20}$$

25. $4x^2 - 27x + 35 = 0$
$$(4x - 7)(x - 5) = 0$$
$$4x - 7 = 0 \qquad \text{or} \qquad x - 5 = 0$$
$$4x = 7 \qquad \text{or} \qquad x = 5$$
$$x = \frac{7}{4}$$

27. $(7 - 5x)^2 = 18$
$$(7)^2 - 2(7)(-5x) + (5x)^2 = 18$$
$$49 - 70x + 25x^2 = 18$$
$$25x^2 - 70x + 49 - 18 = 0$$
$$25x^2 - 70x + 31 = 0$$

$$a = 25, \, b = -70, \, c = 31$$

$$x = \frac{-(-70) \pm \sqrt{(-70)^2 - 4(25)(31)}}{2(25)}$$

$$x = \frac{70 \pm \sqrt{4900 - 3100}}{50}$$

$$x = \frac{70 \pm \sqrt{1800}}{50}$$

$$x = \frac{70 \pm \sqrt{900}\sqrt{2}}{50}$$

$$x = \frac{70 \pm 30\sqrt{2}}{50}$$

$$x = \frac{10(7 \pm 3\sqrt{2})}{50}$$

$$x = \frac{7 \pm 3\sqrt{2}}{5}$$

29. $3z^2 - 7z = 12$
$$3z^2 - 7z - 12 = 0$$
$$a = 3, \, b = -7, \, c = -12$$

$$z = \frac{-(-7) \pm \sqrt{(-7)^2 - 4(3)(-12)}}{2(3)}$$

$$z = \frac{7 \pm \sqrt{49 + 144}}{6}$$

$$z = \frac{7 \pm \sqrt{193}}{6}$$

31. $x = x^2 - 110$
$$0 = x^2 - x - 110$$
$$0 = (x - 11)(x + 10)$$
$$x - 11 = 0 \qquad \text{or} \qquad x + 10 = 0$$
$$x = 11 \qquad \text{or} \qquad x = -10$$

33. $\frac{3}{4}x^2 - \frac{5}{2}x - 2 = 0$

$$4\left(\frac{3}{4}x^2 - \frac{5}{2}x - 2\right) = 4(0)$$

$$3x^2 - 10x - 8 = 0$$
$$(3x + 2)(x - 4) = 0$$
$$3x + 2 = 0 \qquad \text{or} \qquad x - 4 = 0$$
$$3x = -2 \qquad \text{or} \qquad x = 4$$
$$x = -\frac{2}{3}$$

35. $x^2 - 0.6x + 0.05 = 0$
$(x - 0.1)(x - 0.5) = 0$
$x - 0.1 = 0$ or $x - 0.5 = 0$
$x = 0.1$ or $x = 0.5$

37. $10x^2 - 11x + 2 = 0$
$a = 10,\ b = -11,\ c = 2$
$$x = \frac{-(-11) \pm \sqrt{(-11)^2 - 4(10)(2)}}{2(10)}$$
$$x = \frac{11 \pm \sqrt{121 - 80}}{20}$$
$$x = \frac{11 \pm \sqrt{41}}{20}$$

39. $\dfrac{1}{2}z^2 - 2z + \dfrac{3}{4} = 0$
$$4\left(\frac{1}{2}z^2 - 2z + \frac{3}{4}\right) = 4(0)$$
$2z^2 - 8z + 3 = 0$
$a = 2,\ b = -8,\ c = 3$
$$z = \frac{-(-8) \pm \sqrt{(-8)^2 - 4(2)(3)}}{2(2)}$$
$$z = \frac{8 \pm \sqrt{64 - 24}}{4}$$
$$z = \frac{8 \pm \sqrt{40}}{4}$$
$$z = \frac{8 \pm \sqrt{4}\sqrt{10}}{4}$$
$$z = \frac{8 \pm 2\sqrt{10}}{4}$$
$$z = \frac{2(4 \pm \sqrt{10})}{4}$$
$$z = \frac{4 \pm \sqrt{10}}{2}$$

41. $\dfrac{AB}{AC} = \dfrac{AC}{CB}$
$\dfrac{x}{1} = \dfrac{1}{x-1}$
$x(x - 1) = 1(1)$
$x^2 - x = 1$
$x^2 - x - 1 = 0$
$a = 1,\ b = -1,\ c = -1$

$$x = \frac{-(-1) \pm \sqrt{(-1)^2 - 4(1)(-1)}}{2(1)}$$
$$x = \frac{1 \pm \sqrt{5}}{2}$$
$$AB = \frac{1 \pm \sqrt{5}}{2}$$
Since length cannot be negative, $AB = \dfrac{1 + \sqrt{5}}{2}$.

43. Answers may vary.

45. $\sqrt{104} = \sqrt{4}\sqrt{26} = 2\sqrt{26}$

47. $\sqrt{80} = \sqrt{16}\sqrt{5} = 4\sqrt{5}$

49. width $= x$ length $= x + 6$
$A = lw$
$391 = (x + 6)(x)$
$391 = x^2 + 6x$
$0 = x^2 + 6x - 391$
$0 = (x + 23)(x - 17)$
$x + 23 = 0$ or $x - 17 = 0$
$x = -23$ or $x = 17$
width $= 17$ in., length $= 23$ in.

Exercise Set 10.5

1. $\sqrt{-9} = \sqrt{-1}\sqrt{9} = i \cdot 3 = 3i$

3. $\sqrt{-100} = \sqrt{-1}\sqrt{100} = i \cdot 10 = 10i$

5. $\sqrt{-50} = \sqrt{-1}\sqrt{25}\sqrt{2} = i \cdot 5\sqrt{2} = 5i\sqrt{2}$

7. $\sqrt{-63} = \sqrt{-1}\sqrt{9}\sqrt{7} = i \cdot 3\sqrt{7} = 3i\sqrt{7}$

9. $(2 - i) + (-5 + 10i) = 2 - 5 + (-i + 10i) = -3 + 9i$

11. $(3 - 4i) - (2 - i) = 3 - 4i - 2 + i = 3 - 2 + (-4i + i)$
$= 1 - 3i$

13. $4i(3 - 2i) = 12i - 8i^2 = 12i - 8(-1) = 12i + 8$
$= 8 + 12i$

15. $(6 - 2i)(4 + i) = 6(4) + 6i - 2i(4) - 2i(i)$
$= 24 + 6i - 8i - 2i^2 = 24 - 2i - 2(-1) = 24 - 2i + 2$
$= 26 - 2i$

17. Answers may vary.

19. $\dfrac{8 - 12i}{4} = \dfrac{4(2 - 3i)}{4} = 2 - 3i$

21.
$$\frac{7-i}{4-3i} = \frac{(7-i)}{(4-3i)} \cdot \frac{(4+3i)}{(4+3i)}$$
$$= \frac{7(4)+7(3i)-i(4)-i(3i)}{(4)^2-(3i)^2}$$
$$= \frac{28+21i-4i-3i^2}{16-9i^2} = \frac{28+17i-3(-1)}{16-9(-1)}$$
$$= \frac{28+17i+3}{16+9} = \frac{31+17i}{25} = \frac{31}{25}+\frac{17i}{25}$$

23. $(x+1)^2 = -9$
$$x+1 = \pm\sqrt{-9}$$
$$x+1 = \pm\sqrt{-1}\sqrt{9}$$
$$x+1 = \pm 3i$$
$$x = -1 \pm 3i$$

25. $(2z-3)^2 = -12$
$$2z-3 = \pm\sqrt{-12}$$
$$2z-3 = \pm\sqrt{-1}\sqrt{4}\sqrt{3}$$
$$2z-3 = \pm 2i\sqrt{3}$$
$$2z = 3 \pm 2i\sqrt{3}$$
$$z = \frac{3 \pm 2i\sqrt{3}}{2}$$

27. $y^2 + 6y + 13 = 0$
$a = 1, b = 6, c = 13$
$$y = \frac{-6 \pm \sqrt{6^2 - 4(1)(13)}}{2(1)}$$
$$y = \frac{-6 \pm \sqrt{36 - 52}}{2}$$
$$y = \frac{-6 \pm \sqrt{-16}}{2}$$
$$y = \frac{-6 \pm 4i}{2}$$
$$y = \frac{2(-3 \pm 2i)}{2}$$
$$y = -3 \pm 2i$$

29. $4x^2 + 7x + 4 = 0$
$a = 4, b = 7, c = 4$
$$x = \frac{-7 \pm \sqrt{7^2 - 4(4)(4)}}{2(4)}$$

$$x = \frac{-7 \pm \sqrt{49 - 64}}{8}$$
$$x = \frac{-7 \pm \sqrt{-15}}{8}$$
$$x = \frac{-7 \pm i\sqrt{15}}{8}$$

31. $2m^2 - 4m + 5 = 0$
$a = 2, b = -4, c = 5$
$$m = \frac{-(-4) \pm \sqrt{(-4)^2 - 4(2)(5)}}{2(2)}$$
$$m = \frac{4 \pm \sqrt{16 - 40}}{4}$$
$$m = \frac{4 \pm \sqrt{-24}}{4}$$
$$m = \frac{4 \pm \sqrt{-4}\sqrt{6}}{4}$$
$$m = \frac{4 \pm 2i\sqrt{6}}{4}$$
$$m = \frac{2(2 \pm i\sqrt{6})}{4}$$
$$m = \frac{2 \pm i\sqrt{6}}{2}$$

33. $3 + (12 - 7i) = 3 + 12 - 7i = 15 - 7i$

35. $-9i(5i - 7) = -45i^2 + 63i$
$$= -45(-1) + 63i = 45 + 63i$$

37. $(2-i) - (3-4i) = 2 - i - 3 + 4i = 2 - 3 + (-i + 4i)$
$$= -1 + 3i$$

39.
$$\frac{15+10i}{5i} = \frac{(15+10i)}{5i} \cdot \frac{(-i)}{(-i)} = \frac{-15i - 10i^2}{-5i^2}$$
$$= \frac{-15i - 10(-1)}{-5(-1)} = \frac{-15i + 10}{5} = \frac{5(-3i + 2)}{5}$$
$$= -3i + 2 = 2 - 3i$$

41. $-5 + i - (2 + 3i) = -5 + i - 2 - 3i = -5 - 2 + (i - 3i)$
$$= -7 - 2i$$

43. $(4-3i)(4+3i) = (4)^2 - (3i)^2 = 16 - 9i^2$
$$= 16 - 9(-1) = 16 + 9 = 25$$

45. $\dfrac{4-i}{1+2i} = \dfrac{(4-i)(1-2i)}{(1+2i)(1-2i)}$

$= \dfrac{4(1)+4(-2i)-i(1)-i(-2i)}{(1)^2-(2i)^2}$

$= \dfrac{4-8i-i+2i^2}{1-4i^2} = \dfrac{4-9i+2(-1)}{1-4(-1)} = \dfrac{4-9i-2}{1+4}$

$= \dfrac{2-9i}{5} = \dfrac{2}{5}-\dfrac{9}{5}i$

47. $(5+2i)^2 = (5)^2 + 2(5)(2i) + (2i)^2$

$= 25+20i+4i^2 = 25+20i+4(-1)$

$= 25+20i-4 = 21+20i$

49. $(y-4)^2 = -64$

$y-4 = \pm\sqrt{-64}$

$y-4 = \pm 8i$

$y = 4 \pm 8i$

51. $4x^2 = -100$

$x^2 = -25$

$x = \pm\sqrt{-25}$

$x = \pm 5i$

53. $z^2 + 6z + 10 = 0$

$a = 1, b = 6, c = 10$

$z = \dfrac{-6 \pm \sqrt{6^2 - 4(1)(10)}}{2(1)}$

$z = \dfrac{-6 \pm \sqrt{36-40}}{2}$

$z = \dfrac{-6 \pm \sqrt{-4}}{2}$

$z = \dfrac{-6 \pm 2i}{2}$

$z = \dfrac{2(-3 \pm i)}{2}$

$z = -3 \pm i$

55. $2a^2 - 5a + 9 = 0$

$a = 2, b = -5, c = 9$

$a = \dfrac{-(-5) \pm \sqrt{(-5)^2 - 4(2)(9)}}{2(2)}$

$a = \dfrac{5 \pm \sqrt{25-72}}{4}$

$a = \dfrac{5 \pm \sqrt{-47}}{4}$

$a = \dfrac{5 \pm i\sqrt{47}}{4}$

57. $(2x+8)^2 = -20$

$2x+8 = \pm\sqrt{-20}$

$2x+8 = \pm\sqrt{-1}\sqrt{4}\sqrt{5}$

$2x+8 = \pm 2i\sqrt{5}$

$2x = -8 \pm 2i\sqrt{5}$

$x = \dfrac{-8 \pm 2i\sqrt{5}}{2}$

$x = \dfrac{2(-4 \pm i\sqrt{5})}{2}$

$x = -4 \pm i\sqrt{5}$

59. $3m^2 + 108 = 0$

$3m^2 = -108$

$m^2 = -\dfrac{108}{3}$

$m^2 = -36$

$m = \pm\sqrt{-36}$

$m = \pm 6i$

61. $x^2 + 14x + 50 = 0$

$a = 1, b = 14, c = 50$

$x = \dfrac{-14 \pm \sqrt{14^2 - 4(1)(50)}}{2(1)}$

$x = \dfrac{-14 \pm \sqrt{196-200}}{2}$

$x = \dfrac{-14 \pm \sqrt{-4}}{2}$

$x = \dfrac{-14 \pm 2i}{2}$

$x = \dfrac{2(-7 \pm i)}{2}$

$x = -7 \pm i$

63. True

65. True

67. $y = -3$

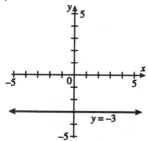

69. $y = 3x - 2$

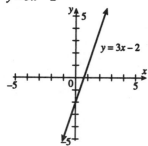

71. $x^2 + 7^2 = 10^2$

$x^2 + 49 = 100$

$\sqrt{x^2} = \pm\sqrt{51}$

$x = \sqrt{51}$ meters or 7.14 meters. Disregard the negative.

Exercise Set 10.6

1. $y = 2x^2$

$y = 2(x - 0)^2 + 0$

vertex (0, 0)

y-intercept   x-intercept

let $x = 0$ let $y = 0$

$y = 2(0)^2$ $0 = 2x^2$

$y = 0$ $0 = x^2$

(0, 0) $0 = x$

(0, 0)

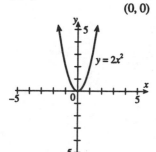

3. $y = (x - 1)^2$

$y = (x - 1)^2 + 0$

vertex (1, 0)

y-intercept   x-intercept

let $x = 0$ $y = 0$

$y = (0 - 1)^2$ $0 = (x - 1)^2$

$y = 1$ $0 = x - 1$

(0, 1) $1 = x$

(1, 0)

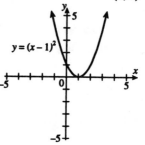

5. $y = -x^2 + 4$

$y = -(x - 0)^2 + 4$

vertex (0, 4)

y-intercept   x-intercept

let $x = 0$ $y = 0$

$y = -0^2 + 4$ $0 = -x^2 + 4$

$y = 4$ $x^2 = 4$

(0, 4) $x = \pm\sqrt{4} = \pm 2$

(2, 0) and (−2, 0)

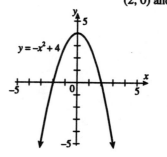

7. $y = \dfrac{1}{3}x^2$

$y = \dfrac{1}{3}(x-0)^2 + 0$

vertex $(0, 0)$

y-intercept

let $x = 0$

$y = \dfrac{1}{3}(0)^2$

$y = 0$

$(0, 0)$

x-intercept

let $y = 0$

$0 = \dfrac{1}{3}x^2$

$0 = x^2$

$0 = x$

$(0, 0)$

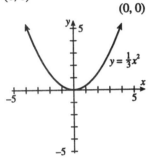

9. $y = (x-2)^2 + 1$

vertex $(2, 1)$

y-intercept

let $x = 0$

$y = (0-2)^2 + 1$

$y = 4 + 1$

$y = 5$

$(0, 5)$

x-intercept

let $y = 0$

$0 = (x-2)^2 + 1$

$-1 = (x-2)^2$

$\pm\sqrt{-1} = x - 2$

$\sqrt{-1}$ **is not a real number.**

No y-intercept

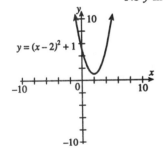

11. $y = -(x+1)^2 + 4$

vertex $(-1, 4)$

y-intercept

let $x = 0$

$y = -(0+1)^2 + 4$

$y = -1 + 4$

$y = 3$

$(0, 3)$

x-intercept

let $y = 0$

$0 = -(x+1)^2 + 4$

$(x+1)^2 = 4$

$x + 1 = \pm\sqrt{4}$

$x + 1 = \pm 2$

$x = -1 \pm 2$

$x = -3, \ x = 1$

$(-3, 0), (1, 0)$

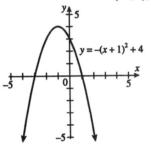

13. $y = -4x^2 + 1$

$y = -4(x-0)^2 + 1$

vertex $(0, 1)$

y-intercept

let $x = 0$

$y = -4(0)^2 + 1$

$y = 1$

$(0, 1)$

x-intercept

let $y = 0$

$0 = -4x^2 + 1$

$4x^2 = 1$

$x^2 = \dfrac{1}{4}$

$x = \pm\sqrt{\dfrac{1}{4}}$

$x = \pm\dfrac{1}{2}$

$\left(\dfrac{1}{2}, 0\right), \left(-\dfrac{1}{2}, 0\right)$

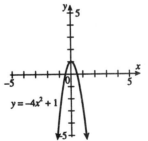

15. $y = -x^2$ F

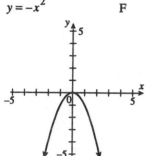

17. $y = (x-2)^2$ A

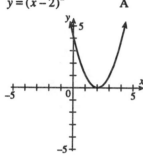

19. $y = (x+3)^2 - 1$ H

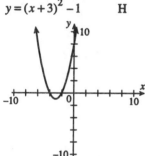

21. $y = 2(x+3)^2$ B

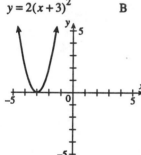

23. $y = -\dfrac{1}{2}x^2 + 1$ D

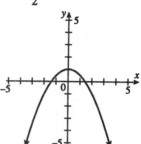

25. $y = x^2 - 4x$

$y = x^2 - 4x + 4 - 4$

$y = (x-2)^2 - 4$

vertex $(2, -4)$

y-intercept,	x-intercept
let $x = 0$	let $y = 0$
$y = 0^2 - 4(0)$	$0 = x^2 - 4x$
$y = 0$	$0 = x(x-4)$
$(0, 0)$	$x = 0$ or $x - 4 = 0$
	$x = 0$ or $x = 4$
	$(0, 0), (4, 0)$

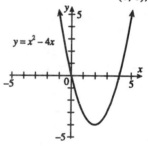

27. $y = x^2 - 2x - 3$

$y = x^2 - 2x + 1 - 1 - 3$

$y = (x-1)^2 - 4$

vertex $(1, -4)$

y-intercept	x-intercept
let $x = 0$	let $y = 0$
$y = 0^2 - 2(0) - 3$	$0 = x^2 - 2x - 3$
$y = -3$	$0 = (x-3)(x+1)$
$(0, -3)$	$x = 3$ or $x = -1$
	$(3, 0), (-1, 0)$

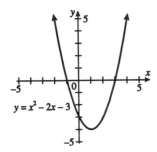

29. $y = x^2 + 2x + 1$

$y = (x+1)^2 + 0$

vertex $(-1, 0)$

y-intercept	x-intercept
let $x = 0$	let $y = 0$
$y = 0^2 + 2(0) + 1$	$0 = (x+1)^2$
$y = 1$	$0 = x + 1$
$(0, 1)$	$-1 = x$
	$(-1, 0)$

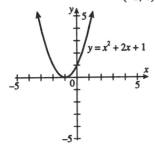

31. $y = x^2 + 7x + 10$

$y = x^2 + 7x + \dfrac{49}{4} - \dfrac{49}{4} + 10$

$y = \left(x + \dfrac{7}{2}\right)^2 - \dfrac{9}{4}$

vertex $\left(-\dfrac{7}{2}, -\dfrac{9}{4}\right)$

y-intercept	x-intercept
let $x = 0$	let $y = 0$
$y = 0^2 + 7(0) + 10$	$0 = x^2 + 7x + 10$
$y = 10$	$0 = (x+5)(x+2)$
$(0, 10)$	$x = -5$ or $x = -2$
	$(-5, 0)$, $(-2, 0)$

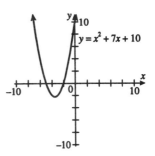

33. $y = x^2 - 6x + 8$

$y = x^2 - 6x + 9 - 9 + 8$

$y = (x-3)^2 - 1$

vertex $(3, -1)$

y-intercept	x-intercept
Let $x = 0$	let $y = 0$
$y = 0^2 - 6(0) + 8$	$0 = x^2 - 6x + 8$
$y = 8$	$0 = (x-4)(x-2)$
$(0, 8)$	$x - 4 = 0$ or $x - 2 = 0$
	$x = 4$, $x = 2$
	$(4, 0)$, $(2, 0)$

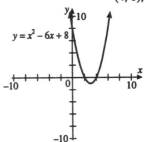

35. Domain: all real numbers
Range: $y \le -4$

37. Domain: all real numbers
Range: $y \ge -2$

39. C

41. A

43. $\dfrac{\frac{1}{7}}{\frac{2}{5}} = \dfrac{1}{7} \cdot \dfrac{5}{2} = \dfrac{5}{14}$

45. $\dfrac{\frac{1}{x}}{\frac{2}{x^2}} = \dfrac{1}{x} \cdot \dfrac{x^2}{2} = \dfrac{x}{2}$

47. $\dfrac{2x}{1 - \frac{1}{x}} = \dfrac{2x}{\frac{x}{x} - \frac{1}{x}} = \dfrac{2x}{\frac{x-1}{x}} = 2x \cdot \dfrac{x}{x-1} = \dfrac{2x^2}{x-1}$

49. $\dfrac{\frac{a-b}{2b}}{\frac{b-a}{8b^2}} = \dfrac{a-b}{2b} \cdot \dfrac{8b^2}{b-a} = \dfrac{a-b}{2b} \cdot \dfrac{8b^2}{-1(a-b)}$

$\qquad = \dfrac{4b}{-1} = -4b$

Chapter 10 - Review

1. $(x-4)(5x+3) = 0$

$\quad x - 4 = 0 \qquad$ or $\qquad 5x + 3 = 0$

$\quad x = 4 \qquad\qquad$ or $\qquad 5x = -3$

$\qquad\qquad\qquad\qquad\qquad\qquad x = -\dfrac{3}{5}$

2. $(x+7)(3x+4) = 0$

$\quad x + 7 = 0 \qquad$ or $\qquad 3x + 4 = 0$

$\quad x = -7 \qquad\qquad$ or $\qquad \dfrac{3x}{3} = \dfrac{-4}{3}$

$\qquad\qquad\qquad\qquad\qquad\qquad x = -\dfrac{4}{3}$

3. $3m^2 - 5m = 2$

$\quad 3m^2 - 5m - 2 = 0$

$\quad (3m + 1)(m - 2) = 0$

$\quad 3m + 1 = 0 \qquad$ or $\qquad m - 2 = 0$

$\quad 3m = -1 \qquad\quad$ or $\qquad m = 2$

$\quad m = -\dfrac{1}{3}$

4. $7m^2 + 2m = 5$

$\quad 7m^2 + 2m - 5 = 0$

$\quad (7m - 5)(m + 1) = 0$

$\quad 7m - 5 = 0 \qquad$ or $\qquad m + 1 = 0$

$\quad \dfrac{7m}{7} = \dfrac{5}{7} \qquad\quad$ or $\qquad m = -1$

$\quad m = \dfrac{5}{7}$

5. $k^2 = 50$

$\quad k = \pm\sqrt{50}$

$\quad k = \pm\sqrt{25}\sqrt{2}$

$\quad k = \pm 5\sqrt{2}$

6. $k^2 = 45$

$\quad \sqrt{k^2} = \sqrt{45}$

$\quad k = \pm\sqrt{9 \cdot 5}$

$\quad k = \pm 3\sqrt{5}$

7. $(x - 5)(x - 1) = 12$

$\quad x(x) + x(-1) - 5(x) - 5(-1) = 12$

$\qquad\qquad\qquad\qquad\qquad\qquad x^2 - x - 5x + 5 = 12$

$\qquad\qquad\qquad\qquad\qquad\qquad x^2 - 6x + 5 - 12 = 0$

$\qquad\qquad\qquad\qquad\qquad\qquad x^2 - 6x - 7 = 0$

$\qquad\qquad\qquad\qquad\qquad\qquad (x - 7)(x + 1) = 0$

$\qquad\qquad x - 7 = 0 \qquad$ or $\qquad x + 1 = 0$

$\qquad\qquad x = 7 \qquad\qquad$ or $\qquad x = -1$

8. $(x - 3)(x + 2) = 6$

$\qquad\qquad x^2 - x - 6 = 6$

$\qquad\qquad x^2 - x - 12 = 0$

$\qquad\qquad (x - 4)(x + 3) = 0$

$\qquad\qquad x - 4 = 0 \qquad$ or $\qquad x + 3 = 0$

$\qquad\qquad x = 4 \qquad\qquad$ or $\qquad x = -3$

9. $(x - 11)^2 = 49$

$\qquad\qquad x - 11 = \pm\sqrt{49}$

$\qquad\qquad x - 11 = \pm 7$

$\qquad\qquad x = 11 \pm 7$

$\qquad\qquad x = 11 + 7 \qquad$ or $\qquad x = 11 - 7$

$\qquad\qquad x = 18 \qquad\qquad$ or $\qquad x = 4$

10. $(x + 3)^2 = 100$

$\qquad\qquad \sqrt{(x+3)^2} = \sqrt{100}$

$\qquad\qquad x + 3 = \pm 10$

$\qquad\qquad x = -3 \pm 10$

$\qquad\qquad x = -13, 7$

11. $6x^3 - 54x = 0$

$\qquad\qquad 6x(x^2 - 9) = 0$

$\qquad\qquad 6x(x + 3)(x - 3) = 0$

$\qquad\qquad 6x = 0 \quad$ or $\quad x + 3 = 0 \quad$ or $\quad x - 3 = 0$

$\qquad\qquad x = 0 \quad$ or $\quad x = -3 \quad$ or $\quad x = 3$

12. $2x^2 - 8 = 0$

$\qquad\qquad \dfrac{2x^2}{2} = \dfrac{8}{2}$

$\qquad\qquad \sqrt{x^2} = \sqrt{4}$

$\qquad\qquad x = \pm 2$

13. $(4p + 2)^2 = 100$

$\qquad\qquad 4p + 2 = \pm\sqrt{100}$

$\qquad\qquad 4p + 2 = \pm 10$

$\qquad\qquad 4p = -2 \pm 10$

$\qquad\qquad p = \dfrac{-2 \pm 10}{4}$

$\qquad\qquad p = \dfrac{-2 + 10}{4} \qquad$ or $\qquad p = \dfrac{-2 - 10}{4}$

$\qquad\qquad p = \dfrac{8}{4} = 2 \qquad\quad$ or $\qquad p = \dfrac{-12}{4} = -3$

14. $(3p+6)^2 = 81$

$\sqrt{(3p+6)^2} = \sqrt{81}$

$3p+6 = \pm 9$

$\dfrac{3p}{3} = \dfrac{-6 \pm 9}{3}$

$p = -5, 1$

15. $x^2 - 10x$

$x^2 - 10x + 25 = (x-5)^2$

16. $x^2 + 16x + 64 = (x+8)^2$

17. $a^2 + 4a$

$a^2 + 4a + 4 = (a+2)^2$

18. $a^2 - 12a$

$\left(\dfrac{-12}{2}\right)^2 = 36$

$a^2 - 12a + 36 = (a-6)^2$

19. $m^2 - 3m$

$m^2 - 3m + \dfrac{9}{4} = \left(m - \dfrac{3}{2}\right)^2$

20. $m^2 + 5m$

$\left(\dfrac{5}{2}\right)^2 = \dfrac{25}{4}$

$m^2 + 5m + \dfrac{25}{4} = \left(m + \dfrac{5}{2}\right)^2$

21. $x^2 - 6x + 7 = 0$

$x^2 - 6x = -7$

$x^2 - 6x + 9 = -7 + 9$

$(x-3)^2 = 2$

$x - 3 = \pm\sqrt{2}$

$x = 3 \pm \sqrt{2}$

22. $x^2 + 6x + 7 = 0$

$x^2 + 6x = -7$

$x^2 + 6x + 9 = 9 - 7$

$(x+3)^2 = 2$

$\sqrt{(x+3)^2} = \sqrt{2}$

$x + 3 = \pm\sqrt{2}$

$x = -3 \pm \sqrt{2}$

23. $2y^2 + y - 1 = 0$

$2y^2 + y = 1$

$\dfrac{2y^2}{2} + \dfrac{1y}{2} = \dfrac{1}{2}$

$y^2 + \dfrac{1}{2}y + \dfrac{1}{16} = \dfrac{1}{2} + \dfrac{1}{16}$

$\left(y^2 + \dfrac{1}{4}\right)^2 = \dfrac{8}{16} + \dfrac{1}{16}$

$\left(y^2 + \dfrac{1}{4}\right)^2 = \dfrac{9}{16}$

$y + \dfrac{1}{4} = \pm\sqrt{\dfrac{9}{16}}$

$y + \dfrac{1}{4} = \pm\dfrac{3}{4}$

$y = -\dfrac{1}{4} \pm \dfrac{3}{4}$

$y = -\dfrac{1}{4} + \dfrac{3}{4}$ or $y = -\dfrac{1}{4} - \dfrac{3}{4}$

$y = \dfrac{2}{4} = \dfrac{1}{2}$ or $y = -\dfrac{4}{4} = -1$

24. $y^2 + 3y - 1 = 0$

$y^2 + 3y = 1$

$y^2 + 3y + \dfrac{9}{4} = \dfrac{9}{4} + 1$

$\left(y + \dfrac{3}{2}\right)^2 = \dfrac{13}{4}$

$\sqrt{\left(y + \dfrac{3}{2}\right)^2} = \sqrt{\dfrac{13}{4}}$

$y + \dfrac{3}{2} = \pm\dfrac{\sqrt{13}}{2}$

$y = -\dfrac{3}{2} \pm \dfrac{\sqrt{13}}{2}$

$y = \dfrac{-3 \pm \sqrt{13}}{2}$

25. $x^2 - 10x + 7 = 0$
$a = 1, b = -10, c = 7$

$$x = \frac{-(-10) \pm \sqrt{(-10)^2 - 4(1)(7)}}{2(1)}$$

$$x = \frac{10 \pm \sqrt{100 - 28}}{2}$$

$$x = \frac{10 \pm \sqrt{72}}{2}$$

$$x = \frac{10 \pm \sqrt{36}\sqrt{2}}{2}$$

$$x = \frac{10 \pm 6\sqrt{2}}{2}$$

$$x = \frac{2(5 \pm 3\sqrt{2})}{2}$$

$$x = 5 \pm 3\sqrt{2}$$

26. $x^2 + 4x - 7 = 0$
$a = 1, b = 4, c = -7$

$$x = \frac{-4 \pm \sqrt{4^2 - 4(1)(-7)}}{2(1)}$$

$$x = \frac{-4 \pm \sqrt{16 + 28}}{2}$$

$$x = \frac{-4 \pm \sqrt{44}}{2}$$

$$x = \frac{-4 \pm \sqrt{4 \cdot 11}}{2}$$

$$x = \frac{-4 \pm 2\sqrt{11}}{2}$$

$$x = -2 \pm \sqrt{11}$$

27. $2x^2 + x - 1 = 0$
$a = 2, b = 1, c = -1$

$$x = \frac{-1 \pm \sqrt{1^2 - 4(2)(-1)}}{2(2)}$$

$$x = \frac{-1 \pm \sqrt{1 + 8}}{4}$$

$$x = \frac{-1 \pm \sqrt{9}}{4}$$

$$x = \frac{-1 \pm 3}{4}$$

$$x = \frac{-1 + 3}{4} \quad \text{or} \quad x = \frac{-1 - 3}{4}$$

$$x = \frac{2}{4} = \frac{1}{2} \quad \text{or} \quad x = \frac{-4}{4} = -1$$

28. $x^2 + 3x - 1 = 0$
$a = 1, b = 3, c = -1$

$$x = \frac{-3 \pm \sqrt{3^2 - 4(1)(-1)}}{2(1)}$$

$$x = \frac{-3 \pm \sqrt{9 + 4}}{2}$$

$$x = \frac{-3 \pm \sqrt{13}}{2}$$

29. $9x^2 + 30x + 25 = 0$
$a = 9, b = 30, c = 25$

$$x = \frac{-30 \pm \sqrt{30^2 - 4(9)(25)}}{2(9)}$$

$$x = \frac{-30 \pm \sqrt{900 - 900}}{18}$$

$$x = \frac{-30 \pm \sqrt{0}}{18}$$

$$x = \frac{-30}{18} = -\frac{5}{3}$$

30. $16x^2 - 72x + 81 = 0$
$a = 16, b = -72, c = 81$

$$x = \frac{72 \pm \sqrt{(-72)^2 - 4(16)(81)}}{2(16)}$$

$$x = \frac{72 \pm \sqrt{5184 - 5184}}{32}$$

$$x = \frac{72 \pm \sqrt{0}}{32}$$

$$x = \frac{72 \pm 0}{32}$$

$$x = \frac{9}{4}$$

31. $15x^2 + 2 = 11x$

$15x^2 - 11x + 2 = 0$

$a = 15, \, b = -11, \, c = 2$

$x = \dfrac{-(-11) \pm \sqrt{(-11)^2 - 4(15)(2)}}{2(15)}$

$x = \dfrac{11 \pm \sqrt{121 - 120}}{30}$

$x = \dfrac{11 \pm \sqrt{1}}{30}$

$x = \dfrac{11 \pm 1}{30}$

$x = \dfrac{11 + 1}{30}$ or $x = \dfrac{11 - 1}{30}$

$x = \dfrac{12}{30} = \dfrac{2}{5}$ or $x = \dfrac{10}{30} = \dfrac{1}{3}$

32. $15x^2 + 2 = 13x$

$15x^2 - 13x + 2 = 0$

$a = 15, \, b = -13, \, c = 2$

$x = \dfrac{13 \pm \sqrt{(-13)^2 - 4(15)(2)}}{2(15)}$

$x = \dfrac{13 \pm \sqrt{169 - 120}}{30}$

$x = \dfrac{13 \pm \sqrt{49}}{30}$

$x = \dfrac{13 \pm 7}{30}$

$x = \dfrac{20}{30}, \, \dfrac{6}{30}$

$x = \dfrac{2}{3}, \, \dfrac{1}{5}$

33. $2x^2 + x + 5 = 0$

$a = 2, \, b = 1, \, c = 5$

$x = \dfrac{-1 \pm \sqrt{1^2 - 4(2)(5)}}{2(2)}$

$x = \dfrac{-1 \pm \sqrt{1 - 40}}{4}$

$x = \dfrac{-1 \pm \sqrt{-39}}{4}$

$x = \dfrac{-1 \pm i\sqrt{39}}{4}$

34. $7x^2 - 3x + 1 = 0$

$a = 7, \, b = -3, \, c = 1$

$x = \dfrac{3 \pm \sqrt{(-3)^2 - 4(7)(1)}}{2(7)}$

$x = \dfrac{3 \pm \sqrt{9 - 28}}{14}$

$x = \dfrac{3 \pm \sqrt{-19}}{14}$

No real solutions.

35. $5z^2 + z - 1 = 0$

$a = 5, \, b = 1, \, c = -1$

$z = \dfrac{-1 \pm \sqrt{1^2 - 4(5)(-1)}}{2(5)}$

$z = \dfrac{-1 \pm \sqrt{1 + 20}}{10}$

$z = \dfrac{-1 \pm \sqrt{21}}{10}$

36. $4z^2 + 7z - 1 = 0$

$a = 4, \, b = 7, \, c = -1$

$z = \dfrac{-7 \pm \sqrt{7^2 - 4(4)(-1)}}{2(4)}$

$z = \dfrac{-7 \pm \sqrt{49 + 16}}{8}$

$z = \dfrac{-7 \pm \sqrt{65}}{8}$

37. $4x^4 = x^2$

$4x^4 - x^2 = 0$

$x^2(4x^2 - 1) = 0$

$x^2(2x + 1)(2x - 1) = 0$

$x^2 = 0$ or $2x + 1 = 0$ or $2x - 1 = 0$

$x = 0$ or $2x = -1$ or $2x = 1$

$x = -\dfrac{1}{2}$ or $x = \dfrac{1}{2}$

38.
$$9x^3 = x$$
$$9x^3 - x = 0$$
$$x(9x^2 - 1) = 0$$
$$x(3x - 1)(3x + 1) = 0$$

$x = 0$ or $3x - 1 = 0$
$$3x = 1$$
$$x = \frac{1}{3}$$

or $3x + 1 = 0$
$$3x = -1$$
$$x = -\frac{1}{3}$$

39.
$$2x^2 - 15x + 7 = 0$$
$$(2x - 1)(x - 7) = 0$$
$2x - 1 = 0$ or $x - 7 = 0$
$2x = 1$ or $x = 7$
$$x = \frac{1}{2}$$

40.
$$x^2 - 6x - 7 = 0$$
$$(x - 7)(x + 1) = 0$$
$x - 7 = 0$ or $x + 1 = 0$
$x = 7$ or $x = -1$

41.
$$(3x - 1)^2 = 0$$
$$3x - 1 = \pm\sqrt{0}$$
$$3x - 1 = 0$$
$$3x = 1$$
$$x = \frac{1}{3}$$

42.
$$(2x - 3)^2 = 0$$
$$\sqrt{(2x - 3)^2} = \sqrt{0}$$
$$2x - 3 = 0$$
$$2x = 3$$
$$x = \frac{3}{2}$$

43.
$$x^2 = 6x - 9$$
$$x^2 - 6x + 9 = 0$$
$$(x - 3)(x - 3) = 0$$
$$(x - 3)^2 = 0$$
$$x - 3 = \pm\sqrt{0}$$
$$x - 3 = 0$$
$$x = 3$$

44.
$$x^2 = 10x - 25$$
$$x^2 - 10x + 25 = 0$$
$$(x - 5)(x - 5) = 0$$

$x - 5 = 0$ or $x - 5 = 0$
$x = 5$ or $x = 5$

45.
$$\left(\frac{1}{2}x - 3\right)^2 = 64$$
$$\frac{1}{2}x - 3 = \pm 8$$
$$\frac{1}{2}x = 3 \pm 8$$
$$x = 2(3 \pm 8)$$
$x = 2(3 + 8)$ or $x = 2(3 - 8)$
$x = 2(11)$ or $x = 2(-5)$
$x = 22$ or $x = -10$

46.
$$\left(\frac{1}{3}x + 1\right)^2 = 49$$
$$\sqrt{\left(\frac{1}{3}x + 1\right)^2} = \sqrt{49}$$
$$\frac{1}{3}x + 1 = \pm 7$$
$$\frac{1}{3}x = -1 \pm 7$$
$\frac{1}{3}x = -8$ or $\frac{1}{3}x = 6$
$x = -24$ or $x = 18$

47.
$$x^2 - 0.3x + 0.01 = 0$$
$$100(x^2 - 0.3x + 0.01) = 100(0)$$
$$100x^2 - 30x + 1 = 0$$
$$a = 100,\ b = -30,\ c = 1$$
$$x = \frac{-(-30) \pm \sqrt{(-30)^2 - 4(100)(1)}}{2(100)}$$
$$x = \frac{30 \pm \sqrt{900 - 400}}{200}$$
$$x = \frac{30 \pm \sqrt{500}}{200}$$
$$x = \frac{30 \pm \sqrt{100}\sqrt{5}}{200}$$
$$x = \frac{30 \pm 10\sqrt{5}}{200}$$
$$x = \frac{10(3 \pm \sqrt{5})}{200}$$
$$x = \frac{3 \pm \sqrt{5}}{20}$$

48. $x^2 + 0.6x - 0.16 = 0$

$100x^2 + 60x - 16 = 0$

$a = 100, b = 60, c = -16$

$x = \dfrac{-60 \pm \sqrt{60^2 - 4(100)(-16)}}{2(100)}$

$x = \dfrac{-60 \pm \sqrt{3600 + 6400}}{200}$

$x = \dfrac{-60 \pm \sqrt{10000}}{200}$

$x = \dfrac{-60 \pm 100}{200} = -\dfrac{160}{200}, \dfrac{40}{200}$

$x = -\dfrac{4}{5}, \dfrac{1}{5}$

49. $\dfrac{1}{10}x^2 + x - \dfrac{1}{2} = 0$

$10\left(\dfrac{1}{10}x^2 + x - \dfrac{1}{2}\right) = 0$

$x^2 + 10x - 5 = 0$

$a = 1, b = 10, c = -5$

$x = \dfrac{-10 \pm \sqrt{10^2 - 4(1)(-5)}}{2(1)}$

$x = \dfrac{-10 \pm \sqrt{100 + 20}}{2}$

$x = \dfrac{-10 \pm \sqrt{120}}{2}$

$x = \dfrac{-10 \pm \sqrt{4}\sqrt{30}}{2}$

$x = \dfrac{-10 \pm 2\sqrt{30}}{2}$

$x = \dfrac{2(-5 \pm \sqrt{30})}{2}$

$x = -5 \pm \sqrt{30}$

50. $\dfrac{1}{12}x^2 - \dfrac{1}{2}x + \dfrac{1}{3} = 0$

$x^2 - 6x + 4 = 0$

$a = 1, b = -6, c = 4$

$x = \dfrac{6 \pm \sqrt{(-6)^2 - 4(1)(4)}}{2(1)}$

$x = \dfrac{6 \pm \sqrt{36 - 16}}{2}$

$x = \dfrac{6 \pm \sqrt{20}}{2}$

$x = \dfrac{6 \pm \sqrt{4 \cdot 5}}{2}$

$x = \dfrac{6 \pm 2\sqrt{5}}{2}$

$x = 3 \pm \sqrt{5}$

51. $\sqrt{-144} = \sqrt{-1}\sqrt{144} = i \cdot 12 = 12i$

52. $\sqrt{-36} = \sqrt{36 \cdot -1} = 6i$

53. $\sqrt{-108} = \sqrt{-1}\sqrt{36}\sqrt{3} = i \cdot 6\sqrt{3} = 6i\sqrt{3}$

54. $\sqrt{-500} = \sqrt{120 \cdot 5 \cdot -1} = 10i\sqrt{5}$

55. $(7 - i) + (14 - 9i) = 7 - i + 14 - 9i$
$= 7 + 14 + (-i - 9i) = 21 - 10i$

56. $(10 - 4i) + (9 - 21i) = 10 - 4i + 9 - 21i = 19 - 25i$

57. $3 - (11 + 2i) = 3 - 11 - 2i = -8 - 2i$

58. $(-4 - 3i) + 5i = -4 - 3i + 5i = -4 + 2i$

59. $(2 - 3i)(3 - 2i) = 2(3) + 2(-2i) - 3i(3) - 3i(-2i)$
$= 6 - 4i - 9i + 6i^2 = 6 - 13i + 6(-1)$
$= 6 - 13i - 6 = -13i$

60. $(2 + 5i)(5 - i) = 10 - 2i + 25i - 5i^2$
$= 10 + 5 - 2i + 25i$
$= 15 + 23i$

61. $(3 - 4i)(3 + 4i) = (3)^2 - (4i)^2 = 9 - 16i^2$
$= 9 - 16(-1) = 9 + 16 = 25$

62. $(7 - 2i)(7 - 2i) = 49 - 14i - 14i + 4i^2$
$= 49 - 4 - 28i = 45 - 28i$

63. $\dfrac{2 - 6i}{4i} = \dfrac{(2 - 6i)(-i)}{4i(-i)} = \dfrac{-2i + 6i^2}{-4i^2} = \dfrac{-2i + 6(-1)}{-4(-1)}$

$= \dfrac{-2i - 6}{4} = \dfrac{2(-i - 3)}{4} = -\dfrac{i}{2} - \dfrac{3}{2} = -\dfrac{3}{2} - \dfrac{i}{2}$

64. $\dfrac{5 - i}{2i} \cdot \dfrac{2i}{2i} = \dfrac{2i(5 - i)}{4i^2} = \dfrac{10i - 2i^2}{-4} = \dfrac{2 + 10i}{-4}$

$= \dfrac{1 + 5i}{-2} = -\dfrac{1}{2} - \dfrac{5i}{2}$

65. $\dfrac{4-i}{1+2i} = \dfrac{(4-i)(1-2i)}{(1+2i)(1-2i)}$

$= \dfrac{4(1)+4(-2i)-i(1)-i(-2i)}{(1)^2-(2i)^2} = \dfrac{4-8i-i+2i^2}{1-4i^2}$

$= \dfrac{4-9i+2(-1)}{1-4(-1)} = \dfrac{4-9i-2}{1+4} = \dfrac{2-9i}{5}$

$= \dfrac{2}{5}-\dfrac{9}{5}i$

66. $\dfrac{1+3i}{2-7i} \cdot \dfrac{2+7i}{2+7i} = \dfrac{2+7i+6i+21i^2}{4-49i^2}$

$= \dfrac{2-21+13i}{4+49} = \dfrac{-19+13i}{53} = -\dfrac{19}{53}+\dfrac{13i}{53}$

67. $3x^2 = -48$

$x^2 = -16$

$x = \pm\sqrt{-16}$

$x = \pm 4i$

68. $\dfrac{5x^2}{5} = \dfrac{-125}{5}$

$\sqrt{x^2} = \sqrt{-25}$

$x = \pm\sqrt{25 \cdot -1}$

$x = \pm 5i$

69. $x^2 - 4x + 13 = 0$

$x = \dfrac{-(-4)\pm\sqrt{(-4)^2-4(1)(13)}}{2(1)}$

$x = \dfrac{4\pm\sqrt{16-52}}{2}$

$x = \dfrac{4\pm\sqrt{-36}}{2}$

$x = \dfrac{4\pm 6i}{2}$

$x = \dfrac{2(2\pm 3i)}{2}$

$x = 2\pm 3i$

70. $x^2 + 4x + 11 = 0$

$a = 1,\ b = 4,\ c = 11$

$x = \dfrac{-4\pm\sqrt{4^2-4(1)(11)}}{2(1)}$

$x = \dfrac{-4\pm\sqrt{16-44}}{2}$

$x = \dfrac{-4\pm\sqrt{-28}}{2}$

$x = \dfrac{-4\pm\sqrt{4\cdot7\cdot-1}}{2}$

$x = \dfrac{-4\pm 2i\sqrt{7}}{2}$

$x = -2\pm i\sqrt{7}$

71. $y = -3x^2$

$y = -3(x-0)^2 + 0$

vertex $(0, 0)$

axis of symmetry $x = 0$

Parabola opens downward.

72. $y = -\dfrac{1}{2}x^2$

vertex $(0, 0)$

axis of symmetry $x = 0$

Parabola opens downward.

73. $y = (x-3)^2$

$y = (x-3)^2 + 0$

vertex $(3, 0)$

axis of symmetry $x = 3$

Parabola opens upward.

74. $y = (x-5)^2$

vertex $(5, 0)$

axis of symmetry $x = 5$

Parabola opens upward.

75. $y = 3x^2 - 7$

$y = 3(x-0)^2 - 7$

vertex $(0, -7)$

axis of symmetry $x = 0$

Parabola opens upward.

76. $y = -2x^2 + 25$

vertex $(0, 25)$

axis of symmetry $x = 0$

Parabola opens downward.

77. $y = -5(x - 72)^2 + 14$

vertex (72, 14)

axis of symmetry $x = 72$

Parabola opens downward.

78. $y = 2(x - 35)^2 - 21$

vertex (35, –21)

axis of symmetry $x = 35$

parabola opens upward

79. $y = -(x + 1)^2$

$y = -(x + 1)^2 + 0$

vertex (–1, 0)

y-intercept,	x-intercept
Let $x = 0$	$y = 0$
$y = -(0 + 1)^2$	$0 = -(x + 1)^2$
$y = -1$	$0 = (x + 1)^2$
(0, –1)	$\pm\sqrt{0} = x + 1$
	$0 = x + 1$
	$-1 = x$
	(–1, 0)

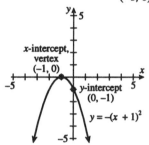

80. $y = -(x - 2)^2$

vertex (2, 0)

x-intercept (2, 0)

y-intercept (0, –4)

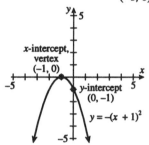

81. $y = (x - 2)^2 + 3$

vertex (2, 3)

y-intercept	x-intercept
Let $x = 0$,	$y = 0$
$y = (0 - 2)^2 + 3$	$0 = (x - 2)^2 + 3$
$y = 4 + 3$	$-3 = (x - 2)^2$
$y = 7$	$\sqrt{-3} = x - 2$
(0, 7)	not a real number
	no x-intercept

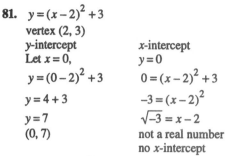

82. $y = (x + 3)^2 - 1$

vertex = (–3, –1)

y-intercept (0, 8)

x-intercepts (–2, 0), (–4, 0)

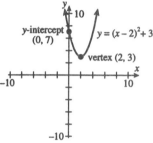

83. $y = x^2 + 5x + 6$

$$\frac{-b}{2a} = \frac{-5}{2}$$

$$y = \left(-\frac{5}{2}\right)^2 + 5\left(-\frac{5}{2}\right) + 6 = -\frac{1}{4}$$

vertex $\left(-\dfrac{5}{2}, -\dfrac{1}{4}\right)$

y-intercept	x-intercepts
Let $x = 0$	Let $y = 0$
$y = 0^2 + 5(0) + 6$	$0 = x^2 + 5x + 6$
$y = 6$	$0 = (x + 3)(x + 2)$
(0, 6)	$x = -3, x = -2$
	(–3, 0), (–2, 0)

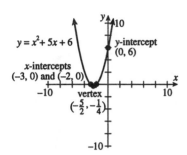

$y = x^2 + 5x + 6$
y-intercept (0, 6)
x-intercepts (−3, 0) and (−2, 0)
vertex $\left(-\frac{5}{2}, -\frac{1}{4}\right)$

84. $y = x^2 - 4x - 8$

$\dfrac{-b}{2a} = \dfrac{-(-4)}{2 \cdot 1} = \dfrac{4}{2} = 2$

$y = 2^2 - 4(2) - 8 = -12$

vertex (2, −12)
y-intercept
Let $x = 0$
$y = 0^2 - 4(0) - 8$
$y = -8$
(0, −8)

x-intercepts
Let $y = 0$
$0 = x^2 - 4x - 8$

$x = \dfrac{-(-4) \pm \sqrt{(-4)^2 - 4(1)(-8)}}{2(1)}$

$x = \dfrac{4 \pm \sqrt{48}}{2}$

$x = \dfrac{4 \pm 4\sqrt{3}}{2}$

$x = 2 \pm 2\sqrt{3}$

$(2 - 2\sqrt{3},\ 0),\ (2 + 2\sqrt{3},\ 0)$

x-intercept $(2 - 2\sqrt{3}, 0)$
x-intercept $(2 + 2\sqrt{3}, 0)$
$y = x^2 - 4x - 8$
(0, −8) y-intercept
(2, −12) vertex

85. $y = 2x^2 - 11x - 6$

$\dfrac{-b}{2a} = \dfrac{-(-11)}{2 \cdot 2} = \dfrac{11}{4}$

$y = 2\left(\dfrac{11}{4}\right)^2 - 11\left(\dfrac{11}{4}\right) - 6$

$y = -\dfrac{169}{8}$

vertex $\left(\dfrac{11}{4},\ -\dfrac{169}{8}\right)$

y-intercept
Let $x = 0$
$y = 2(0)^2 - 11(0) - 6$
$y = -6$
(0, −6)

x-intercept
Let $y = 0$
$0 = 2x^2 - 11x - 6$
$0 = (2x + 1)(x - 6)$
$x = -\dfrac{1}{2},\ x = 6$

$\left(-\dfrac{1}{2},\ 0\right),\ (6,\ 0)$

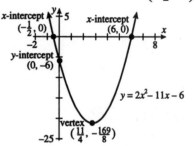

x-intercept $\left(-\frac{1}{2}, 0\right)$
x-intercept (6, 0)
y-intercept (0, −6)
$y = 2x^2 - 11x - 6$
vertex $\left(\frac{11}{4}, -\frac{169}{8}\right)$

86. $y = 3x^2 - x - 2$

$\dfrac{-b}{2a} = \dfrac{-(-1)}{2 \cdot 3} = \dfrac{1}{6}$

$y = 3\left(\dfrac{1}{6}\right)^2 - \dfrac{1}{6} - 2$

$y = -\dfrac{25}{12}$

vertex $\left(\dfrac{1}{6},\ -\dfrac{25}{12}\right)$

y-intercept
Let $x = 0$
$y = 3(0)^2 - 0 - 2$
$y = -2$
(0, −2)

x-intercepts
Let $y = 0$
$0 = 3x^2 - x - 2$
$0 = (3x + 2)(x - 1)$
$x = -\dfrac{2}{3},\ x = 1$

$\left(-\dfrac{2}{3},\ 0\right),\ (1, 0)$

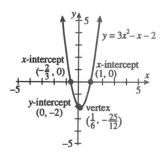

87. 1 solution; $x = -2$

88. 2 solutions; $x = 3$, $x = -1$

89. No real solutions

90. 2 solutions; $x = \pm 2$

Chapter 10 - Test

1. $2x^2 - 11x = 21$
$2x^2 - 11x - 21 = 0$
$(2x + 3)(x - 7) = 0$
$2x + 3 = 0$ or $x - 7 = 0$
$2x = -3$ or $x = 7$
$x = -\dfrac{3}{2}$

2. $x^4 + x^3 - 2x^2 = 0$
$x^2(x^2 + x - 2) = 0$
$x^2(x + 2)(x - 1) = 0$
$x^2 = 0$ or $x + 2 = 0$ or $x - 1 = 0$
$x = 0$ or $x = -2$ or $x = 1$

3. $5k^2 = 80$
$k^2 = 16$
$k = \pm\sqrt{16}$
$k = \pm 4$

4. $(3m - 5)^2 = 8$
$3m - 5 = \pm\sqrt{8}$
$3m - 5 = \pm\sqrt{4}\sqrt{2}$
$3m - 5 = \pm 2\sqrt{2}$
$3m = 5 \pm 2\sqrt{2}$
$m = \dfrac{5 \pm 2\sqrt{2}}{3}$

5. $x^2 - 26x + 160 = 0$
$x^2 - 26x = -160$
$x^2 - 26x + 169 = -160 + 169$

$(x - 13)^2 = 9$
$x - 13 = \pm\sqrt{9}$
$x - 13 = \pm 3$
$x = 13 \pm 3$
$x = 13 + 3$ or $x = 13 - 3$
$x = 16$ or $x = 10$

6. $5x^2 + 9x = 2$

$\dfrac{5x^2}{5} + \dfrac{9}{5}x = \dfrac{2}{5}$

$x^2 + \dfrac{9}{5}x = \dfrac{2}{5}$

$x^2 + \dfrac{9}{5}x + \dfrac{81}{100} = \dfrac{2}{5} + \dfrac{81}{100}$

$\left(x + \dfrac{9}{10}\right)^2 = \dfrac{40}{100} + \dfrac{81}{100}$

$\left(x + \dfrac{9}{10}\right)^2 = \dfrac{121}{100}$

$x + \dfrac{9}{10} = \pm\sqrt{\dfrac{121}{100}}$

$x + \dfrac{9}{10} = \pm\dfrac{11}{10}$

$x = -\dfrac{9}{10} \pm \dfrac{11}{10}$

$x = -\dfrac{9}{10} + \dfrac{11}{10}$ or $x = -\dfrac{9}{10} - \dfrac{11}{10}$

$x = \dfrac{2}{10} = \dfrac{1}{5}$ or $x = \dfrac{-20}{10} = -2$

7. $x^2 - 3x - 10 = 0$
$a = 1,\ b = -3,\ c = -10$

$x = \dfrac{-(-3) \pm \sqrt{(-3)^2 - 4(1)(-10)}}{2(1)}$

$x = \dfrac{3 \pm \sqrt{9 + 40}}{2}$

$x = \dfrac{3 \pm \sqrt{49}}{2}$

$x = \dfrac{3 \pm 7}{2}$

$x = \dfrac{3 + 7}{2}$ or $x = \dfrac{3 - 7}{2}$

$x = \dfrac{10}{2} = 5$ or $x = \dfrac{-4}{2} = -2$

8. $p^2 - \dfrac{5}{3}p - \dfrac{1}{3} = 0$

$3\left(p^2 - \dfrac{5}{3}p - \dfrac{1}{3}\right) = 3(0)$

$3p^2 - 5p - 1 = 0$

$a = 3, b = -5, c = -1$

$p = \dfrac{-(-5) \pm \sqrt{(-5)^2 - 4(3)(-1)}}{2(3)}$

$p = \dfrac{5 \pm \sqrt{25 + 12}}{6}$

$p = \dfrac{5 \pm \sqrt{37}}{6}$

9. $(3x - 5)(x + 2) = -6$

$3x(x) + 3x(2) - 5(x) - 5(2) = -6$

$3x^2 + 6x - 5x - 10 = -6$

$3x^2 + x - 10 + 6 = 0$

$3x^2 + x - 4 = 0$

$(3x + 4)(x - 1) = 0$

$3x + 4 = 0 \qquad$ or $\qquad x - 1 = 0$

$3x = -4 \qquad$ or $\qquad x = 1$

$x = -\dfrac{4}{3}$

10. $(3x - 1)^2 = 16$

$3x - 1 = \pm\sqrt{16}$

$3x - 1 = \pm 4$

$3x = 1 \pm 4$

$x = \dfrac{1 \pm 4}{3}$

$x = \dfrac{1 + 4}{3} \qquad$ or $\qquad x = \dfrac{1 - 4}{3}$

$x = \dfrac{5}{3} \qquad$ or $\qquad x = -\dfrac{3}{3} = -1$

11. $3x^2 - 7x - 2 = 0$

$a = 3, b = -7, c = -2$

$x = \dfrac{-(-7) \pm \sqrt{(-7)^2 - 4(3)(-2)}}{2(3)}$

$x = \dfrac{7 \pm \sqrt{49 + 24}}{6}$

$x = \dfrac{7 \pm \sqrt{73}}{6}$

12. $x^2 - 4x + 5 = 0$

$a = 1, b = -4, c = 5$

$x = \dfrac{-(-4) \pm \sqrt{(-4)^2 - 4(1)(5)}}{2(1)}$

$x = \dfrac{4 \pm \sqrt{16 - 20}}{2}$

$x = \dfrac{4 \pm \sqrt{-4}}{2}$

$x = \dfrac{4 \pm 2i}{2}$

$x = \dfrac{2(2 \pm i)}{2}$

$x = 2 \pm i$

13. $3x^2 - 7x + 2 = 0$

$(3x - 1)(x - 2) = 0$

$3x - 1 = 0 \qquad$ or $\qquad x - 2 = 0$

$3x = 1 \qquad$ or $\qquad x = 2$

$x = \dfrac{1}{3}$

14. $2x^2 - 6x + 1 = 0$

$a = 2, b = -6, c = 1$

$x = \dfrac{-(-6) \pm \sqrt{(-6)^2 - 4(2)(1)}}{2(2)}$

$x = \dfrac{6 \pm \sqrt{36 - 8}}{4}$

$x = \dfrac{6 \pm \sqrt{28}}{4}$

$x = \dfrac{6 \pm \sqrt{4}\sqrt{7}}{4}$

$x = \dfrac{6 \pm 2\sqrt{7}}{4}$

$x = \dfrac{2(3 \pm \sqrt{7})}{4}$

$x = \dfrac{3 \pm \sqrt{7}}{2}$

15. $2x^5 + 5x^4 - 3x^3 = 0$

$x^3(2x^2 + 5x - 3) = 0$

$x^3(2x - 1)(x + 3) = 0$

$x^3 = 0 \qquad$ or $\quad 2x - 1 = 0 \quad$ or $\quad x + 3 = 0$

$x = 0 \qquad$ or $\quad 2x = 1 \qquad$ or $\quad x = -3$

$x = \dfrac{1}{2}$

16. $9x^3 = x$

$9x^3 - x = 0$

$x(9x^2 - 1) = 0$

$x(3x - 1)(3x + 1) = 0$

$x = 0$ or $3x - 1 = 0$ or $3x + 1 = 0$

                  $3x = 1$ or $3x = -1$

                  $x = \dfrac{1}{3}$ or $x = -\dfrac{1}{3}$

17. $\sqrt{-25} = \sqrt{-1}\sqrt{25} = i \cdot 5 = 5i$

18. $\sqrt{-200} = \sqrt{-1}\sqrt{100}\sqrt{2} = i \cdot 10\sqrt{2} = 10i\sqrt{2}$

19. $(3 + 2i) + (5 - i) = 3 + 2i + 5 - i$
$= 3 + 5 + (2i - i) = 8 + i$

20. $(4 - i) - (-3 + 5i) = 4 - i + 3 - 5i$
$= 4 + 3 + (-i - 5i) = 7 - 6i$

21. $(3 + 2i) - (3 - 2i) = 3 + 2i - 3 + 2i$
$= 3 - 3 + (2i + 2i) = 4i$

22. $(3 + 2i) + (3 - 2i) = 3 + 2i + 3 - 2i$
$= 3 + 3 + (2i - 2i) = 6$

23. $(3 + 2i)(3 - 2i) = (3)^2 - (2i)^2 = 9 - 4i^2$
$= 9 - 4(-1) = 9 + 4 = 13$

24. $\dfrac{3 - i}{1 + 2i} = \dfrac{(3 - i)(1 - 2i)}{(1 + 2i)(1 - 2i)}$

$= \dfrac{3(1) + 3(-2i) - i(1) - i(-2i)}{(1)^2 - (2i)^2} = \dfrac{3 - 6i - i + 2i^2}{1 - 4i^2}$

$\dfrac{3 - 7i + 2(-1)}{1 - 4(-1)} = \dfrac{3 - 7i - 2}{1 + 4} = \dfrac{1 - 7i}{5} = \dfrac{1}{5} - \dfrac{7}{5}i$

25. $y = (x - 1)^2 + 2$
vertex $(1, 2)$
y-intercept   x-intercept
Let $x = 0$ Let $y = 0$
$y = (0 - 1)^2 + 2$ $0 = (x - 1)^2 + 2$
$y = 3$ $-2 = (x - 1)^2$
$(0, 3)$ $\sqrt{-2} = x - 1$
                  Not a real number;
                  no x-intercept

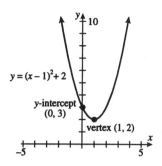

$y = (x - 1)^2 + 2$
y-intercept $(0, 3)$
vertex $(1, 2)$

26. $y = -2(x + 3)^2$
vertex $(-3, 0)$
y-intercept   x-intercept
Let $x = 0$ Let $y = 0$
$y = -2(0 + 3)^2$ $0 = -2(x + 3)^2$
$y = -2(9)$ $0 = (x + 3)^2$
$y = -18$ $0 = x + 3$
$(0, -18)$ $-3 = x$
                  $(-3, 0)$

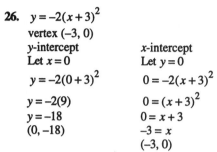

vertex and x-intercept $(-3, 0)$
$y = -2(x + 3)^2$
y-intercept $(0, -18)$

27. $y = x^2 - 7x + 10$

$\dfrac{-b}{2a} = \dfrac{-(-7)}{2(1)} = \dfrac{7}{2}$

$y = \left(\dfrac{7}{2}\right)^2 - 7\left(\dfrac{7}{2}\right) + 10$

$y = -\dfrac{9}{4}$

vertex $\left(\dfrac{7}{2}, -\dfrac{9}{4}\right)$

y-intercept   x-intercepts
Let $x = 0$ Let $y = 0$
$y = 0^2 - 7(0) + 10$ $0 = x^2 - 7x + 10$
$y = 10$ $0 = (x - 2)(x - 5)$
$(0, 10)$ $x = 2, x = 5$
                  $(2, 0), (5, 0)$

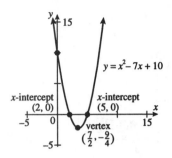

$$y = x^2 - 7x + 10$$

x-intercept (2, 0)   x-intercept (5, 0)

vertex $\left(\frac{7}{2}, -\frac{9}{4}\right)$

Chapter 10 - Cumulative Review

1. $y + 0.6 = -1.0$
$y + 0.6 - 0.6 = -1.0 - 0.6$
$y = -1.6$

2. $8(2 - t) = -5t$
$16 - 8t = -5t$
$16 - 8t + 8t = -5t + 8t$
$16 = 3t$
$\dfrac{16}{3} = \dfrac{3t}{3}$
$\dfrac{16}{3} = t$

3. Let x = liters of 40% solution
$12 - x$ = liters of 70% solution
$0.4x + 0.7(12 - x) = 0.5(12)$
$0.4x + 8.4 - 0.7x = 6$
$-0.3x = -2.4$
$x = 8$
$12 - x = 4$
8 liters of 40% solution
4 liters of 70% solution

4. $3x + y = 12$

  a. $3(0) + y = 12$
$0 + y = 12$
$y = 12$
$(0, 12)$

  b. $3x + 6 = 12$
$3x = 6$
$x = 2$
$(2, 6)$

  c. $3(-1) + y = 12$
$-3 + y = 12$
$y = 15$
$(-1, 15)$

5. $m = \dfrac{2 - 7}{2 - (-1)} = -\dfrac{5}{3}$

answer: $\dfrac{3}{5}$

6. a. $3^0 = 1$

  b. $(ab)^0 = 1$

  c. $(-5)^0 = 1$

  d. $-5^0 = -1$

7. $(3a + b)^3 = (3a + b)(9a^2 + 6ab + b^2)$
$= 27a^3 + 27a^2b + 9ab^2 + b^3$

8.
$$\begin{array}{r}
2x + 4 \\
3x - 1 \overline{\smash{)}6x^2 + 10x - 5} \\
\underline{6x^2 - 2x} \\
12x - 5 \\
\underline{12x - 4} \\
-1
\end{array}$$

Answer: $2x + 4 - \dfrac{1}{3x - 1}$

9. $r^2 - r - 42 = (r - 7)(r + 6)$

10. $x^2 + 12x + 36 = (x + 6)(x + 6) = (x + 6)^2$

11. $y^3 - 27 = (y - 3)(y^2 + 3y + 9)$

12. $64x^3 + 1 = (4x + 1)(16x^2 - 4x + 1)$

13. $(5x - 1)(2x^2 + 15x + 18) = 0$
$5x - 1 = 0$ or $2x^2 + 15x + 18 = 0$
$5x = 1$ or $(2x + 3)(x + 6) = 0$
$x = \dfrac{1}{5}$ or $2x + 3 = 0$ or $x + 6 = 0$
                $2x = -3$ or $x = -6$
                $x = -\dfrac{3}{2}$

14. $\dfrac{2x^2 - 2xy + 3x - 3y}{2x + 3} = \dfrac{2x(x - y) + 3(x - y)}{2x + 3}$
$= \dfrac{(2x + 3)(x - y)}{2x + 3} = x - y$

15. $\dfrac{3.79}{14} = 0.271$

$\dfrac{4.99}{18} = 0.277$

14 oz. for $3.79

16. a. $g(x) = \dfrac{1}{x}$
Domain: all real numbers except 0

   b. $f(x) = 2x + 1$
Domain: all real numbers

17. $2x + y = 10$
$x = y + 2$

$2(y + 2) + y = 10$
$2y + 4 + y = 10$
$3y + 4 = 10$
$3y = 6$
$y = 2$

$x = y + 2$
$x = 2 + 2$
$x = 4$
$(4, 2)$

18. a. $\sqrt{36} = 6$

   b. $\sqrt{64} = 8$

   c. $-\sqrt{25} = -5$

   d. $\sqrt{\dfrac{9}{100}} = \dfrac{3}{10}$

   e. $\sqrt{0} = 0$

19. a. $(\sqrt{5} - 7)(\sqrt{5} + 7) = (\sqrt{5})^2 + 7\sqrt{5} - 7\sqrt{5} - 7^2$
$= 5 - 49 = -44$

   b. $(\sqrt{7x} + 2)^2 = (\sqrt{7x})^2 + 2\sqrt{7x} + 2\sqrt{7x} + 2^2$
$= 7x + 4\sqrt{7x} + 4$

20. a. $\dfrac{2}{1+\sqrt{3}} \cdot \dfrac{(1-\sqrt{3})}{(1-\sqrt{3})} = \dfrac{2(1-\sqrt{3})}{1-3} = \dfrac{2(1-\sqrt{3})}{-2}$
$= -1(1-\sqrt{3}) = -1 + \sqrt{3}$

   b. $\dfrac{\sqrt{5}+4}{\sqrt{5}-1} \cdot \dfrac{(\sqrt{5}+1)}{(\sqrt{5}+1)} = \dfrac{5+5\sqrt{5}+4}{5-1} = \dfrac{9+5\sqrt{5}}{4}$

21. $(x-3)^2 = 16$

$\sqrt{(x-3)^2} = \pm\sqrt{16}$
$x - 3 = \pm 4$
$x = 3 \pm 4$
$x = -1, 7$

22. $2y^2 + 6y = -7$
$2y^2 + 6y + 7 = 0$
$a = 2, \ b = 6, \ c = 7$

$y = \dfrac{-6 \pm \sqrt{6^2 - 4(2)(7)}}{2(2)}$

$y = \dfrac{-6 \pm \sqrt{36 - 56}}{4}$

$y = \dfrac{-6 \pm \sqrt{-20}}{4}$

$y = \dfrac{-6 \pm 2i\sqrt{5}}{4}$

$y = \dfrac{-3 \pm i\sqrt{5}}{2}$

23. $m^2 = 4m - 5$
$m^2 - 4m + 5 = 0$
$a = 1, \ b = -4, \ c = 5$

$m = \dfrac{-(-4) \pm \sqrt{(-4)^2 - 4(1)(5)}}{2(1)}$

$m = \dfrac{4 \pm \sqrt{16 - 20}}{2}$

$m = \dfrac{4 \pm \sqrt{-4}}{2}$

$m = \dfrac{4 \pm 2i}{2}$

$m = 2 \pm i$